PHalarope Books are designed specifically for the amateur naturalist. These volumes represent excellence in natural history publishing. Most books in the PHalarope series are based on a nature course or program at the college or adult education level or are sponsored by a museum or nature center. Each PHalarope book reflects the author's teaching ability as well as writing ability. Among the books:

The Amateur Naturalist's Handbook
Vinson Brown

At the Sea's Edge: An Introduction to Coastal Oceanography for the Amateur Naturalist
William T. Fox/illustrated by Clare Walker Leslie

Botany in the Field: An Introduction to Plant Communities for the Amateur Naturalist
Jane Scott

Discover the Invisible: A Naturalist's Guide to Using the Microscope
Eric V. Gravé

Exploring the Oceans: An Introduction for the Traveler and Amateur Naturalist
Henry S. Parker

Exploring Tropical Isles and Seas: An Introduction for the Traveler and Amateur Naturalist
Frederic Martini

Nature with Children of All Ages: Activities and Adventures for Exploring, Learning, and Enjoying the World Around Us
Edith A. Sisson, the Massachusetts Audubon Society

Owls: An Introduction for the Amateur Naturalist
Gordon Dee Alcorn

The Plant Observer's Guidebook: A Field Botany Manual for the Amateur Naturalist
Charles E. Roth

Pond and Brook: A Guide to Nature Study in Freshwater Environments
Michael J. Caduto

Thoreau's Method: A Handbook for Nature Study
David Pepi

The Wildlife Observer's Guidebook
Charles E. Roth, Massachusetts Audubon Society

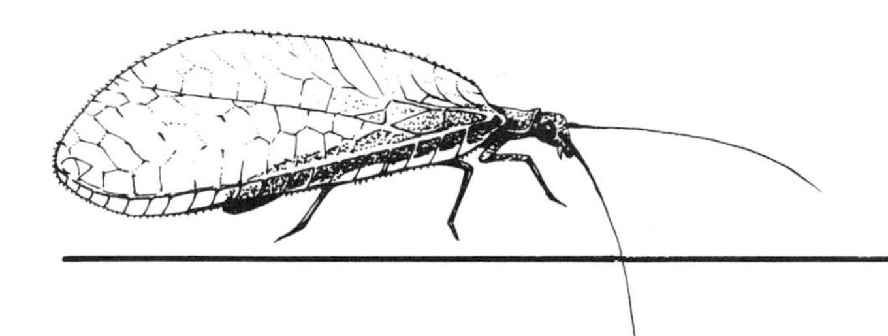

ROSS H. ARNETT, Jr. received his Ph.D. from Cornell University. He is now a research taxonomist at the Florida State Collection of Arthropods and a professor of Entomology at the University of Florida. Dr. Arnett is a member of the board of directors of the American Entomological Institute, and the author of 22 books.

RICHARD L. JACQUES, Jr. received his Ph.D. in entomology from Purdue University. He is a professor of biology at Fairleigh Dickinson University, New Jersey, and has taught entomology and zoology for over ten years.

Professors Arnett and Jacques are also co-authors of the best-selling *Simon & Schuster's Guide to Insects*.

Insect Life:

A Field Entomology Manual for the Amateur Naturalist

Ross H. Arnett, Jr., Ph.D.

Research Taxonomist
Florida State Collection of Arthropods, Gainsville, Florida

Richard L. Jacques, Jr., Ph.D.

Professor of Biology
Fairleigh Dickinson University, Rutherford, New Jersey

Illustrated by Adelaide Murphy

A SPECTRUM BOOK

Prentice-Hall, Inc., Englewood Cliffs, N.J. 07632

Library of Congress Cataloging in Publication Data

Arnett, Ross H.
Insect life.

"A Spectrum Book."—T.p. verso.
Bibliography:
Includes index.
1. Entomology—Field work. I. Jacques, Richard L.
II. Title.
QL464.A76 1985 595.7'00723 85-562
ISBN 0-13-467259-3
ISBN 0-13-467242-9 (pbk.)

Production coordination: Fred Dahl
Cover design: Mike Freeland
Cover illustration: Adelaide Murphy
Manufacturing: Frank Grieco

This book is available at a special discount when ordered in bulk quantities. Contact Prentice-Hall, Inc., General Publishing Division, Special Sales, Englewood Cliffs, N. J. 07632.

A SPECTRUM BOOK

10 9 8 7 6 5 4 3 2 1

Printed in the United States of America

ISBN 0-13-467259-3

ISBN 0-13-467242-9 {PBK.}

Prentice-Hall International (UK) Limited, *London*
Prentice-Hall of Australia Pty. Limited, *Sydney*
Prentice-Hall Canada Inc., *Toronto*
Prentice-Hall of India Private Limited, *New Delhi*
Prentice-Hall of Japan, Inc., *Tokyo*
Prentice-Hall of Southeast Asia Pte. Ltd., *Singapore*
Whitehall Books Limited, *Wellington, New Zealand*
Editora Prentice-Hall do Brasil Ltda., *Rio de Janeiro*
Prentice-Hall Hispanoamericana, S.A., *Mexico*

Dedicated to J. Chester Bradley of Cornell
(1884–1975)
A man who understood taxonomy

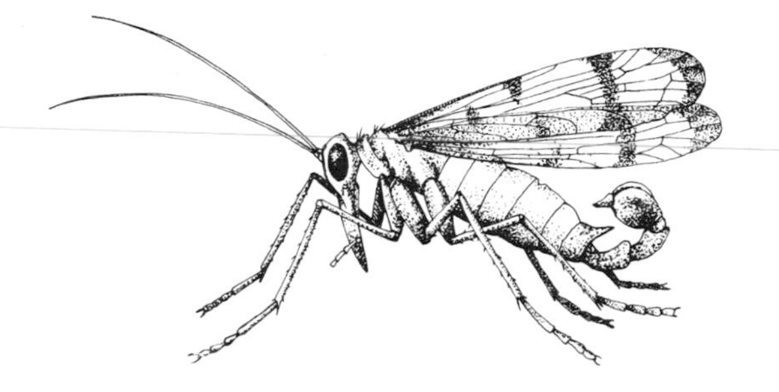

"Soon after dark they arrived, and gave me a most cordial welcome. Jupiter, grinning from ear to ear, bustled about to prepare some marsh-hens for supper. Legrand was in one of his fits—how else shall I term them?—of enthusiasm. He had found an unknown bivalve, forming a new genus, and, more than this, he had hunted down and secured, with Jupiter's assistance, a Scarabaeus *which he believed to be totally new, but in this respect he wished to have my opinion on the morrow."*

Edgar Allan Poe, *The Gold Bug*

Contents

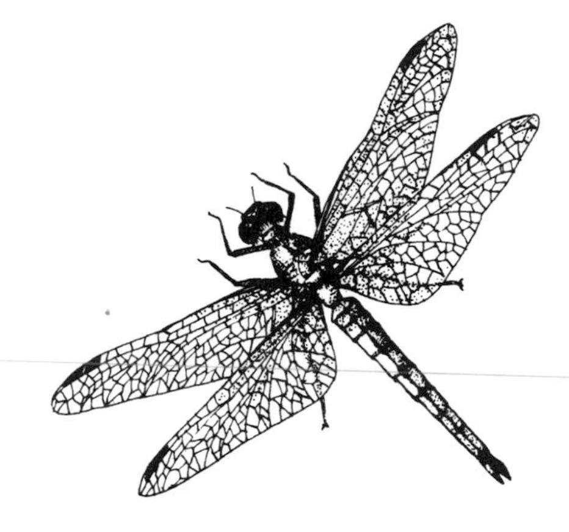

Preface

Full comprehension of the complexity of biological systems comes only as one becomes aware of the fantastic array of kinds of organisms, each interlinked with the other in one way or another throughout the world.

"Insects are far more than just names, but to record data about insects, or any other creature for that matter, it is necessary to start with a name." Thus began Professor Bradley's lecture. He pointed out how to collect, record observations, and make nothing less than accurate identifications. He also pointed out that knowledge of this sort is wholly superficial, until one gains an insight into the extent and complexity of the "world of insects." Each specimen represents part of a natural system, some of the same systems of which we are also a part. The way to gain this insight is the purpose of this book.

Not one of these systems is easily comprehended. One must start with those species termed "common," because they are easy to find and, therefore, easy to study. For example, one well-studied insect is the "common" fruit fly, *Drosophila melanogaster.* Yet all the detailed knowledge now available about this fly is meaningless unless one realizes the size, range, and complexity of the group to which it belongs and *its relationship to all other organisms.* The process this species uses to obtain energy from the sun via fruit, to circulate this energy through the members of its community, and then topass it on to other organisms is just one example. Knowing as much as we do about this tiny creature's genes probably tells us the most we can know about the process of evolution, but nothing about its role in an ecosystem.

The combined weight of all insects exceeds all other terrestrial animals. This alone might place them in the forefront as creatures to know and understand. That most persons regard insects as pests to be eradicated with all possible haste is a tragedy, yet we continue to train our children to scream and step on any insect encountered or, at least, to run in fear of being bitten or stung. Almost no parent or relative encourages them to observe and learn.

"Little Miss Moffett sat on a tuffet, eating her curds and whey. Along came a spider [really not an insect], and sat down beside her, and frightened Miss Moffett away." This is a double tragedy, because poor Miss Moffett was the daughter of the sixteenth-century entomologist, Thomas Moffett (1553–1604), who dissected and illustrated insect parts. We can only surmise that daughter Moffett, as do daughters today, really didn't "cope" with her father's idiosyncrasy. Children show a natural curiosity about insects, but too often they are not encouraged to explore this particular realm of living. If, however, these young people persist in satisfying their curiosity, they soon run into a blank wall. Without the help of some older person trained, or at least experienced, in entomology, they turn hopelessly away from the study of insects because of the staggering number of species they see and are unable to comprehend.

This book is designed to bridge the gap between the many popular articles and books about insects and the somewhat more specialized literature. Instead of the usual family treatment and discussion of economic or common species, you will find here what is needed to collect, preserve, rear, and identify insects found nearby, no matter where you live, and we tell you how to properly study these species. This guide may also get many persons over the fear of insects—fear of bodily harm and fear of the presumed austerity of the subject. We believe that insect study offers as much, if not more, satisfaction as bird study. We, frankly, want to impress would-be bird watches that they can add to their fun by becoming insect watches.

If we can envision the task ahead, the work seems less—we see a light at the end of a tunnel. In this book we show you exactly what insect study is all about: what is easy, what is hard, what is enjoyable, and what is tedious. Most of all, we show you how to do it right.

We quarrel with those persons who believe that knowledge

of nature is unnecessary or who condescendingly insist one first needs to understand DNA and other aspects of molecular biology. College students forced to make a small sampling of the local flora and fauna as a part of classroom study think that they then are capable of doing biological research or teaching with understanding. Their rejection of an indepth study of insect life, by labeling it too specialized, denies them the pleasure of a very rewarding pursuit. School and college curriculum directors might well consider requiring the study of biology, including entomology and botany, of *every* student. To think otherwise is like living back in the days of little Miss Moffett. This unfortunate perspective of the life and role of humans on this earth is akin to the lack of understanding shown by those industries known to pollute the environment for the sake of profit. These minds do not connect ecosystems with daily life any more than people in the Dark Ages could know that malaria was transmitted by mosquitoes, instead of "bad air." An absurd parallel? Not at all! A large department store chain took bamboo baskets infested with wood-boring beetles off the market, not because they thought there was a danger of spreading a pest, which there wasn't, but because they feared that purchasers would think that those wood-boring beetles would transmit a disease to those who purchased the baskets. Now we ask you, if the beetles are busy transmitting diseases, why do they bore into wood instead of people?

To merely begin to understand our world, every student should read and study this book. But, you say, this is the harangue of a specialist. But, shouldn't you know what you are paying for when a plane flies over spraying chemicals, even in minute quantities, for mosquito or fire ant control? Or shouldn't you understand something about the fruit fly problem, pesticide use and misuse, or threats of disease transmission by insects? Reading the newspaper merely gives you the score, but you don't know anything about the game.

To get everyone to read this book is impossible; the need to do so is no less real. We hope you will read it and spread your knowledge of the principles you have absorbed. Without this kind of information, many wrong decisions have been and continue to be made by government and industry. It is said that "without old maids the British Empire would fall, for without their cats, the mice would eat the bumble bees, and there would be no pollenators for the clover that feeds the cattle that feeds the British Navy"! Something to think about, isn't it?

Several friends read this book as it developed. Their suggestions greatly improved our message. We particularly thank Drs. Philip S. Callahan, U.S. Department of Agriculture; Eugene J. Gerberg, Insect Control and Research, Inc.; and

Robert E. Woodruff, Florida Department of Agriculture, for their suggestions. Dr. James E. Lloyd, University of Florida, provided information about fireflies. Thomas and Paula Bartlett helped find obscure passages likely to confuse high school readers and suggested ways to clarify the text. Thanks go to Robert P. Magda for reading the first draft and for his helpful comments, and to Dr. Robert E. Wall, Provost of the Rutherford Campus of Fairleigh Dickinson University, for providing funds for the art work.

Most of the illustrations were provided by the authors or specially drawn for this book by Adelaide Murphy. Some were provided by friends, the use of which is gratefully acknowledged.

Finally, we particularly wish to acknowledge the help of Mary Anne Arnett Held, who pointed out places that needed clarification, thus helping weed out the usual ambiguities that creep into any manuscript.

We thank all of those who helped us and, naturally, hope that they approved of our work. All errors are our own, for which we beg your forgiveness.

1

The World of Insects

"And the poor beetle, that we tread upon,
In corporal sufference finds a pang as great
As when a giant dies."

William Shakespeare (1564–1616), *Measure for Measure*

A tiny ant hurries down a scent trail on the trunk of a giant tropical tree. Suddenly an equally tiny mosquito straddles the trail directly in front of the ant. With its curved, somewhat-swollen proboscis, the mosquito prods the ant between its front legs. This causes the ant to open its mouth and regurgitate ant honey from its crop, whereupon the mosquito plunges its proboscis into the ant's mouth and sucks up a quantity of the sweet fluid.

Little is known about this strange mosquito, a member of the genus *Malayia*. They live in Java, Sri Lanka, and Africa, where some of the many species of the ant genus *Cremastogaster* also occur. This seldom-observed, unique interaction exists between these ants and these mosquitoes, and it is established to be a natural phenomenon, only one of many examples of symbiosis. No one yet knows how this complex relationship came about thousands of years ago. Even more remarkable is the size of these creatures: neither is over 3 mm in length. Obviously, whatever correlation there is between size and complexity of behavior in animals, it isn't apparent here.

What happens in this tiny world?

We read about and see films of big game animals being forced, because of water shortage, to gather around an isolated water hole during the dry season on the plains of Africa. We all know about prey-predator food chain pyramids (fig. 1.1) in these

Figure 1.1. *Food-chain pyramid. Note the decreasing number of organisms at each trophic level; this reflects the different energy demands.*

grasslands. Lions and other cats take advantage of this concentration of food and prey on antelope and other thirsty animals coming there for a drink.

Seldom pictured, however, are thirsty insects at the same pool. Tiger beetles, the cats of the insect world, frequent these water holes in large numbers (fig. 1.2). Sometimes these hungry insects emerge from their pupal cells in such numbers that the prey-predator balance is destroyed. Tiger beetle larvae develop in moist soil around a water hole or a stock tank. Most any unfortunate insect coming to drink will be torn to pieces in a matter of seconds and eaten by these tiny "tigers." Moths, butterflies, even wasps, are subjected to their ravenous appetites. Yet, we have seen net-winged beetles (Lycidae) walk through a pack of tiger beetles and receive only a few nips from members of the mob. After such nips, tigers quickly back off and don't try to attack again. Studies show net-winged beetles are protected by secreting a powerful acrid juice. One may well wonder why these few insects are so fully protected, while this advantage, and hence their resulting survival, escaped so many species. We know of no comparable cases among big game animals, but find similar protection in the disdained skunk. Note, however, the order of magnitude of the two food chains, big game animals, and tiny "tigers." Tiger beetles and net-winged beetles are less than 25 mm in length, while a full-size tiger measures up to 2.5 m in length, or 100 times as large. Even so, the struggle for life is as intense for insects as it is for mammals.

Figure 1.2. *Tiger beetles at a water hole in the Arizona (Sonoran) desert.*

Our own backyard

Let's leave behind for a moment the plains of Africa and giant trees of Java and look at our own backyard. Any summer morning we may spread apart tops of some tall flowering timothy and observe activity at stem bases. Hurrying along a faint path, tiny red ants stop here and there to pick up a bit of organic matter, then off again, possibly taking these back to their subterranean nest. Their world is one of giant grass "trees." Their 3 mm size compares to the 10 mm diameter stem, as our size compares to a great lumber tree. They wander in a forest of grass stems as we do in a dense pine stand.

Jonathan Swift imagined such a world as the one just discussed. Only, he called it Lilliput, and, through the magic of words, let us follow his surgeon-seaman Lemuel Gulliver on a visit to that remarkable world. Swift kept his character in Lilliput from November 5, 1699, to April 13, 1702, a much longer time than Darwin spent in the Galapagos Islands. Darwin's observations led to a unifying theory of evolution. Perhaps Gulliver's observations in Lilliput excited the French naturalist, Henri Fabre, to search for Lilliput in his backyard bramble patch, for, surely, the words from Fabre's pen are as much a story of Lilliput as those of Swift. Fabre, among the first to realize that much more goes on in the tiny world of insects than just feeding and egg-laying, wrote many volumes describing their lives.

Suppose for a moment that our travels lead us to the beautiful land of Lilliput. To us their tallest trees would seem nothing more than blades of grass. A man of extraordinary visual imagination, Swift quickly realized the appeal of a Gulliver in such an exciting land as Lilliput and planned a visit to a land of reverse order of magnitude. This should be equally fascinating, he reasoned. Hence, not many weeks after his return to England, Gulliver set sail for Brobdingnag. Here Gulliver felt like a Lilliputian, the ant feeding the mosquito, the tiger beetle at the water hole, or the ant in the grass forest, while the trees were but grass stems to the gigantic king and people of Brobdingnag!

It's a matter of perspective

Many human beings are Brobdingnagians when it comes to their reactions to insects. Because "bugs" are so small, they are either to be ignored or destroyed. Depending on our previous experience, our reactions to insects are seldom positive. We must learn, as did Gulliver through his travels, exactly what a difference size makes in one's approach to and appraisal of something. The level of magnitude colors our thinking about many things. We stand in awe as we watch a tiger bring down its prey, while we are oblivious to a tiger beetle and its prey. Although many people may be extremely fond of cats and dogs, and view lions and elephants with respect, they feel no sense of remorse at stepping on an ant or swatting a fly. Seldom do we hesitate to leave a squashed moth weakly vibrating on the alley pavement, feeling no sadness at its pain!

Order of magnitude

We talked about the order of magnitude between the world of the lion and the world of tiger beetles, a difference of approximately 100 times. Let's make some comparisons, Gulliver style. A house fly is about 8 mm in length. Suppose this fly were enlarged only 10 times (fig. 1.3). This would be a difference of the same magnitude as between six-foot Gulliver and 60-foot Brobdingnagians, or six- or seven-inch Lilliputians. Enlarged flies would then be the size of small mice. We still would feel little remorse over killing a mouse, although we believe many of us feel sorry for dead furry mice in traps. Multiply again by only two, and the fly becomes the size of a guinea pig, now something of a size that a child can care for. If this fly were then covered with fur, instead of spines, it, too, would at least attract attention.

Return now to tiger beetles at water holes. Using the same order of magnitude and multiplying first by ten, we would see a beetle the size of a playful kitten. Multiply again by two, and our beetle becomes the size of a small bobcat, as fearful a predator as the tiger beetle. These size relationships become apparent only when one has a chance to live outdoors and experience their problems by firsthand observation.

For many summers we camped in southern Arizona collecting insects and carrying on research on population dynamics of several species of beetles. One night in camp we discussed the problems facing human beings, cows, and small insects in finding mates in broad Arizona valleys. In days before automobiles, a man needed a horse to ride into town or to another ranch to find a mate. A bull finds his mates on the range by fighting and staking out his territory. He fathers calves and protects mothers. In similar low density population situations, man does much the same: stakes out his property and protects his family from aggression.

We then compared the size of a cow—about 1.8 m in length—with the size of beetles under study. These averaged about 10 mm in length. Mating problems for beetles, we pointed out, were about 180 times as great as for cows in the same area. Once this was established, we looked at our beetle population problem with an entirely new insight. To these Lilliputian-sized creatures, finding a mate is not easy. In fact, it is about 180 times

Figure 1.3. *Gulliver reviews the passing "Lilliputian"-size insects.*

more difficult for them than for a man or a cow in the same region.

We then began to ask questions about how they find mates, instead of taking it for granted that tiger beetles and other insects, as well, breed to the maximum extent environment allows. We should not assume all insect populations are excessively abundant and must be stopped at all cost. Only some specialized process allows these creatures to breed and maintain their species in the vast area of their habitat. Man became a social animal, built towns, had parties and dances, all ultimately useful for mate-finding. Most insects developed wings, and some evolved special glands to secrete sex attractants, both aiding the mating processes.

With these comparisons in mind, we began to relate data previously gathered, but seemingly of little significance. Thinking in proper order of magnitude puts an entirely new slant on responses we make to insect activity.

Think little

To this point we have tried to show how differences in size provoke different humans responses. What we may call the Gulliverian syndrome plays a major part in generally unfavorable responses of humans to the sight of insects. To overcome this we must "think little" instead of "thinking big."

Other thoughtless human behavior patterns play a significant role in most of our reactions to insects. Many parents set a bad example for their children by showing signs of fear, dislike, or even disgust in the presence of any insect, harmful, indifferent, or beneficial. Often parents dispatch any bug on sight, usually with much commotion and violent swatting with rolled newspapers, shoes, or any handy object. This prevents a child from seeing an insect as a living creature or as anything but a nauseous glob of semi-homogenized tissue. Because this happens from early childhood, children acquire an almost automatic response, which is rarely overcome in later years. School curricula only recently provide and encourage a more enlightened view of biological subjects.

Readers may object, believing this aversion to insect life natural and protective. Few moderns know what to fear; that is, they do not know what can harm them. Most human diseases are spread by other humans, or by domestic animals. We let our dogs lick our faces and spread worm eggs that they have just licked from their own anuses. At the same time, we imagine we can get a disease from eating the worm in an apple! Anyone who thinks these reactions "natural" or instinctive, instead of learned from misinformed adults, are unaware that millions of people in the past and many still today eat insects and depend on these high protein sources for a major portion of their diet. Those who believe that insect protein is used only as a last resort should consider the fact that termites are 68% protein, while a delicious, expensive steak, so full of cholesterol, is only 15% protein!

Because of their small size, many insects tend to look much alike on first sight. Hence, remarkable differences between approximately one million species go unappreciated. Here, particularly, order of magnitude is important. Suppose every insect fell within the size range of mammals (20 millimeters to 200 meters) (fig. 1.4). Each species would then be as apparent to us as each mammal species. The difference between cows and horses, lions and tigers, elephants and buffalo, and so on becomes easy to determine. Small-sized rodent species may look alike, but they are still usually large enough to identify, at least when dead in a trap. Many of us easily adjust our perspective (usually with the aid of field glasses) to study bird life. Birds, of course, are often brightly colored, sing, and attract our attention. Even so with insects, and many of these have songs, if only we would take the time to listen (see chap. 8). Since birds and insects overlap their size ranges, it should not be difficult to make the next observational adjustment—perhaps aided by a hand lens or low power microscope—and peer into our Lilliputian world of insects.

Figure 1.4. *Enlarge a fly to the size of a Guinea pig, and you will see how different each species really is.*

Many kinds and many adaptions

Estimates of the number of species of insects vary widely. One late figure, nearly one million, includes only those known and described. Every year hundreds of new species are found, and these new names fill pages of many entomological journals. Some entomologists believe that there are as many as ten million species, while others are more conservative and feel this figure probably will stabilize at two million species. Reasons for these divergent opinions are simple: too many insects and too few entomologists to study them. Naming insect species is an unusually complex procedure, one which has little to do with insects themselves. Naming is merely the beginning of an information storage system and certainly not a true indication of the fascinating activities of insects themselves. It begins our study, not ends it.

Why so many species?

Insect species equal in number all other kinds of organisms together. Less than one-fourth as many species comprise the entire plant kingdom. Insects outnumber all other animals by two to one. No one really knows why so many kinds of insects or why one kind is so much more abundant than another kind. Entomologists say that, because insects are able to adapt to many different habitats, and, in so doing become modified for that habitat, they readily speciate. Adaptation results in speciation. But why do insects outnumber other organisms in such an unbalanced proportion to other living organisms? Perhaps, if we look closely at their many adaptations for life on this earth, we will find the answer to this question. This we will do in the following chapters.

Preview

Many of the biological phenomena we have come to know were learned through a study of higher animals, particularly mammals, and especially humans. Historically these phenomena first developed in insects and other lower animals, although, independently, perhaps, of lines of development leading to vertebrates. Adaptive features appear time and time again, without genealogical connections, since they are adaptive. For example, paddles enhance the ability of an animal to move in water. Therefore, "paddles" appear in many completely unrelated aquatic animals. It seems obvious that many principles of biology may be best illustrated by insects, because of the great diversity in this group. This is, indeed, the case. Examples of these principles are briefly described below and illustrated in detail in text to follow.

Some biological principles illustrated by insects

The biological principles of living organisms are much the same. Differences are primarily adaptive. This is to say that all living creatures have much the same basic features, but each species and each group of species express these various features of life in widely different ways. Thus, we define life as cellular, composed of one or many cells, basic units carrying on processes necessary to continue life. All living things reproduce in some manner. Food-getting permits growth and provides energy for reproduction. Organisms are usually manifested as individuals. Some are interconnected in a manner permitting them to share processes. These form a particular kind of colony found among certain types of marine invertebrates, particularly Coelenterates, the corals, and similar organisms. All biological processes eventually cease in the individual, after which it enters the state known as death. Before death, the individual is able to maintain its distinctness from the environment. After death, the individual merges into the environment through the process of decay. Decay is brought about by other organisms, such as fungus and bacteria, or through the process of being eaten and digested by another organism.

It is apparent that some individuals are able to survive, because they are more efficient in the way they live. We say that they are better "adapted." Better adaptation results in longer survival. If these adaptations are genetically determined and the individual reproduces, genetic traits may be passed on to its offspring. When this happens, a slight amount of evolution takes place. When changes are great, we say the species evolved. We believe this process takes place in all species, and, therefore, evolution is a basic principle of life.

All other biological processes and functions are manifestations of these basic principles. Diversity of form and process results in the great array of species. Since most (but not all) of this diversity came about through adaptation to the environment, we must conclude that evolution is the result of adaptation. It is, of course, much more complex than these simple statements. The evolutionary process is described in more detail in chapter 4.

Each chapter of this book is written to show how these basic principles are illustrated by insect life. Thus chapter 5 describes food-getting processes necessary to sustain life. Chapter 6 outlines complex reproductive cycles and how they vary from group to group. Much of survival depends on behavior—how an insect reacts to its environment. Chapters 7, 8, 9, and 10 give more details on insect survival.

Remaining chapters of the book consider in some detail relationships of insects to man. Both insects and humans have the same biological requirement, because, obviously, they are both biological organisms. Overlapping the needs of these two kinds of organisms—insects and man—causes war between humans and some insect species. Actually, the species in conflict with man are few compared to the total number of species (see chaps. 12, 13, and 14).

Particular attention is given to the way insects are studied, because we are eager to have you, the reader, share with us the joy of observing insects. To do this, you must be informed about proper ways to handle insects. Therefore, we have added several chapters (15, 16, 17, and 18) to introduce you to these procedures.

Who studies insects?

Those interested in insect life, other than students, fall into two major groups: amateurs and professionals. Many more amateurs than professionals study insects, since we generally define amateurs as those who are not earning their living by the study of entomology and professionals as those who are paid to practice the profession as entomologists. The line between the two groups is fuzzy, and it is unimportant to make a clear-cut distinction. It should be emphasized that distinctions between the two groups are not based on knowledge of insect life. Much of the development of the field was and still is the result of studies by amateurs.

Professional entomologists generally specialize in one of the following branches of entomology.

Industrial and home pest control. Application of pesticides and other means for controlling insects, arachnids, rodents, and other pests of buildings and homes concerns these professionals and their technicians. Persons entering this branch are generally called pest control operators (PCOs). Professionals are employed by private industry or teach in universities.

Applied entomology. Entomologists studying various methods for control of insect pests comprise a high percentage of professional entomologists. Many of them are employed by commercial companies, particularly chemical and petroleum industries. Others are professors at universities and employees of state and federal experiment stations.

Biological control. Use of parasites, pathogens, and predators for control of insect pests is a spinoff from applied entomology. Applied entomologists involved in this kind of research are usually employed by agencies of state and federal governments, although some commercial companies are rearing insects for release as parasites and predators of crop pests.

Pest management. When control practices involve using ecological principles, parasites, pathogens, and predators, as well as pesticides, to control insect populations, particularly when combined to prevent pollution and habitat destruction, these procedures are termed "pest management." Much of current research in applied entomology is now concerned with methods of managing insect pest populations involving all aspects of their biology. Professionals concerned with this are employed by commercial firms, universities, and various state and federal agencies.

Forest entomology. Because of the specialized nature of forest management, some entomologists specialize in this branch of entomology. Insect pests of forest trees differ considerably from species found on field crops. Because of their different biologies, their control involves wholly different techniques. These entomologists are employed by lumber companies and governmental agencies.

Medical entomology. Some entomologists specialize in the study of insects harmful to humans and domestic animals. As with forest entomology, species involved are different and, therefore, special knowledge is needed to control them. Medical entomologists are associated with some commercial companies, county, state, and federal agencies, and universities.

Academic entomology. Because background education is needed by professional entomologists, universities need professors who have specialized in various branches of entomology. Some teach applied aspects of the field; others, basic subjects: insect physiology, insect morphology, insect ecology, and insect systematics. Vast literature stores on this subject require university courses on the use of entomological literature. Also, because insects are involved in almost every ecological system, every entomologist must have a broad background in biology, geology, chemistry, and physics. To become an academic entomologist takes years of university study.

Apiculture. Rearing honey bees to get honey for food, a special branch of entomology, involves both individuals without training in entomology and professional entomologists specializing in the art of beekeeping. Most professional apiculturalists are associated with private companies, university experiment stations, or agricultural extension offices.

Systematic entomology. As does apiculture, insect systematics attract professional and amateur entomologists. Our discussion of this branch has been left for last, because it leads into the subject of this book. Professional systematic entomologists identify and classify insects of the world. They are employed to make identifications of insect pests, in addition to other kinds. These entomologists work for universities, the federal government, and various agencies of state governments. Their identifications make possible the study of literature on species, to find answers to various questions besides, "What is the name of this bug?" For example, if an insect is a pest, by knowing its correct name, all literature on the biology and control of the species becomes available. If not a pest, often one may find out where the insect lives, interesting facts about its biology, and so on. Insects are intercepted at ports of entry in baggage, and in products being brought in. When specimens are identified, entomologists may ascertain whether a species threatens crops or produce in this country, and then advise officials about steps to prevent their entry into the country.

Besides applied aspects of insect systematics, many systematists name new species and place these species in the correct order, family, and genus. Most of this work is done at universities maintaining collections of insects or at museums operated by various governmental agencies. Readers must understand that the insect study collections referred to here are not those on display for visitors to the museum, but the "behind the scene" collections used for scientific studies. These reference collections contain by far the largest number of specimens. Those on display are meager representatives of those used for scientific study.

Amateur entomologists. Insect collectors, professional and amateur, find insect collecting and arranging of collections a fascinating study. Both make observations on insect biology, besides identification. Amateur entomologists are almost entirely systematists.

Amateur systematic entomologists outnumber professional systematic entomologists. In the past, the work of amateurs provided most of the basic classification. Until the beginning of the twentieth century, almost all systematic work on insects was carried on by amateurs, usually physicians, the only persons with much training in natural history. Thus, names of most insects described before 1900 were given to the species by amateurs. Many amateurs today name and classify insects, but it should be clear in the minds of readers that these "amateurs" are not beginners. They may have the same background and experience in systematic entomology as the professional. They do not just start out naming new species. Classification would be chaotic, if this were the case.

Who are the amateurs?

One would not expect a banker to be a collector of insects. Yet David Rockefeller, whose lifetime career has been in banking, still adds to his collection of beetles whenever an opportunity occurs. As a boy, he became interested in collecting beetles. While a student at Harvard, he took courses in entomology and has continued this interest by maintaining a large private collection of beetles.

The famous novelist, Vladimir Nabokov, an amateur collector of butterflies, named the famous Karner Blue Butterfly (*Lycaeides melissa samuelis* Nabokov), a rare species known only from the pine barrens in eastern New York State and a few other localities. He named several other kinds of butterflies, as well.

Television stars, sports figures, mechanics, truck drivers, and travelling sales representatives are among those who collect insects. Many of these individuals do not engage in original research, but simply enjoy the physical activity of collecting and the fun of identifying and arranging specimens in a collection. The exchange and sale of specimens is a part of these activities. (See chap. 14 for more details on the value of insect collections.)

Those interested in systematic entomology, amateur and professional, are characterized by their long hours at work. Most of these individuals, even though they may be employed professionally as systematists, continue their study of insect classification into evening hours, weekends, and in fact, never retire from this work. The day after their retirement parties, they are back hard at work on some systematic problem.

On the following pages, we will tell you how to emulate these individuals, and we hope that you, too, will find for yourself joy and fascination in the study of insects.

2 Where and When to Study Insects

"At eve the beetle boometh
Athwart the thicket lone."

Alfred Lord Tennyson (1809–1892), *Claribel*

The best place to start is at home. Youngsters capture insects visiting garden flowers or attracted to porch lights at night. When these are pinned, labelled, and displayed, parents, neighbors, and teachers are amazed at the variety of "everyday" insects that went previously unnoticed. Unfortunately, as pointed out in the previous chapter, interest soon wanes through lack of encouragement and proper instruction. Few young people continue to keep up their collections. They pass up what could be a lifetime hobby or even a rewarding profession.

What to expect around home

Suburban or country "flies" are among the first insects noticed because they are pests. Don't assume every household "fly" is the same. Dozens of species are attracted to household odors. Some are the common house fly; others are carrion and dung flies, syrphid flies (some of which are easily mistaken for bees), and miscellaneous other kinds. When you find out from chapters 15 and 16 how to collect and store insects, capture all these and examine them with a good hand lens. Their structure and diversity of form are remarkable and their parts truly beautiful. They are difficult to identify to species and, for now, they may be put aside for future study when you have learned more entomology.

Several species of ants commonly invade homes. Try to locate their nests and study their contents. Some, such as the carpenter ant, are destructive, while others merely snatch a little food now and then and really do little harm.

Certain household pests, including the much maligned cockroaches and the pesky, but delightful, fruit flies, find their way in, unwelcome though they may be. Fleas (if there are pets in the house), silverfish, and book lice join the crowd and expand our growing list of things to add to the collection. Most people, not realizing different species may or may not be pests, spray with insecticides without finding out exactly whether the insect in the house is a problem, or just an incidental wanderer that may be booted out the door along with stray dogs and uninvited traveling salespersons (too bad we can't get rid of unwanted phone sales pitches as easily). A thorough survey of household insects and an assessment of these species through accurate identification make an exciting first project for the novice entomologist (and it will save money and worry).

Everytime a door opens, more insects enter (here, that is, where we are writing in sunny Florida, but you will have to wait until spring, summer, and fall if you are "up north"). Houses make excellent traps, but why not go outside where many species abound? Bees and beetles of many kinds visit flowers; hence, here is a good place to start. Dozens of species feed on the leaves of plants growing in the garden, and more cling to trees and bushes. A single day in July will produce at least one hundred species, surely a pleasing sight once mounted, identified, and properly arranged in boxes. A second project for you to do (and which will also save money and worry) is to learn to identify garden pests and distinguish them from the casual visitor and the absolutely necessary pollenators. Why spray if there are no pests? You don't ward off elephants unless you live in Africa or India, the only places these animals are dangerous.

Other local places in which to collect

Moving on to nearby fields or wild areas near one's home or on a college campus, park, or similar area, you will find hundreds of additional species, because of the greater diversity of the habitats. For example, if a creek flows through the field and empties into a pond, aquatic bugs, beetles, dragonflies, damselflies, stoneflies, mayflies, and many other "aquatics" compete for space in these habitats. If you lack a pond, shut off the water filter and stop adding chlorine to your swimming pool! (If you don't have a pool, try digging a small garden pond.) In a few weeks you will have your own pond, and you can call it "ye ol' swimming hole." Once you have a good culture of aquatic insects, you can introduce bullfrogs and enjoy a good supply of frogs' legs for the dinner table. Almost all 30 orders of insects are represented in or near wild water.

Nighttime collecting

Many insects are nocturnal, and, for reasons not entirely clear, hundreds of nocturnal species are attracted to the area of lights at night. Dr. Philip S. Callahan of the U.S. Department of Agriculture has shown, through a series of sophisticated observations and experiments, that flickering lights and moth antennae (and, presumably, beetle, bug, and other insect antennae, as well) resonate, and that this is why insects come to light, not because they see the light.

Now this makes sense. We have always wondered why it is that moths, well known to be nocturnal, come to light, the same light that they shun all day long. Common sense would tell you that, if light causes moths to hide during the day, they would not be attracted to lights at night. Establishing these facts, it is equally evident that the areas near lights abound with insects. Most people feel that this is unfortunate and go to the expense of screening porches and patios to prevent insects from entering. (They also purchase the completely useless "bug zappers," which attract everything *but* the specific insect pests that they are trying to get rid of. In fact, they attract far more pesky insects than ordinary porch lights would, because the flickering of the UV black light is generally more efficient as an insect attractor. But, mosquitoes and flies are attracted to people and household odors, not to UV light, thereby completely defeating the purpose of the zapper, unless one enjoys smelling burning insect flesh.)

For entomologists, however, this phenomenon is a great asset, because it is a way to get more specimens for a collection.

Moths and beetles comprise the greatest number of species attracted to lights. Some flies and the adults of a variety of aquatic insects will come to these lights, along with a great variety of bugs.

To sum up local collecting areas, convince your neighbors that you are using a UV light to lure away their insects, and that the weeds in your lawn are really rare flowers that you are cultivating instead of grass. If you are successful, you will not only get more insects for your collection, but you will have more time and money to spend on collecting, because of the time not spent on lawn mowing and the money not spent on lawn mowers and the gasoline to run them. And—you will have a great oxygen supply around your yard.

But local collecting is finite and, after a while, you will exhaust the chance of finding additional species to add to your collection. It is then time to travel to other areas.

Local preserves

Most all regions throughout the country have a preserved area ranging from a National Park, a National Forest, a State Park, down to a small area set aside by a local natural history society. Because these regions are not disturbed by farming or industry, a variety of habitats provide suitable places for collecting and observing insect life. Some of these regions are off limits for collecting natural history specimens, except when special permission has been obtained (see chapter 17). Remember that picnic areas in parks are sprayed with insecticides to kill "pest" species. The sprays are not selective; hence, if mosquitoes and flies are not present in these areas, chances are that dragonflies, butterflies, and other beautiful insects also will not be present, making the area useless as a collecting area, permission or not. You, now an insect explorer, will have to search hard for a place to collect additional specimens.

Habitat classification

After a little experience at collecting, it will be evident that certain species are always found in more or less the same place. One soon realizes that insects select a particular habitat. Then one sees that there are woods species, field species, aquatics, and so on.

The purpose of habitat classification is to provide you with a comprehensive view of the environment successfully invaded by insects. Insects are an integral part of each habitat. They are living, functioning creatures, very definitely related to, and restricted in their activities by, the surrounding environment. Specimens in an insect box are representives of these habitats. Thus, when the name *Cicindela sexguttata* conjures in the mind of the reader an alert, great-eyed beetle flashing brilliant green in the sunlight as it warily watches from a few feet down a country path; when the name *Pompilus* brings up the image of a nervous, steel-blue wasp, jerkily hunting around into every nook and cranny, hoping to find a ground spider to capture and bear away to its nest; when "cicada" means hot summer days and shrill, but elusive, calls from sufficiently ponderable creatures on the branches of trees, perhaps the sudden frantic and desperate shrilling of one caught in a net, or held between thumb and forefinger; when the technical names of these and other insects paint such word pictures in your mind, you have acquired, or are fast acquiring, a true love of nature. No longer is a specimen merely a set of morphological characters delineated in a key to the species, but it represents a part of a living, ever-changing ecosystem.

In fitting an insect into any of the categories used here, two tests should be applied: first, the habitat must be one regularly and necessarily occupied, not incidentally or accidentally visited; second, the environment should be connected with some fundamental life-function, such as a source of food or the site of the immature stages. Most adult insects are freely moving creatures and may search out new habitats with temporary visits, but don't become established (that is, do not breed) in these areas.

An insect's ecological relationships may change with the stage of its life cycle. During the nuptial flight, an ant or a honey bee may be truly a part of an aerial environment, for that environment becomes a primary one for a brief period and necessary for the accomplishment of a life function. During the rest of their lives, the ants and the queen bees of the nuptial flight definitely abjure the air, while, to even worker bees, the air is but an incident of transport from food-collecting rounds to the hive, both of which may be looked on as their true habitat. The larva of a cat or dog flea inhabits detritus (disintegrated debris) in the home, but its life as an adult is in a different habitat, usually on the cat or dog. Hence, it may be necessary in relegating an insect to its true habitat in the following scheme to state at what period or stage it belongs there; and, even in a given phase of the life cycle, it may belong in more than one category as its activities or functions change.

There follows behind each major category in table 2.1 a technical term that may be used with some precision in stating the habitat of a species. However, one must realize that many different classification schemes have been published, and that this is only one, one that may be modified to serve any special purpose.

Table 2.1. *Habitat classification.*

1. Aquatic (Hydrobiota)
    - 1.1 Dwelling on the surface of the water (Ex. water striders, whirligig beetles)
    - 1.2 Living submerged in the water
        - 1.21 On still water (Lentic)
            - 1.211 Free-swimming and suspended forms (Pelagic) (Ex. some fly larvae)
            - 1.212 Bottom-dwelling forms (Benthic)
                - 1.2121 Burrowing forms (Ex. certain mayfly and dragonfly naiads)
                - 1.2122 Bottom-resting forms (Ex. water bugs and beetles)
                - 1.2123 Forms that climb among aquatic vegetation (Ex. naiads of damselflies and certain mayflies)
        - 1.22 Running water forms (Lotic)
            - 1.221 Under stones (Ex. naiads of many mayflies and of stoneflies and caddisfly larvae)
            - 1.222 Fastened to surface of submerged objects (Ex. waterpennies, certain fly larvae and pupae, including black fly larvae and pupae)
2. Ground-dwelling insects (Geobiota)
    - 2.1 Dwelling in the soil (Subterranean)
        - 2.11 In mud flats (Hygrophilous) (Ex. various subaquatic beetles)
        - 2.12 In normal soil, neither arid nor wet (Mesophilous) (Ex. tiger beetle larvae; nest of many ants, wasps, and bees)
    - 2.2 Beneath stones (Subsaxean)
        - 2.21 In wet places (Hygrophilous) (Ex. certain carabid beetles)
        - 2.22 In neither wet nor dry situations (Mesophilous) (Ex. many ants, beetles, and beetle larvae)
        - 2.23 In desert and extremely dry soil (Arenicolous) (Ex. various beetles)
    - 2.3 In leaf mold (Humicolous) (Ex. springtails, minute beetles)
    - 2.4 On surface of the ground (Terrestrial)
        - 2.41 On mud flats and sand bars (Hygrophilous) (Ex. grouse locusts, toad bugs, pygmy mole crickets, shore bugs, some rove beetles, ground beetles, and minute semi-aquatic beetles)
        - 2.42 On normal soil (Agrarian) (Ex. many ground beetles)

- 2.43 On dry sand and desert (Arenicolous) (Ex. various tiger beetles, some grasshoppers, velvet ants, ants, and various flies)
- 2.44 On exposed rock surface (Saxicolous) (Ex. some grasshoppers)

3. Plant-inhabiting insects (Phytobiota: see also the following discussion of the natural regions, or biotas, of the world)
   - 3.1 Living externally on a plant
     - 3.11 Feeding on leaves and petals (Phyllophagous) (Ex. caterpillars, beetle larvae and adults, sawfly larvae, etc.)
     - 3.12 Sucking sap (Galactopoietic) (Ex. aphids, scale insects, leafhoppers, tree hoppers, etc.)
     - 3.13 On flowers (Anthophilous) sucking nectar or eating pollen (Ex. bees, wasps, various flies, various beetles, etc.)
     - 3.14 On roots (Rhizocolous) (Ex. certain aphids, beetle larvae, etc.)
   - 3.2 Living internally on a plant
     - 3.21 Boring in living tree trunks, branches, and stems (Lignicolous) (Ex. larvae of various beetles and moths)
     - 3.22 Dwelling in natural cavities of plants (Cavicolous) (Ex. many tropical ants)
     - 3.23 Causing gall formation or dwelling in galls (Cecidicolous) (Ex. many flies and wasps, some aphids)
     - 3.24 Inhabiting fungi (Mycocolous) (Ex. larvae of fungus gnats, many beetles)
4. Woodland and savannah species (Sylvan) (Ex. scorpionflies, many kinds of flies, wasps, bees, ants, etc.)
5. In open field (Campestrian) (Ex. many treehoppers, spittle bugs, beetles, wasps, ants, bees, butterflies, flies, etc.)
6. Parasitic on or infesting animals (Zoobiota)
   - 6.1 External parasites (Ectoparasites) and visitors
     - 6.11 Permanent parasites (Ex. biting lice, chewing lice, certain beetles, a few species of moth larvae)
     - 6.12 Temporary visitors (Ex. various bloodsucking insects, such as mosquitoes, horse flies, black flies, punkies, fleas, etc.)
7. Dwellers in dead plant and animal matter (Necrobiota)
   - 7.1 Living in dead wood (Xylophilous)
     - 7.11 Under bark (Ex. flat bugs, histerid beetles, certain cucujids, and other beetles, beetle larvae, and some bugs)
     - 7.12 In rotten logs (Ex. bess beetles, tenebrionid beetles, certain elaterid larvae, etc.)

7.2 Living in decaying vegetation (saprozoic) (Ex. certain beetles, some fly larvae, etc.)
7.3 Living in dead animal tissue or excrement (Saprophilous)
  7.31 Living in excrement (Scatophilous) (Ex. histerid and rove beetles, some scarabaeids, various fly larvae, etc.)
  7.32 Living in carrion (Saprophilous) (Ex. silphid beetle larvae and adults, hister beetles, rove beetles, hide beetles, larvae of flies, etc.)

Many, if not most, insect species are restricted to specialized habitats. Some are confined to a single habitat—that is, they never leave their habitat, while many others are found in a wide variety of places, that is, they are polyphagous. A great many of the described species have no recorded habitats. Recently collected, newly described species often are based solely on specimens collected at lights, and, therefore, nothing is known about their habitats. Older species are based on specimens that were collected in certain habitats, but where was not recorded. Indeed, many of these old specimens merely recorded the country, or at most, the state or province. This explains why it is important that detailed records be kept on even the most common species. Detailed records may yield volumes of scientific data.

Biomes of the world

The biological habitats as outlined below, when taken all together for major areas of the world, form biomes. Usually biomes are characterized mostly by the dominant plant cover (fig. 2.1). Insect distribution coincides only indirectly with these biomes, mainly because many species are dependent on plant life. They are either directly herbivorous, or indirectly, as predators feeding on herbivores. Thus, their distribution in the biome is not directly dependent upon the physical factors that determine the plant cover, but, rather, by those selective factors that coupled them with their plant host. Insects often occur in more than one of these regions, but so do plant species. Nevertheless, biomes give us a convenient way to refer to distribution areas, instead of using changing political divisions of the earth. Unfortunately, the preceding statement is not entirely true, because, again, we know so little about most species, and it is impossible to flatly state that the species lives in one or another biome or range of biomes. Still, each of these biomes has a characteristic assemblage of insect species. The biomes of the world

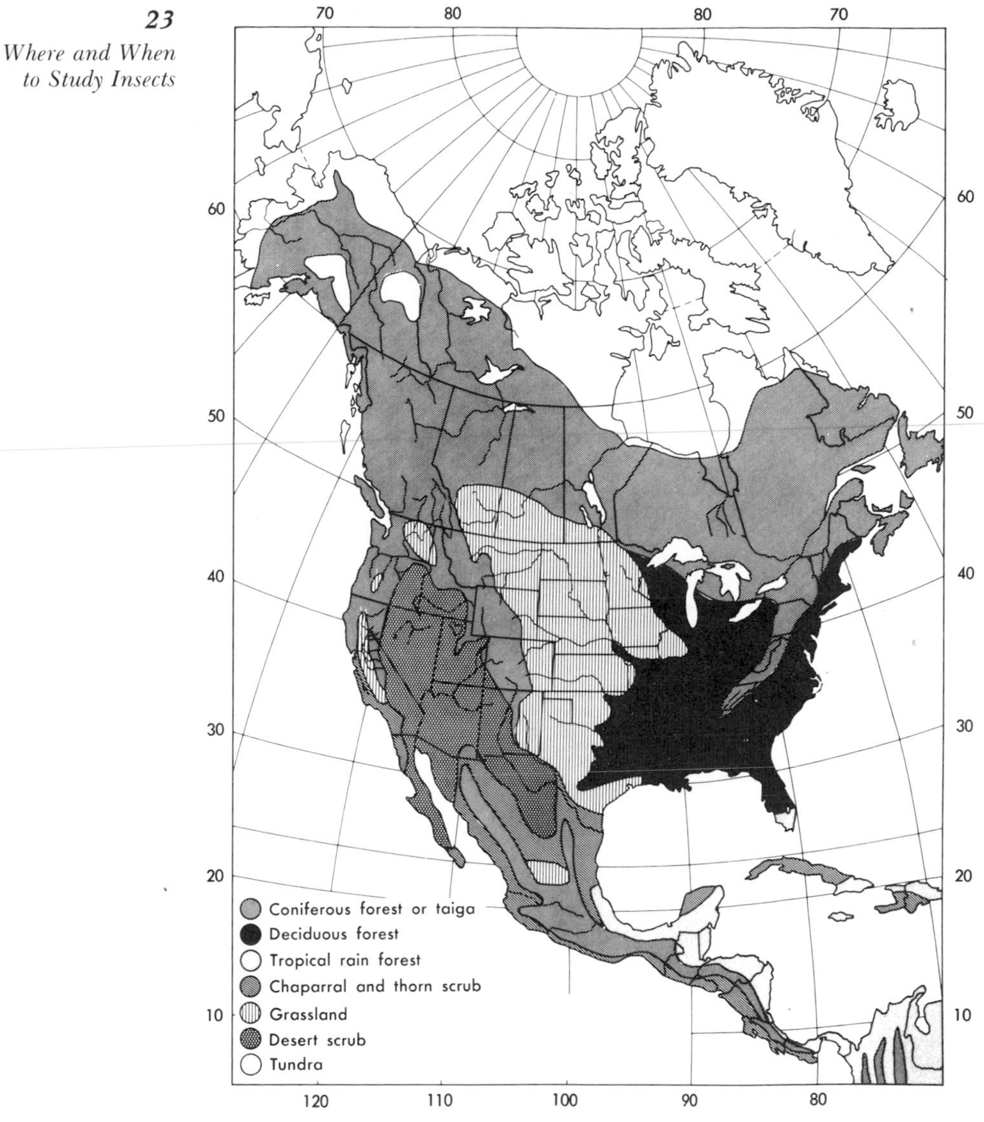

Figure 2.1. *Map showing the biomes of North America.*

are briefly described, and some of the characteristic insects are mentioned as follows.

Ocean. Only one kind of insect lives its entire life in the ocean. These are the only truly oceanic species. The bug, *Halobates sericeus* Eschscholtz (Hemiptera), is found off the coast of California and ranges across the Pacific Ocean. This species comprises the entire catalog of oceanic insects. The reason for this is unknown. Every other major group of animals is represented in the diverse marine habitats of the world. Various shore (littoral zone) insects are represented along the coasts, living in algae in the intertidal zone or on the rack along the beach. Among these insects are springtails, bristletails, bugs, caddisflies, and beetles of several families. None of these are able to live alone on the ocean—only the one species cited.

Oceanic islands. Hawaii is politically, not ecologically, a part of North America. Technically, therefore, we have as part of our American insect fauna, a great number of oceanic island species, and, fortunately, the Hawaiian chain is a splendid example of such islands. Insects, like man, are attracted to the balmy breezes and even temperatures of these beautiful isles. Even so, this doesn't seem to count as a top tourist attraction for its visitors. Island invasions are a vast subject, the topic of many volumes. The size of the island determines the variety of its insect fauna. Large islands with high mountain ranges will have good representations of the majority of insect families and most of the orders. Smaller islands, those remote from the mainland, will have a restricted fauna. Although certain mainland pest species occur, many, if not most, of the species are endemic—that is, they have evolved on the islands from some early invader shortly after the island was formed. This means that young islands have fewer species than old islands.

Tundra. Wind-swept northern polar regions, tops of certain high mountains, and a few islands of the southern hemisphere have a flora composed of cold-resistant plants in the tundra biome. These cold plains are devoid of trees, or, at most, very small, stunted trees only a few inches high. Very few insect species occur in these regions, but those that do may be extremely abundant. Mosquitoes and biting flies appear in vast hordes during the short summer. Butterflies are common. Springtails abound in the soil above the permafrost and on the sparse vegetation. Only a few species of wasps, bees, and beetles have been recorded from these barren regions.

Coniferous forests (taiga). Forests composed of coniferous trees, particularly pines, spruce, and some other species, dominate much of the northern part of North America, Europe, and Asia. These quiet regions, where only the sound of the wind in the trees and the mournful song of the doves detract from the beauty of the vast stands of a single, or at most a few, species, range from the very cold regions at the edge of the tundra, into the temperate zone, on to the tops of mountains, to the tropics. Although its insect fauna is richer than that of the tundra, the diversity of species is not great, because of the limited habitats.

Species are largely limited to wood-boring beetles, needle-feeding moth caterpillars, plant-feeding wasp caterpillars, a variety of parasitic wasp and fly larvae and their adults. Open marshy areas and lakes provide breeding grounds for flies, especially mosquitoes, while black flies inhabit the streams and rivers. Collecting in these areas requires beating vegetation, using UV lights, and erecting Malaise traps (see chap. 15). Newly cut trees attract many beetles. Even forest fires attract certain species of buprestid beetles and others. The area disturbed by logging interfaces with standing trees, providing diverse habitats, invaded by many species attempting to establish new colonies of their kind.

Deciduous (temperate) forests. Temperate forests are dominated by deciduous trees, but are interspersed with coniferous trees. This diverse plant life attracts a richer insect fauna than conifers alone. Expect to find many leaf-eating moths and butterflies, leaf-mining moths and flies, aphids, ants, and leaf beetles, along with an equal variety of herbaceous plants and a variety of birds. Complex ecosystems intertwine throughout the habitats of these beautiful forests. Lakes, ponds, swampy areas and meandering streams complement this biome and add to its rich fauna and flora. Special communities formed from the leaf litter found on the forest floor provide habitats for many very small beetles, fly larvae, and other insects. Perhaps more species live in these forests than in any other biome, except the tropical rain forest. The transition area between forest and grassland is doubly rich in insect life, because the species of the forest and those of the grassland meet as each tries to invade the other's territory.

Grassland. Great stretches of land in midcontinents are nearly devoid of trees, not because they have been cut down, but because of a lower rainfall in these areas and because of underground streams flowing through subterranean limestone deposits. Except along the margins of the few streams and along river beds, the only trees present are those planted as wind barriers. Grasses of many kinds form the dominant plant cover. Unfortunately for the naturalist, most of the natural grasslands have been converted into fields of grain, forage crops, or are overgrazed by cattle. The natural insect fauna is almost overlooked, because of the abundance of pest species that accompany civilization in these regions. Hordes of grasshoppers seem to dominate the insect fauna. As their common name suggests, these orthopterans are grass feeders. Their destruction of forage and grain plants is a natural result of a change of food plants from those previously growing "wild" in this biome to the crop species now comprising their food. But this is nothing new. The world's literature, including the Bible, gives poetic descriptions of grasshopper damage, as if these usually gentle insects were the special envoys of the devil. Other insect forms join the grasshoppers in the tasty meals provided by man's skill at cultivation. Among them are the infamous armyworms, caterpillars of noctuid moths, leafhoppers, harvester ants, and predators of many kinds, the latter attacking these hordes of phytophagous insects. In terms of individuals, the population density of grassland species is very high, but the diversity of species may not be as great as that of some of the other biomes.

Deserts. "Arid" regions of the world vary considerably; many regions look to be deserts, but, instead, they are overgrazed by cattle and other livestock. A true desert is a region with no more than 25 cm of rainfall each year. Many desert regions of the world are in "rain shadows" of mountains, while others are probably man-made. The American western arid regions are a combination of true deserts and greatly overgrazed land. The strangeness of the desert landscape with a backdrop of mountains, with the simplicity of its flora and fauna, combine to form an excitingly beautiful habitat. Large trees are absent, but a flora of uniquely specialized plants offers a diverse landscape. Characteristically, the plants have long root systems designed to reach deeply into the soil to tap subterranean collections of water. Leaves are small or absent, a feature that reduces water loss. Some typical plants include cactus, creosote bush, palo verde, ocotillo, desert willow, and, the invader from Mexico, mesquite.

An amazing variety of animals, including insects, are adapted for life in these hot areas where water is always the limiting factor. The families of insects represented in the desert regions are much the same throughout the world; only the genera and species differ. Obviously, certain groups were preadapted for life in the desert, but, as always, there are exceptions. In plants, for example, the cactus family is restricted to the New World. The Old World counterparts are species of the plant family Euphorbiaceae. Tenebrionid and buprestid beetles, syrphid flies, and many other insect families are typical. Many insects are attracted to flowers that appear shortly after the brief rains. Some deserts have a rainy period each year, usually in the summer; others may have additional winter rains, while some may go for two or more years without any rain at all.

The use of UV lights to attract insects is exceedingly productive during the rainy periods. At the same time, beating branches and searching the desert floor with a light at night helps round out a collection of desert species.

Chaparral and thorn scrub are not true desert areas, and such regions are loosely defined. These regions, usually adjacent to true desert, are characterized by the same limited, seasonal rainfall, but enough rain falls to support certain species of oaks and thorny shrubs. These areas have a fauna similar to that of the desert. Many areas throughout the west have been overgrazed by cattle to the point of causing a desert. When grazing is stopped or at least controlled, grasses take over again, interfacing the desert with true grassland and prairie.

Tropical rain forests. A tropical rain forest is, in a sense, the opposite of a coniferous forest. The former is hot and humid; the latter, cool with low humidity. More diagnostic, however, is the composition of the flora. The coniferous forest has large stands of a single species. The tropical forest is so diverse that it is difficult to find two trees of the same species within sight of each other. With such a diversity of plants, it is no surprise that animal life is equally diverse. Group for group, the number of species of most families is greater by many times in the tropical rain forest. It is here, of course, that the greatest number of yet undescribed species live.

True tropical rain forests exist in a limited portion of southern Florida. There was once a tropical rain forest in the region of Brownsville, Texas, but, alas, the trees have disappeared and there is little to remind us of this once-beautiful area. We must now go into Mexico to find tropical rain forests, and, thence, on south through many regions of Central and South America. Old World tropical rain forests are found in central Africa, Indomalaysian Region, southeastern Asia, and northern Australia. These rain forests are frequently confused with "jungles." A jungle is characteristic of highland forests in a tropical region. They have sparse vegetation in the lower story, but instead are characterized by high trees with many vines—"Tarzan" country. Rainfall is extraordinarily heavy in the lowland forests, plentiful, but less, in the higher jungles. The highest annual rainfall of the world occurs in these forests, as high as 1600 cm in India, for example.

Tropical insects, well known for their brilliant colors, fantastic shapes, and giant size, often stray into the extreme southern part of the United States. A few of the species are established there, thus making it possible for us to collect some tropical species without leaving continental United States. If possible, go farther south. There is no end to the beauty of tropical insects.

Rain forests are not restricted to the tropics. A few temperate rain forests occupy extensive area in the northern hemisphere. One, that found on the Olympic Peninsula in Washington state, is a good example, but it harbors a sparse insect fauna.

Mountains. Insects of high mountains are much more numerous than one might imagine. Some unique insects reside here. For example, the rare rock crawlers (Grylloblattodea) are found high above the tree line among the rocks at the edges of glaciers and in the sparse vegetation of these tundra-like areas. Numerous species of butterflies flit across the alpine meadows during the few weeks of summer. The typically alpine Old World genus *Parnassius* is found on the top of mountains in Alaska, British Columbia, Idaho, Colorado, and California. These beautiful insects are related to the swallowtail butterflies, but lack the "tails." Some mountain butterflies (Lycaenidae) have a complex relationship with ants and lupines, the plants on which their larvae feed. The ants tend the larvae and protect the pupae during their metamorphosis to the adult stage, all in exchange for the secretions of the butterfly larvae, which are favored by the ants. Springtails are at home in this country and form an important food source for other insects in the alpine community.

Northern isolated mountaintops seem to be extensions of the arctic biome. Mountains in desert regions are interesting "island" habitats. In Arizona, for example, several small mountain ranges are isolated by the surrounding desert. The insect fauna of these mountaintops, although not alpine or subalpine in most places, harbor many species not found at lower elevations. Sometimes the species are northern extensions of the range of species living in mountainous regions in Mexico and Central America. For example, the beetle genus *Sisenes* (Oedemeridae) is represented by fourteen species in Mexico and Central America, all in the higher elevations of their mountains. Only one of these species, *S. championi*, reaches the United States, and it is confined to the high elevations of Mount Lemon in Arizona, completely isolated by the surrounding desert.

Inland waters. The aquatic insects of our lakes, ponds, pools, streams, swamps, marshes, and bogs include all the species of whole orders (for example, Odonata and Ephemeroptera), entire families (Coleoptera: Dytiscidae), and one to several species of particular families of other orders (Lepidoptera: some pyralid moths). Almost every order of insects has some aquatic members, with their immature stages modified for aquatic life. Adult females of many species enter water for egg-laying and some even for feeding. Mayflies and stoneflies, with few exceptions, as well as the damselflies and dragonflies (Odonata) have all of their species aquatic in the naiad (immature) stage. Aquatic insects are the special subject of several books, one of which is listed in appendix II.

Aquatic habitats, by the sheer nature of various bodies of water, provide a special habitat for insect life. Insects evolved from terrestrial ancestors; hence, it is believed that their aquatic life is a secondary specialization derived from terrestrial ancestors. Factors affecting the insect fauna are the size of the body of water, depth, rate of flow, if any, temperature, rate of evaporation, chemical composition, including pollutants, and plant life. The selection process results in adaptations to the varied habitats caused by the various combinations of these factors. Many aquatic insects are plant feeders, either on living plants or detritus formed by dead plant life. Predatory insects feed on phytophagous insects, and they, in turn, are eaten by fish and other aquatic animals. The ecosystems in each of the bodies of water are exceedingly complex. This subject interests many wildlife biologists and amateur anglers.

Symbioses in insect associations. Parasites, commensals, and inquilines (see glossary) are abundant among hexapods. Many species live in close association with other species, both as ectoparasites or endoparasites, and others as plant gall formers. True mutualism is rarely clearcut. Some of these later relationships have been carefully studied. Although these habitats do not really form a separate biome in the sense used for the previously listed habitats, this subject must be considered, if we are to account for all the species. Three entire orders are ectoparasites, Mallophaga, Anoplura, and Siphonaptera, the lice and fleas. Other ectoparasites show up here and there among the other orders. Many species of Hymenoptera and Diptera are endoparasitic (parasitoids) of other insects, and some are internal parasites of birds and mammals. How these relationships came about is the subject of a newly defined study, coevolution. A special vocabulary has been devised to explain these phenomena. Table 2.2 lists and defines these terms. Details about this subject are discussed throughout this book.

Table 2.2. *Symbiotic insect associations.*

Symbiosis: Two species living in intimate and constant association during at least part of their life cycles, sometimes with mutual relationships, assuring them of reciprocal benefits, but also as parasites (Ex. Bark beetles and fungus spores; see parasitism below.)

Commensalism: Two species associating regularly, but without one living on the other. One species usually benefits from the other by protection or nutrition without the other species gaining any advantage, or both may benefit (Ex. Ants and aphids).

Inquilinism: The regular localization of one species in the nest of another as a permanent parasite, or, within the interior of another, the guest may feed on the host or their brood.

Facultative inquilines: Species often encountered as ant guests, but less often than elsewhere (Ex. Certain spiders, centipedes, etc).

Synechthry: The host is hostile to the guests (synechthrans), because the latter prey on them or on their brood (Ex. Certain beetles in ant nests).

Synoecy: An indifferent relationship exists between host and guests (synoeketes) (Ex. Certain staphylinid beetles and ant nests).

Symphily: The relationship between the host and its true guests (symphiles), i.e., those of their inquilines that they appreciate or even cultivate because of some service or substance rendered (Ex. Certain pselaphid beetles with trichome that secrete substances attractive to ants).

Episoites: One species living on the surface of another, but not at the expense of the host (Ex. Fungus and lichens living on the surface of beetles or on the fur of sloths).

Parasitism: An association of two species, one living entirely at the expense of the other, at least during part of the life cycle. Several special relationships between host and parasite are separately defined in the following three entries.

Ectoparasites: Parasites living only on the surface of the host, feeding either on the integument or sucking blood as a regular source of their food; may or may not breed on the host (Ex. Mallophaga, Anoplura, Siphonaptera, and others).

Endoparasites: Parasites that live entirely within the body of the host for at least a part of the life cycle of the parasite. The parasite generally does not kill the host and never completely consumes it (Ex. Certain fly larvae in mammals; rarely scarabaeid beetle larvae in the rectum of mammals).

Parasitoids: Parasites that live during their immature stages within the host, destroy the host through feeding, and emerge from the host as adults; distinguished from endoparasitism, because the parasite kills its host and completely consumes it. (Note: this term is not generally accepted by entomologists; it is applied, however, to the endoparasitism found among certain flies and wasps.)

Faunal regions

Several classifications of regions of North America are based on the fauna and flora. One system, devised by C. Hart Merriam and several of his colleagues many years ago (table 2.2), is widely accepted and appears in many books and articles on animals, including those on insects. The terms they used for certain geographical regions are shown on maps, but they are not easily recognized either from a study of the flora and fauna of the regions or in lists of species. Superficially, the Merriam system seems to coincide with the map of the biomes previously discussed. It differs principally because Merriam divided the eastern deciduous forest into three fauna areas. The northern Alleghenian comprises the northern part of eastern United States and extreme southeastern Canada. The upper sonoran consists mostly of eastern Nebraska east to parts of Pennsylvania, New Jersey, eastern Kansas, Missouri, east to Virginia, Tennessee, and the higher portions of Georgia, North Carolina, and South Carolina. The remaining portion of southeastern United States is lower sonoran. All of the Merriam regions, zones, and faunas are compared with the biomes discussed previously (fig. 2.1). For other classifications, see one proposed by Dice in 1943 (fig. 2.2). These many regions are defined in Dice's book. Unfortu-

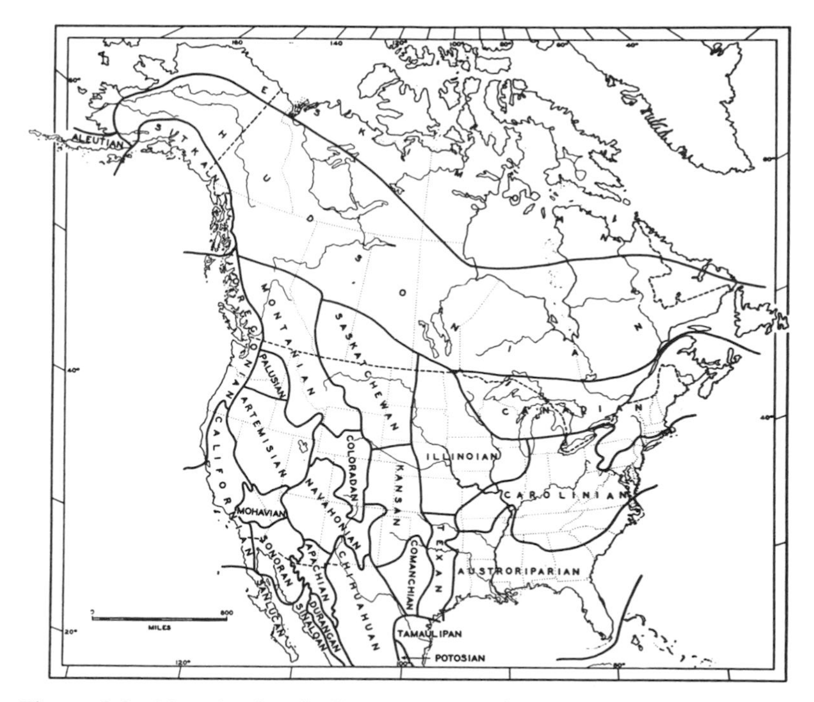

Figure 2.2. *Map showing the biotic provinces of part of North America (Veracruzian province not shown). (Courtesy of University of Michigan Press)*

nately, this classification has not been widely used by entomologists, probably because it was developed primarily for the use of mammalogists, but it is extremely useful for the description of the habitats of populations. It should be consulted before field trips. Finally, an extremely detailed map and description of the floral regions of the United States was published in 1964 by A. W. Kuechler. Each region, 116 in all, is described, giving the physiognomy, dominant plants, and samples of other plants; all 116 are shown on a large colored map.

Seasonal distribution

Unlike many birds and some mammals, few insects migrate in the true sense of the word (chap. 10). Even though you do not see them, almost all insect species are present in some stage in their own habitat throughout the entire year. They hibernate in various ways. Many species pass the winter as adults in cold regions, but this is not usually the way for most species. One may, however, see a mourning cloak butterfly leave its hiding place and, on a warm February day, even in the northern snow belt, fly around for a few hours when the sun strikes its body and raises the temperature of the wing muscle to about 15° C. Adults are present in tropical regions almost the year around. Adult activity throughout most of the U.S. and Canada is confined only to certain times of the year. One may expect to collect adults in abundance during the periods most favorable for reproduction. For most species, egg-laying takes place when food is likely to be abundant, although sometimes eggs are laid in the late summer or fall and overwinter in that stage. Variations in length of the summer season and life span of the adult must be in synchrony. Some leaf beetle adults, for example, appear early in the spring and remain active well into the fall. Others are around only for a short time.

If you are searching for a particular species, you must know the dates that the adults will first appear and when they will disappear. This may be determined by examining the locality labels on museum specimens. Then visit the area at the proper time. If you are interested in finding out what species occur in a particular area, you must collect throughout the season, because some species will emerge early, others later, and so on.

Each species has its own pattern of overwintering. Aphids usually pass the winter as eggs. Many moths and most butterflies do likewise. Some moths and butterflies overwinter as adults concealed behind loose bark of trees or under the snow beneath loose leaves. These adults seldom make good specimens for a collection, so leave them to lay their eggs in the spring. The process of overwintering causes them to lose wing scales, disrupting their color pattern. However, since the females are probably carrying fertile eggs, they may be captured for rearing purposes, if you are sure that there is food available for the caterpillars (see chap. 18). Other insects spend cold months as nymphs or larvae, protected under bark, in rotten logs, and similar hiding places. Certain butterflies and moths, various beetles, and other groups of insects spend the winter in the pupal stage. Moth collectors gather these cocoons in the fall, keep them outdoors throughout the winter, protected from birds, but exposed to the snow and rain. They are brought inside in the spring; adults emerge, and beautiful specimens are obtained.

Some species are active throughout the winter. Winter stoneflies, previously mentioned, emerge from the naiad stage during the winter, and adults may be found on snow, tree trunks, fences, and bridges near their emergence sites. Snow fleas (Collembola) are numerous on melting snow banks in early spring.

3

Getting Acquainted

"And the Lord God having formed out of the ground all the beasts of the earth, and all the fowls of the air, brought them to Adam to see what he would call them: for whatsoever Adam called any living creature, the same is its name. And Adam called all the Beasts by their names, and all fowls of the air, and all the cattle of the field."

Genesis 2:19–20

Several years ago a large oil company ran an advertisement in a popular magazine. They said, "If you don't know an intake valve from a *butterfly* valve, we have a simple offer for you." The ad showed a *moth*, not a butterfly, perched on an automobile valve.

"Didn't make much difference, though," you say. "No big deal!"

Well, how about another ad by the same company in a home magazine claiming to tell you, the reader, "how to recognize and prevent bug damage"? The chemical company branch even offered a free color folder on garden insects. The only trouble with their expensive, full-color spread was that the picture of the "codling *moth*" turned out to be the larva of the cabbage butterfly feeding on cabbage, a food no self-respecting codling moth larva would be caught eating!

Still one more example out of many, a well-known book company sent out a flyer on their authoritative "encyclopedia" on "insects," described as discussing the "vast and incredibly varied world of insects—from tiny daphnia water fleas . . . to fantastic scorpions and giant hairy spiders." Not an insect in the bunch! Well, these are the kinds of mistakes ad people and editors make, and, of course, they were soon corrected after the next mailbag full of letters from entomologists arrived.

The way to name insects, as a part of the information storage process, that is, the recording of data about insects, requires considerable knowledge. It is the name that is the key or guide to references that will give further information about a particular species of insect. For example, once specimens are identified, there are what one might call dictionaries to the available information about insects, and the name is the code word necessary to use these reference works. What these are will be described later in this chapter. Correct identification is important for several reasons:

1. Every species has its own habits, habitats, its place in time and space, and a distinct Latin name
2. Right names associate right data
3. New data must be associated with the right name for the transmission of correct data
4. To retrieve correct data, the correct name must be used
5. Correct names may retrieve incorrect data based on previously incorrect identifications

For these reasons, you must be careful how you use names. To understand the full meaning of these statements, we must examine the naming process first and then the identification process. These two steps are distinct. Failure to make a distinction between the two steps has led to many mistakes and has clouded a significant portion of our literature. Mistakes are made by all, whether amateur or professional. To correct mistakes, one must know how they happen.

Professional and amateur entomologists

In chapter 1, we discussed briefly the differing roles of amateurs and professionals and their contributions to the science. Since this book is primarily for amateurs (but only because the so-called professional has supposedly been exposed to all of this information during the process of becoming a professional, which may or may not be true), we will confine our discussion of insect systematics to those aspects of greatest interest to the amateur. Many, if not most, of us are amateur insect taxonomists, because we are paid to teach, and we do our taxonomic research in our "spare" time. Spare time is hard to apply to taxonomic work, because our work week usually is far more than forty hours. Generally, however, we think of an amateur entomologist as someone who studies insects when not at work and whose work is entirely different, in no way connected with entomology, as we pointed out in chapter 1.

Our concern here is not to distinguish between these two groups of entomologists, but to point out some of the requirements for work in taxonomic entomology. Unlike most branches of science, taxonomy has been in the past, and is today, practiced by nonprofessionals. Almost all the species described prior to 1900 were named by amateurs, usually physicians. Many of the museum curators after 1900 have been amateurs turned professional, because of their special interests. Specialization took place shortly after 1800, after the passing of the great masters, Linnaeus, Fabricius, Olivier, Say, and others. Many of the 19th century insect taxonomists specialized in moths, butterflies, and beetles. The classification of other groups developed more slowly, because they were less colorful, often smaller, and more difficult to study.

For a while after World War II, many professional insect taxonomists were able to find jobs. This, in turn, resulted in the publication of hundreds of new species each year, and many major monographs were published. In fact, nearly as much has been published on most groups of insects since 1945, as during the time from 1758, the starting date of zoological nomenclature, to 1945. However, for the past few years, job openings for taxonomists have almost ceased, and graduate students are turning to pest control, particularly biological control of insects and certain aspects of the study of the environment. This has resulted in the return of the field to amateurs and the renewed interest in entomology by amateurs.

Many new societies were started at the end of World War II. Most of these continue to publish journals, even with the scarcity of professionals. During the past decade, special interests have resulted in the publication of more than sixty newsletters or similar informal publications devoted to a single genus, a single family, or a single order. These publications take the place of the local meetings that have long been the way of life for the older societies. Less than ten of the entomological societies in the United States (see appendix I) hold meetings. The largest annual meeting is held by the Entomological Society of America (ESA), the major professional entomological society in the United States. Some of the smaller, specialized societies meet at the same time as the ESA. No society in this country requires professional status for membership, although some of the major European ones do; or, at least they require substantial publications of a professional nature, for example, the requirements for membership in the Royal Entomological Society of London. Even here, there is a subtle difference, because members of that society may be amateurs in the true sense. On the other hand, the American Register of Professional Entomologists (ARPE), affiliated with the ESA, does require professional status. New members are required to take an expensive and extensive test in the branch of entomology in which they wish to be registered. The tests are administered by those entomologists designated as capable of screening applicants. All of this is designed to elevate the field of entomology, particularly the control aspects of the field, to a professional state comparable to that of engineering and other professional "trades." Unfortunately, size of the membership in this registry is lower than might be expected. It does not have the support of many professionals, particularly those in the systematic branch of the field. It has little to offer to teachers and museum curators, for example.

How does this pertain to the amateur readers of this book? We are eager to have amateurs help with the chore of describing and cataloging the insect fauna of the world, besides having their support of the societies and publications devoted to systematic entomology. But, frankly, not all amateurs have been an asset to the field. Without the proper background, some amateurs have published descriptions of new species and genera and created many tangles in the literature. Observing, collecting, and recording data in private notebooks is one thing. Formal publication is another matter. Before amateurs publish (and professionals, as well) they must have the following qualifications:

1. A broad view of biology, which includes knowledge about the general classification of all organisms; a basic understanding of the cellular nature of organisms, including the nature of DNA, laws of heredity, and the basics of the evolution process; knowledge about the process of speciation; nature of isolating mechanisms; and basic ecology and biogeography.
2. A thorough knowledge of the International Rules of Zoological Nomenclature
3. The theory of phylogenetics, that is, how to assess characteristics and how to use these to interpret evolutionary relationships of species
4. Knowledge of the literature on the special groups of interest and the rules for cataloging
5. How to write scientific papers

Once this background is attained, it will still be necessary to enlist the aid of experienced systematists to review your work.

Throughout this book, we have introduced you to some of the background information needed. We cannot overemphasize that this book contains only some of the basics. You must go on from here. For example, we have not treated insect morphology in any detail, yet a thorough knowledge of the structure of insects is absolutely necessary before original research can be undertaken.

The remainder of this chapter is to tell you how to identify, but, before you can do this, you will need more literature. The bibliography at the end of this book will start you on the way to finding this literature.

Insect names

Every species of insect has, as we have indicated throughout the book, a Latin name composed of two words, while, unlike birds, only a few species have common English names. Common names in English follow certain rules informally set down by professional entomologists. These rules differ somewhat from ordinary English usage. All of the common names used in this book follow these rules. Briefly, a common name is usually, but not always, composed of two words, one descriptive, and one indicative of the order of insects to which the species belongs. For example, the name "house fly" is two words, which tells you that the insect is a true fly (order Diptera). "Dragonfly" is one word because these insects are not true flies. "Bed bug" is two words because it is a true bug (order Hemiptera). "Honey bee" is often written as one word, but this insect is a true bee (order Hymenoptera). "Firefly" is one word, but "fire beetle" is two words; in these cases, the insects are beetles (Coleoptera). Throughout this book numerous other cases will be found. One particularly troublesome example is "antlion," which most writers would spell "ant lion," but a moment of reflection will tell you that if it were written as two words, this would indicate that the insect is some kind of lion—a mammal—not an insect. The application of the scientific (Latin) name is controlled by the International Rules of Zoological Nomenclature (known as the "code"). This states briefly that there can be only one correct name, that is, the first name published. Further, the name must meet the criteria of publication as defined by the code. Names are Latin words, or Greek, or other language words Latinized. The rules for the formation and use of these names follow the rules of Latin grammar. Every species' name is composed of at least two parts, the generic name and the specific name. These combined comprise the scientific name.

The code specifies that the generic name must be unique in the animal kingdom. If the name is applied to more than one group of species, the more recent name must be replaced. The two names are homonyms, and only the oldest name is valid. To determine what names have been used for genera, it is necessary to refer to published lists of genera.

Species are assigned to various categories that comprise the classification of organisms. These are shown in table 3.1. To understand this table, you must be able to distinguish between a taxon (pl. taxa) and a category. A taxon is the group of individuals that are named. The category is the hierarchal rank given to these names. Thus, species are assigned to a genus, genera to a family, families to an order, orders to a class, classes to a phylum, and so on. Various subdivisions of the categories are used, principally, subspecies, subgenera, tribes, and subfamilies. Other subdivisions are used in table 3.1, showing the arrangement of the orders. Note that all family names end in "idae," all subfamily names in "inae," and all tribes in "ini." Because a species is always composed of two names, a species cannot be referred to without including the genus. Thus, we cannot refer to *sapiens* in reference to humans, but always *Homo sapiens*. Nor can we refer to *Homo* in reference to humans, because that is a generic name and has more than one species assigned to it. It is possible to refer to *Homo* sp., if it is uncertain just which species is being considered, or *Homo* spp., if two or more species are being considered. (Note that the word "species" is both singular and plural, and the only way to indicate which is meant is to use the abbreviation "sp." for singular and "spp." for plural. The word "specie" applies only to money.) Names assigned as taxa refer to real organisms, whereas the categories to which the taxa are assigned is an artificial ranking of groups of species.

Table 3.1. *Categories and taxa.*

Category	***Taxon***
Kingdom	Animalia
Phylum (pl. phyla)	Arthropoda
Class	Insecta
Order	Coleoptera
Family	Oedemer*idae*
Subfamily	Oxycop*inae*
Tribe	Oxycop*ini*
Genus (pl. genera)	*Oxycopis*
Specific name	*mcdonaldi*
Species and author	*Oxycopis mcdonaldi* (Arnett)
Common name	Florida false blister beetle

Further information about the code, its application, and the many rules to follow may be found in several recent textbooks. It is necessary to know these rules, before any original taxonomic work can be attempted.

Scientific names are pronounced much the same as English words, instead of as the classical pronunciation taught in schools. However, certain words need to be explained.

1. All syllables are pronounced. For example, Wes-tri-ni-a-na, which is accented on the next-to-the-last syllable. In words of more than two syllables derived from Latin words with a short vowel, the accent falls on the third from the last syllable, for example, Chry-so-mél-i-dae and Xy-lo-ste-a-na. Since there is no simple way to determine this vowel quality, it is best to pronounce the word in the way it is easiest to say and has the most pleasant sound, for example, Co-rýd-a-lus, instead of Co-ry-dal-us.

2. Consonants are pronounced as in English, except for the following: "ch" is always pronounced as "k"; "th" as in "*th*ough," rather than in "*th*is; "ct" as "t" in "ten." For example, Ctenophora is pronounced Te-nóf-or-a. For words beginning with "Gn," the "g" is silent; with "Mn," the "m" is silent; with "Ps," the "p" is silent; and with "pt," the "p" is silent.

3. Diphthonged "ae" and "oe" both have the sound of long "e"; other diphthongs follow English. For example, Oedemeridae is pronounced e-de-mér-i-dae.

4. Vowels are almost always pronounced as in English.

5. There is no letter "w" in the Latin alphabet, but certain Latinized English words contain it, for example, "Walesiomór-phus." In those special cases, the "w" is pronounced as in English, not given the German "v" sound.

Identification of insects

The term "character" refers to structural features used to identify and classify organisms. These features differ from group to group. Those used for insects require a special vocabulary and a basic knowledge of insect morphology. We are not able to go into details about this topic in this book. Many books are available for this purpose (see appendix II). All we intend to do here is to explain the steps to be taken when making an identification.

Characters generally are of two types. Those used for identification are selected to separate or distinguish one species from another. These are analytical characters, and they may be qualitative, as well as quantitative—number of parts, size, shape, color, features of the body wall, and so forth. For example, in the identification of beetles, we are concerned about the number of tarsal segments, the shape of antennal segments, and similar anatomical features. To identify moths, you must know how the wing veins are arranged and the color patterns of the wings as well as other anatomical features.

When classifications are made (as opposed to identifications), it is necessary to use synthetic characters, features used to group species, rather than to separate them. These involve such basic features as segmentation, neuroganglia, muscular arrangement, location of spiracles, type of mouthparts, and features of the life cycle. As more is learned about groups of insects, classifications are changed. The discovery of new features will often require the reclassification of the group having these features. Therefore, over the many years (approaching 250 years) that insects have been classified according to the system we use today (table 3.2), many changes have been made in the arrangement of the species, genera, families, and orders. More changes may be expected as these studies continue.

Table 3.2. *Insect orders.*

Superclass HEXAPODA		
Extant Orders	***Common Names of Orders***	***Extinct Orders***
Class ENTOGNATHA		
1. Collembola	Springtails	
2. Protura	Proturans	
3. Entotrophi	Entotrophs	
Class INSECTA		
Subclass Archaeognatha		
4. Microcoryphia	Bristletails	"Monura"
Subclass Dicondylia		
Superorder Thysanura		
5. Thysanura	Silverfish and allies	
Superorder Pterygota		
Division Palaeoptera		
6. Ephemeroptera	Mayflies	Paleodictyoptera
7. Odonata	Dragonflies, damselflies	Megascoptera
		Diaphanopterodea
		Protodonata
Division Neoptera		
8. Plecoptera	Stoneflies	
9. Embiidina	Webspinners	
10. Phasmatodea	Walkingsticks	

11. Orthoptera	Grasshoppers, katydids, crickets	Protorthoptera Caloneurodea
12. Grylloblattodea	Grylloblattids	
13. Dermaptera	Earwigs	
14. Dictyoptera	Mantids, cockroaches	Protelytroptera
15. Isoptera	Termites	
16. Zoraptera	Angel insects	
17. Psocoptera	Barklice, booklice	
18. Mallophaga	Chewing lice	
19. Anoplura	Sucking lice	
20. Hemiptera	Bugs	
21. Homoptera	Cicadas, leafhoppers, whiteflies, aphids, scale insects	
22. Thysanoptera	Thrips	
"Holometabola"		
23. Neuroptera	Fishflies, lacewings, snakeflies, antlions	Glosselytrodea
24. Coleoptera	Beetles, weevils, stylopids	
25. Hymenoptera	Wasps, ants, bees	
26. Trichoptera	Caddisflies	
27. Lepidoptera	Moths, skippers, butterflies	
28. Mecoptera	Scorpionflies and allies	
29. Diptera	Flies, keds	
30. Siphonaptera	Fleas	

Steps for making identifications

As we have explained, the names obtained through the identification process are the code or index words necessary for the storage and retrieval of information about a species. Obviously, the identification must be accurate to retrieve the correct information. The sort of information the collector is interested in is generally to find out where the species fits into the collection and whether this fills a gap in completing the collection.

Remember, only individuals are identified. It is impossible to identify a species, a family, or an order. This paradox is easily explained. One can identify only an individual specimen as a member of a population of a species, and that species is assigned to a family and an order. These assignments, although based on our present knowledge of the evolution of the group, are still arbitrary assignments made for ease in handling large amounts of information and for synthesis of information. As your work progresses, keep in mind that orders and families are learned through the association of these species and that it is only individuals you are studying. Any individual specimen can be studied, and different specimens by different students anywhere in the world will yield the same results. This seems ample proof of the existence of species.

To identify most insects to species, one must have a good hand lens or a low-power stereoscopic, dissecting microscope. Of course, reference books are needed for keys and illustrations. Most insects can be identified to order without much effort. All except the smallest specimens can be sorted to order with no more than a hand lens. Beyond that, things become more difficult. We have included in this chapter a simple key to the orders, but we go no further.

Keying insects to order

Look at the pictures used to illustrate the key at the end of this chapter. Compare these with the table 3.2. Notice that we are keying adults only. Immature stages are difficult to identify, and they are not considered here. To use this key, follow these instructions:

A key is composed of "couplets" (rarely "triplets"). All are numbered, and each half describes alternative insect features. Look at the specimen to be identified and decide which of the two alternatives corresponds to its features. The number that appears at the end of the alternative that fits the specimen indicates the next couplet to be consulted. Repeat the process for this new couplet and for all subsequent couplets, until an alternative is reached that has an order name at the end. The process of identification is then complete. Note that from the second couplet on, a number in parentheses appears after the couplet number, e.g., 6(3). The number in parentheses indicates the previous couplet read, thereby making it easy to backtrack, in case an error was made or one is suspected.

Making accurate identifications

Many reference works are needed to make identifications. Any means possible may be used—comparing specimens with labelled material, asking experts, and so on—if they lead to accuracy. Actually, most identifications are made in two stages, sometimes at different times. The first stage is the preliminary identification, useful in sorting and storing specimens until the final identification can be made. The second stage is the final verification of identification. This is the chore of the specialists.

Keys for identification. Keys are shortcuts giving the pertinent information from the descriptions of the groups keyed. Because of their brief nature, they do not cover every feature, and, therefore, they do not always work. Names obtained from keys must be checked further. Keys and descriptions are found in a variety of reference manuals and in scientific articles and monographs.

The use of field guides. Field guides to insects are now available for many insects found in the United States and Canada. Compared to those available for birds, insect guides are extremely limited, but, of course, you must remember that, instead of dealing with a few hundred species of birds, there are about 90,000 species of insects in the United States and Canada. New guides are being produced as more people become interested in insects. It is expected that before much longer, guides will be available for all species (see appendix II for suggested guides). Many of the new guides illustrate some of the common species in full color, and these colors rival those of birds. With field books, one can identify insects to order, to many families, and to a few of the strikingly distinct species. They are never detailed enough for accurate identifications, mainly because they do not tell you what is missing. Separate books for some orders of insects, particularly beetles, butterflies, moths, true bugs, grasshoppers and their allies are among those now available with others being prepared.

Regional manuals. England and some European countries have regional manuals for the identification of insects. These manuals are similar to the "floras" used by botanists. With the aid of these books, it is possible to identify all species of the region covered. Few works of this nature cover the insects of the United States and Canada. Some have been published, and others are planned. Manuals comparable to those available for plant identification are badly needed. We hope that more will be written soon.

Revisions. In place of regional identification manuals, most American workers depend on family revisions. A revision is a comprehensive work that describes all species of a group for a specified area. For example, the work may treat the species of an entire family for Canada and the United States. Families with a great number of species are divided into subfamilies and tribes, making it possible to produce revisions in parts by covering one tribe or subfamily at a time. Unfortunately, there is no comprehensive plan to cover all insects; these revisions appear at irregular intervals. They are published in scattered journals and serials, which makes it impossible to gather a complete collection of these works for all insects. Certain insects are more popular than others. These are usually the large species, often brightly colored, making them simpler to identify.

Revisions of these groups are easier to find. Revisions include keys to all taxa covered. The keys refer to complete descriptions of the species in the body of the work. Often the features used in the keys are illustrated, particularly features of the male genitalia useful for species recognition. The descriptions are followed by lists of specimens examined, repeating the label data. This provides the user with information about the distribution of the species and data on their ecology and biology.

Monographs. A large comprehensive revision is termed a monograph. It differs little from a revision, except that it presumably includes all known information about the group covered. Of course, these are extremely useful to the specialist. It becomes his "bible" for the group, if it is current. But, as with all systematic works, it is not permanent. Sooner or later, a new monograph is needed and will be written.

Collections for comparison. Perhaps the easiest way to identify specimens is by comparing the unknown with accurately identified specimens housed in large collections or collections of specialists. Of course, to examine specimens in these collections, one must have learned all the rudiments of systematic entomology to assure the proper handling of the specimens and to know how to make these comparisons. Amateurs must be well advanced, before they can expect to receive permission to study the research specimens housed in these collections.

Identification by specialists. Equally helpful as a way to get accurate descriptions is the aid given by specialists. All systematic entomologists eventually specialize in certain groups of insects, those they find most interesting, and usually those in most need of study. They are able to make accurate identifications of the species of their speciality. Although some specialists resent being asked to make identifications for amateurs, most realize that material submitted by these collectors is likely to contain new data helpful in learning the distribution of the species, and, sometimes, new species are collected, and these are recognized and described by the specialist.

It is important that the amateur follow the rules for submitting specimens for identification as given in table 3.3. You can learn who the specialists are by studying the recent literature on the group and by consulting recent editions of the *Naturalists' Directory and Almanac (International)* in your local library.

Table 3.3. *Rules for submitting specimens for identification.*

1. Do not assume that specialists desire more specimens for their own sake; usually they are interested only in establishing new records or additional species. The specialists are doing you a favor by spending valuable time to make identifications. At consultant wages, this could be expensive. Therefore, a request for identification is equivalent to asking for a substantial donation.

2. Never ship specimens for identification without making prior, detailed arrangements with the specialist, including:
   a. return shipping costs
   b. time that specimens are to be returned
   c. specimens to be retained as identification fee
   d. disposition of type specimens, if any

3. All specimens submitted for identification must be:
   a. properly prepared and preserved
   b. provided with exact locality data
   c. sorted to group

4. Never send bulk collections or unsorted collections with a request that the material be picked over to find things of interest.

5. Remember that a refusal to make identifications of large masses of material does not necessarily indicate a lack of interest, but sometimes just a lack of time or facilities.

6. Under certain circumstances, a specialist may request a fee for providing an identification. This practice has become nearly universal for court cases, commercial activities, such as pest control operations, or environmental impact studies. Make sure that both you and the specialist agree on a fee, if any, before the identifications are made.

Becoming a specialist. Once your general collection is arranged by sorting your specimens to order, most of them to family, and you have made preliminary identifications of the common species, you may ask: "What next?" You can go on enlarging your general collection, but soon you will see that it is impossible for you to identify and arrange all the insects you can collect in your own area. A general collection with these preliminary identifications is a valuable teaching aid. A collection of this kind should be present in every county. It might be at the local high school, boy and girl scout troops, or at a community college or local historical museum. With such a collection, others can find answers to their questions, and you can become the local "bug" expert.

For your own satisfaction, you probably will begin to specialize without even thinking about it. You will find that you concentrate your collecting more on one group and less on others. You are now becoming a specialist, and you might dig deeper into the literature to find out more about your special group. Once you do this, you will want to know how to make a verified identification. When you know how to do this, you will probably begin to make identifications for others and realize the importance of the rules given in table 3.3.

Verification of identifications. Catalogs of the species of many groups—all groups, actually—are available, but these may be old and out-of-date. Regardless, the first step toward making accurate, verified identifications is to check a catalog of the species of your group. If none is available, you may have to make your own. The reason for this is that all catalogs are immediately out-of-date when they are published, because, even while the work is in press, new species are being discovered, and new information is being published. Your special catalog should be on file cards, looseleaf notebook sheets, or in your computer; the last is the most versatile and easiest way to keep the catalog in shape. A catalog is a working tool to be added to as details are discovered by the study of specimens or uncovered in the literature. You can extract from this developing catalog a working checklist for quick reference, and you may even wish to duplicate this for use by your colleagues. As you add information to this catalog, you will find that you will have a descriptive catalog. This can be the source for data to use in a revision of certain groups or can even be developed into a monograph. It all depends on how far you want to go into this endless field of study.

Comparing specimens with original descriptions. As your work progresses, you will outgrow the keys you are using and find it necessary to compare your specimens with the original descriptions of the species or, in extremely old species, with the descriptions and notes in subsequent revisions. If there is already a comprehensive revision or monograph of the group, you will still make the comparisons as a recheck of the work of previous students (none of us is infallible). Also, you must do this when the next step, the study of type specimens, is undertaken.

Comparing specimens with types. Every species has a type specimen, the actual specimen on which the original description was written. This specimen, the holotype, represents the populations of the species. It is the standard measure for the name. Do not make the mistake of thinking that this specimen is "typical" of the species. It may not be. For example, suppose the type specimen of *Homo sapiens*, humans, was Napoleon. Was he typical of our species? Certainly not, but he certainly can be considered to be a sample of one population of humans. The holotype, then, is the starting point for any accurate, positive, identification. Whenever we put a name label on a specimen, we are, in effect, saying that the specimen agrees with the holotype, meaning that we feel that the labelled specimen belongs to the same species as the species the holotype represents. Look around you and compare all those people you know with Napoleon and decide whether you, your friends, and those you see on TV and read about in magazines and books are members of the same species as the one to which Napoleon belonged. The holotype is not typical, but represents the species. Compare the human holotype with the holotype of the gorilla. Any problem in separating the two species? Certainly not. The same is true when you compare other species with other holotypes. It may not always be easy to do this, but the principle is sound.

Not all holotypes are available for study. The major museums of the world harbor these specimens, but many of the types of the older species have been lost or destroyed. In their place are designated neotypes, a new specimen selected to represent the lost type. Or, in other cases, the old species never had a single specimen designed as the holotype. The species is represented by a series of two or more specimens. Subsequent revisors have selected one of these specimens as a lectotype, a substitute for the undesignated holotype. Note the difference between a neotype and a lectotype. The first is any specimen selected, because there is no longer an original type series, that is, any of the specimens used for the writing of the original description. In the other case, the lectotype is selected from the original series and, therefore, is exactly the same as a holotype. Neotypes could be, and sometimes are, through errors made by the designation of the neotype, a different species from the one described originally. Of course, this could happen in the selection of a lectotype, but this would be rare.

Before you go any further, it is necessary to know where these types are located and that there is a type for every named species. This applies to all those species considered to be synonyms, that is, a second or later name given to a species in error. We have left this until now, because, once you understand what a type is, you can then understand that two species can be declared synonyms *only* if you, the revisor, believe that the type specimens of the two species are the same, that they both represent the same species. This means, then, that the valid name is the one described at the earliest date, as required by the law of priority of the International Rules of Zoological Nomenclature.

Type specimens are found in museums, university collections, and in some private collections. The last are usually large collections, which may be considered to be equivalent to a museum collection. Therefore, in order to compare your specimens with types, you must visit these collections and study the specimens. Obviously, these valuable specimens are not available for study to anyone other than recognized specialists in the group; professional or amateur, they must be able to demonstrate that they are qualified to make these studies.

Once the comparisons are made, the specimens compared with the types are labelled to show that this has been done, and a description of these specimens is made to show how they agree, and any difference noted. This specimen then becomes an "unofficial" substitute for the type. Mistakes can be made. These "compared-with-type" specimens have no official status. They are, however, convenient working tools. These certainly are reasonable substitutes to use when it is impossible to study types.

Type species. Considerable misunderstanding about genera is prevalent among nontaxonomists (and even some taxonomists). A genus is a subjective unit consisting of one or more species. No rules have been devised to guide the systematist in the selection or rejection of genera. One seems to develop a "generic sense" and work from there. The number of species in a genus for the insects of the United States and Canada averages just under seven. This varies from 4.8 species per genus in the Lepidoptera, to 7.5 in the Coleoptera, 7.6 in the Diptera, and 8.9 in the Trichoptera. The less popular and less known orders seem to have more species per genus—an indication of the subjective nature of the taxon. Therefore, to provide objectivity to the nomenclature of the category, the code requires each genus to have a type species. A type species of a genus is, in a way, easy to determine. First, the species that becomes the type must be a species that was originally included in the genus when the generic name was first proposed. Second, the species chosen must be one that is validly described. Note that there is no requirement that the species be "typical" of the genus. To think otherwise leads to considerable confusion about the generic names of insects.

Before any sound taxonomic work can be done, the type species of all genera of the group must be determined. Three ways, and only three ways, account for the designation of type species. First, if only one species is assigned to the genus when the generic name is proposed, and that species is a validly described one, that species, and only that species, is the type. This type designation is termed "monobasic."

Second, if more than one validly described species is included in the original proposal of the generic name, and the author selects one of these, that species becomes the type by "original designation." These two ways of type selection are simple and, generally, clearcut. Occasionally, there may be some question about the author's intentions, but these may be objectively settled by applying the International Rules.

The third method, "by subsequent designation," has certain inherent problems. The later designation, that is, designation of a type species for genera that are neither monobasic nor have author-designated types, follows the code's requirement of priority. The first valid subsequent designation must stand. But it is not always easy to find the earliest designation, and, sometimes, the manner in which an author designated the type is obscure. Early describers did not have rules of nomenclature to follow. Therefore, even if they realized the need for this objective tool, they didn't always designate a type species that meets present-day requirements of the code. For example, they may have designated as type a species that was not included in the original proposal of the generic name. Many other technical details must be considered when making a list of genera and their types.

Note that we did not say that the genus must be described, but only that the name must be proposed with at least one validly described species included. This means that generic names can be validated by listing them in a catalog of species. This was often done in the older literature; sometimes, however, the species listed had not yet been described. To determine the validity of the generic name, a thorough study of all old literature must be made, not to find generic characters, but to arrive at the correct type species. Actually, it makes no difference what is said about the genus. The name rests on the characters of specimens of the described type species. This is what makes it so important to have the right species. This species may have none of the features originally attributed to the genus. That makes no difference; it is up to the systematist to study the type species and groups of other species according to these findings. The first, the discovery of the correct type species, is a purely arbitrary, objective, search of the literature. The second, once the first is properly done, is a matter of interpretation of the zoology of the problem. These two *must* be kept separate. On the proper understanding of this concept rests the permanence of names. Most of the name-changing that occurs today is the result of earlier mistakes in the association of species in a genus. This we have discussed in more detail later in this chapter.

We need to emphasize that to do work of this nature, two things are required. First, an interest in these details and a willingness to see them through. Second, the availability of a major entomological library. No single library will contain all literature needed to make these studies, but one must start at a leading institution.

How misidentifications are made. We cannot conclude this brief discussion of naming insects without pointing out how errors of identification are made. Obviously, when these errors occur, the wrong information is retrieved about the species. Misidentifications all stem from a lack of information about the type specimen of the species. If the description of the species in a revision or monograph is not of that represented by the type, a misidentification results. To correct this error, new information must be published that applies the correct name to the species. This results in a name change in the literature. When this happens to a pest species, considerable confusion results, because the literature on the species uses two or more names. Sometimes the correction of the name is due to the strict application of the code's "law of priority," that is, the rule that the oldest name must be used. This could cause so much confusion that the rule is discarded. Instead, a ruling is made to conserve the name. The correct name is not applied, and the familiar name is retained. This is, of course, a monument to an error. It satisfies the current generation; later generations would not mind using the correct name, because they never would have had to learn the incorrect name. To tie together the old literature with the new (that which uses the correct name) is a simple matter of cataloging.

Other times, misidentifications are made simply because of the identifier's lack of knowledge. That person believes the identification is correct, but it is not, although this fact may go undetected until a revision or monograph is published and used to correct the identification in collections.

One way many mistakes are made is by the so-called "spot identification" method. Often it is assumed by nontaxonomists that a taxonomist should be able to recognize and give names to any specimen collected in the local area, say a state or a county, where the taxonomist is employed. These individuals are required to name all species brought in by applied entomologists. As a result, a small reference collection is assembled for "quick naming" of these species. Almost never is this correctly done. Even if the specimens in the reference collection are correctly named, it is not possible for any single individual to identify correctly all species as required in such cases. To do so results in many mistakes. Fortunately for the taxonomist, most of these errors are of little importance at the time, except, of course, for the mar on the integrity of the individual involved. Once in a while, a species of great significance slips by, and the error later comes to rest on the back of the systematist forced to work under these pressures. The only defense is to refuse to make an identification under pressure. Give the requesting person the name as close as possible, but do not give a specific name without checking the specimen carefully. It is *always* best to ask for assistance of a specialist, if there is any possible doubt, regardless of the immediate need for a name. There is *no* need for an incorrect name.

World revisions. Even though we decry the lack of regional manuals in entomology, we realize that the ultimate need is for world revisions. At the generic level, this is frequently possible. As the number of species increases, the addition of genera to form tribes, subfamilies, and even families, the task of doing the necessary research and writing a world monograph greatly increases.

A study of the life works of various entomologists sheds some light on the magnitude of this task. For example, the late Professor Charles P. Alexander of the University of Massachusetts was able to describe over 10,000 species of crane flies, but it took a lifetime of squeezing in this work while teaching and doing administrative work as a dean. Unfortunately, the time needed to describe these species did not leave time to monograph them. His long life (1889–1981) ran out before the task was completed.

Maurice Pic (1866–1957), a French naturalist, described about 20,000 species of beetles. Again, he too was unable to monograph any of the groups he studied, and his descriptions were extremely short. It has taken two entomologists, Professor E. G. Linsley and John A. Chemsak, since before 1961 to monograph the longhorned beetles of the United States and Canada, and they are not yet finished.

How names change. Probably the most annoying thing facing the amateur insect taxonomist is coping with name changes. We get letters from readers of our various books telling us that we used such and such a name, but another book (older, of course) uses a different name. How come? It seems fitting, therefore, to explain this to you with the hope that you will understand, and learn, not only to expect name changes, but to accept them as an advance in the field and to be happy that these improvements are being made.

The reasons for name changes are finite. These are listed here, somewhat in the order of their frequency.

1. *Synonymy.* A species is described and the name is used in, say, the United States. The new species was treated along with other known species that occur in the United States. As more work is completed, it is discovered that the species in question was described many years ago in Mexico. The two species are synonyms, and, henceforth, the older of the two names must be used. We have already pointed out the difficulty of completing world revisions, the only way to prevent this kind of synonymy. Even then, there is no guarantee. Sometimes old species are assigned to the wrong family, and synonyms are discovered when the species is correctly assigned.

2. *Misidentifications.* Failure to study the holotypes of all species of a group may result in the misidentification of a species, and the consequent name change is necessary when this error is discovered.

3. *Type species of genera.* Much research has been published without determining the type species of the genera involved. Then, when these are worked out, it is learned that different generic names have to be used for the species treated, which, of course, results in name changes. Frequently, names applied to species of economic importance are involved, much to the annoyance of the applied entomologists.

4. *Wrong generic assignment.* Name changes related to the type species of genera are misassignment of species. It often happens that species are originally assigned to genera "borrowed" from the Old World literature. It was natural for the early workers in this country to use keys to the genera written for European species. Our species were then assigned to these genera. As work progressed, it became apparent that our species are not often congeneric with the Old World species, congeneric in the sense that our species agree with the type species of the Old World genus. New generic names are needed for our species, resulting, sometimes, in dozens of name changes. The most annoying thing that happens, then, is that the Old World generic name is dropped from our catalogs and checklists, making it difficult to determine the new assignments of the species or to relate the old literature to these new names. The only way to avoid this is to learn how to catalog carefully, an art not generally practiced. The mundane chore of cataloging is not widely appreciated by entomological colleagues, some of whom fail to appreciate fully the need for lists of type species and complete catalogs.

5. *Homonyms.* Because of the vast number of names for insects and the slowness of cataloging them, the same name may be used twice in the same genus, or the same generic name may

be used twice in the animal kingdom. This is, of course, not permitted by the "Rules." Frequently, the specific name "bicolor" is used, because it is an easy name to construct, as well as being descriptive of many insects. Or, sometimes genera are synonymized, bringing together two names in the same genus, also against the rules of nomenclature. Whenever homonyms are detected, the youngest name must be replaced by a new name.

6. *Species complexes.* As systematic work progresses, particularly when biological studies are made, it is often learned that what was supposed to be a single species is actually a complex of species. A notable example of this is the complex of species formerly known as "the tree cricket," *Oecanthis niveus*, which turns out to be a complex of several species distinguished best by differences in their songs. Making name changes to comply with findings such as these keeps collection curators busy!

Identification by computers

The sudden availability of comparatively inexpensive computers has made possible new methods for systematists. We have already mentioned the help computers provide in keeping catalogs up to date. (See under "Verification of identifications," p. 48.) Several major catalogs are now stored in computerized files. Although updates are theoretically possible on a daily basis, as yet, none of these are operating to their full potential, usually due to administrative delays.

Several computer-assisted insect identification software packages and files are now available. One advantage of these keys is the ease with which the file can be searched. As with all data that are computer managed, the identifications obtained from the file are no better than the data added. At this writing, no file is suitable for research-quality identifications. Used by students as study guides, these programs work, given their parameters. This is not to say that programs can't be written suitable for research. In fact, one has been announced, called "Autokey," created by Professor Ronald A. Hellenthal of Notre Dame University. It is clear, however, that it will take computers with greater capacities than those currently available as personal computers to run these programs.

The future promises to be exciting for systematics as these tools are developed and refined. Perhaps this will help to make up for the current lack of personnel.

Table 3.4. *Key to the orders of insects.*

NOTE: Insects have three pairs of legs; other invertebrates are sometimes confused with insects. Spiders, scorpions, whipscorpions, ticks, and mites cannot be identified to order with the following key.

1. Common insect orders 2
   Less common insect orders (those that are rarely collected) 28

2(1). Without wings at all stages 3
   Without wings at certain stages (immatures)........... 10
   Always with wings as adults.. 12

3(2). Free living, not attached to another animal for feeding 4
   Ectoparasites, attached to another animal for feeding, or in nests; not free living.. 7

4(3). Abdomen with three tails 5
   Abdomen without tails.. 6

5(4). Compound eyes always present, large, and contiguous, or nearly so, middorsally; body strongly convex dorsally (bristletails) (fig. 3.1)......... MICROCORYPHIA
   Compound eyes absent or small and widely separated dorsally (silverfish and their allies) (fig. 3.2).. THYSANURA

6(4). Antennae moderate to long; abdomen often with a jumping mechanism (springtails and their allies) (fig. 3.3) .... COLLEMBOLA
   Antennae very short, often only one-segmented; legs absent; mouthparts sucking, but hidden from view by oval or somewhat elongate, scalelike body (scale insects only) (fig. 3.4) .... HOMOPTERA

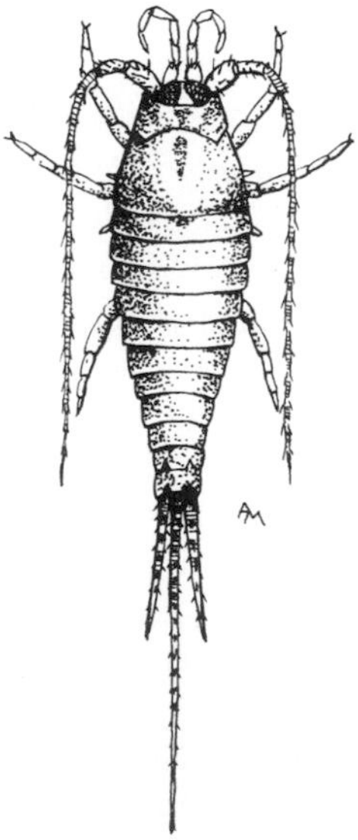

Figure 3.1.
Bristletail (Microcoryphia).

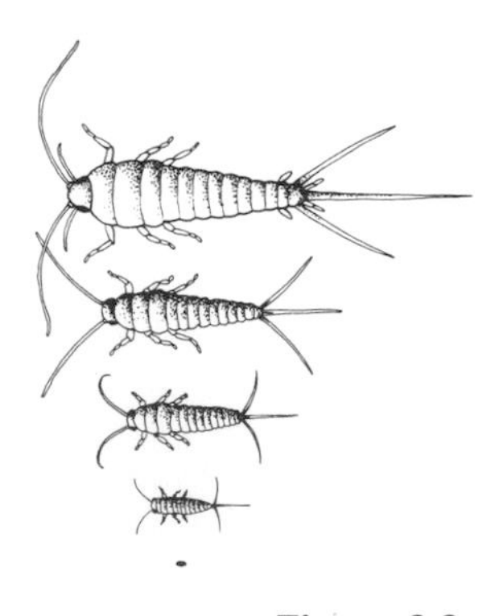

Figure 3.2.
Silverfish (Thysanura).

Figure 3.3.
Springtail (Collembola).

Figure 3.4. *Scale insect (Homoptera). (Courtesy of M. C. Williams)*

NOTE: Many other adults of free-living insects lack wings, but these all resemble the winged adults of the same order and, generally, are not confused with these forms. For example, ants usually are wingless, although there are winged stages, but they would never be confused with any of the preceding. Therefore, if the specimen does not agree with any of the preceding couplets, continue reading this key.

7(3). Body flattened 8
Body compressed; brownish or red, with jumping hind legs; on mammals and birds, in their nests, or rarely on man (fleas) (fig. 3.5) .... SIPHONAPTERA

8(7). Antennae long, easily visible; body color dark red or brown (bed bugs) (fig. 3.6) .... HEMIPTERA
Antennae very short; body whitish 9

Figure 3.5.
Flea (Siphonaptera).

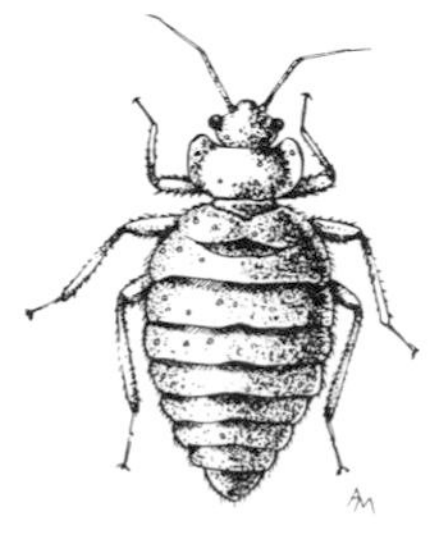

Figure 3.6.
Bed bug (Hemiptera).

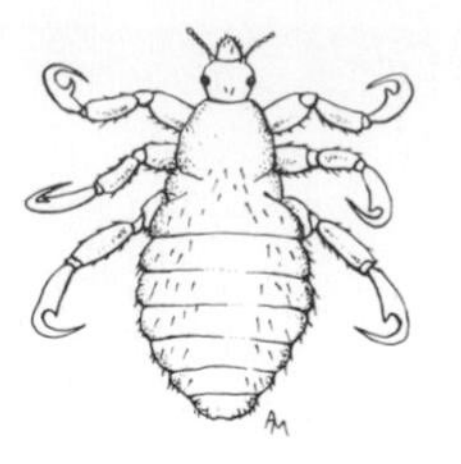

Figure 3.7.
Body louse (Anoplura).

9(8). Sucking mouthparts; live on mammals and man (lice) (fig. 3.7) .... ANOPLURA
Chewing mouthparts; on mammals and birds and in their nests (chewing lice) (fig. 3.8) MALLOPHAGA

10(2). Nymphs, naiads, larvae, pupae, and eggs do not have wings, but are the immature stages of winged adults of the orders that follow; some resemble adults; others (caterpillars, grubs, maggots, and pupae) are very different from the adult stage
Adults without wings, except for certain stages, and certain species 11

11(10). Resemble wasps; black, brown, or yellow; abdomen with one or two humped segments near thorax (ants) (fig. 3.9) .... HYMENOPTERA

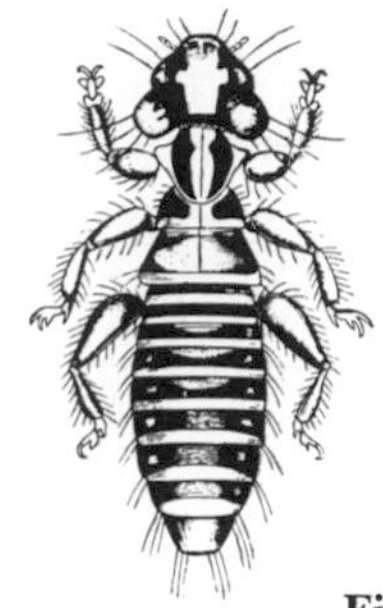

Figure 3.8.
Biting louse (Mallophaga).

Figure 3.9.
Ant (Hymenoptera).

Resemble lice; body large; head black or yellow; body white to light brown; live in wood or soil, not exposed to the sun (termites) (fig. 3.10) .... ISOPTERA

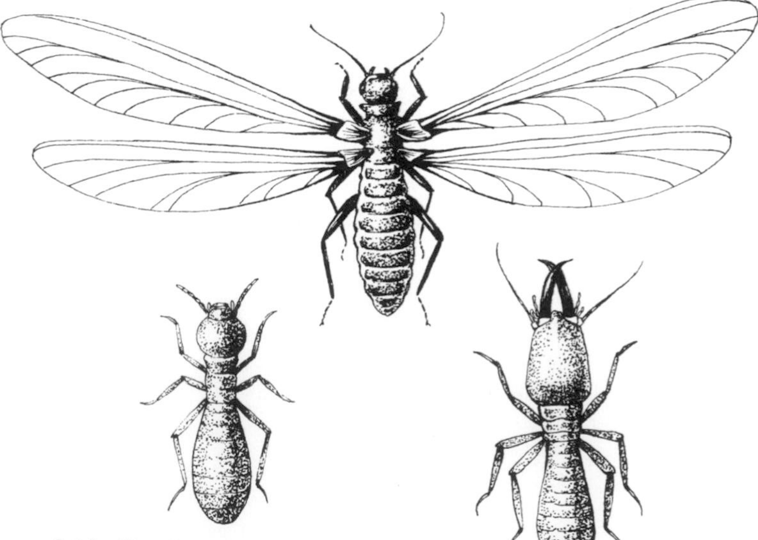

Figure 3.10. *Termite worker (Isoptera).*

Resemble sticks or twigs; body brownish, greatly elongated; legs long, very slender (walkingsticks) (fig. 3.11) .... PHASMATODEA

12(2). With one pair of wings only, hind pair reduced to small, clublike organs or absent (true flies) (fig. 3.12) .... DIPTERA

With two pairs of wings 13

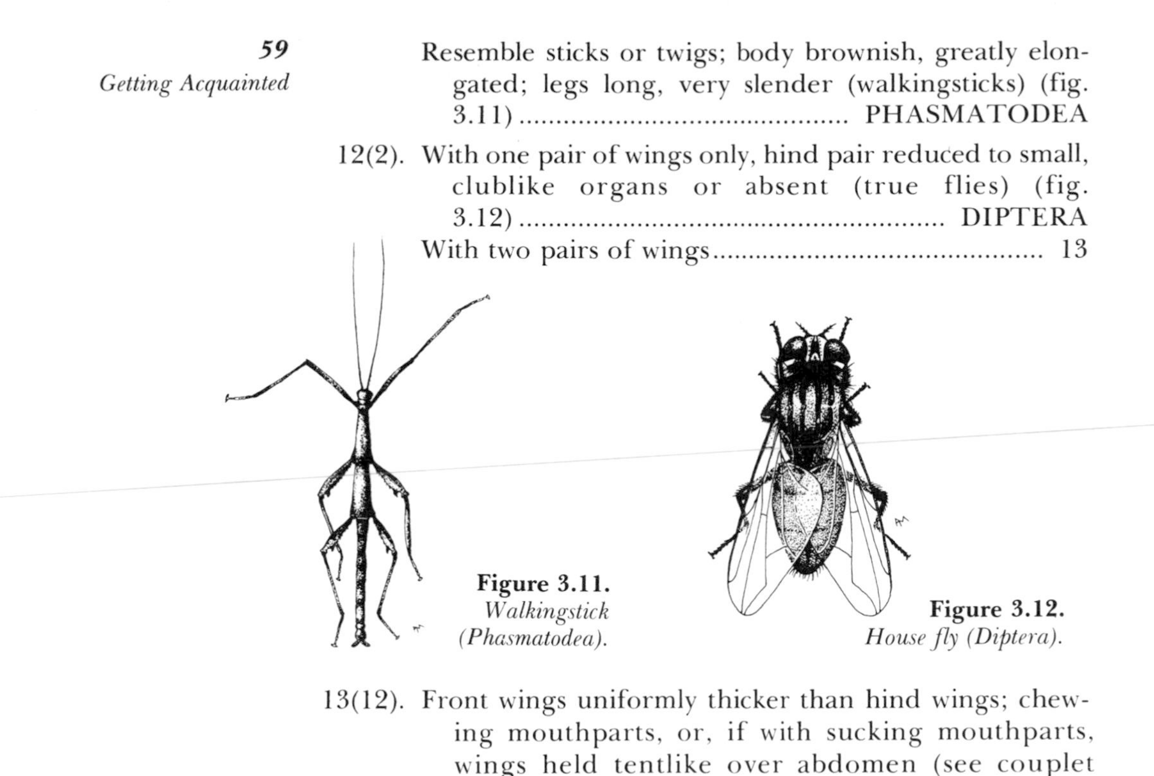

Figure 3.11.
Walkingstick (Phasmatodea).

Figure 3.12.
House fly (Diptera).

13(12). Front wings uniformly thicker than hind wings; chewing mouthparts, or, if with sucking mouthparts, wings held tentlike over abdomen (see couplet 16) 14

Front wings with basal portion thicker, apical portion thin, membranous, transparent; mouthparts sucking; body broad, or narrow and elongate; wings cross apically when held in repose; usually with a large triangular portion between base of front wings (true bugs) (fig. 3.13) .... HEMIPTERA

Both pairs of wings membranous, usually transparent, unless pigmented or covered with scales 20

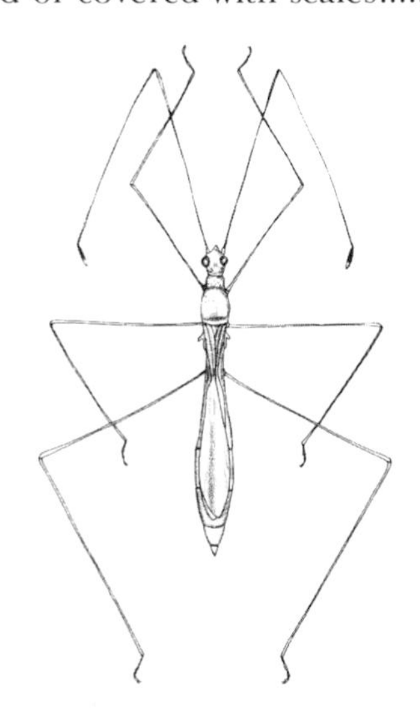

Figure 3.13.
Stilt bug (Hemiptera).

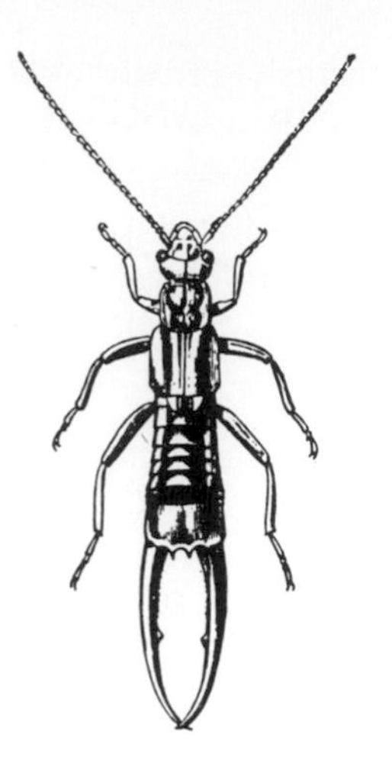

Figure 3.14.
Earwig (Dermaptera).

14(13). Abdomen with terminal, forceplike pinchers (earwigs) (fig. 3.14).. DERMAPTERA
Abdomen without terminal pinchers........................ 15

15(14). Hind wings fold beneath front wing covers (elytra), which usually extend to the end of the abdomen, but, if short, abdomen lacks terminal pinchers (beetles, weevils, and stylopids) (fig. 3.15)..... COLEOPTERA
Hind wings, if folded, the folds are lengthwise only... 16

16(15). Hind wings fold lengthwise under leathery front wings.. 17
Hind wings not folded; both pairs of wings held tentlike over abdomen; long sucking mouthparts projecting

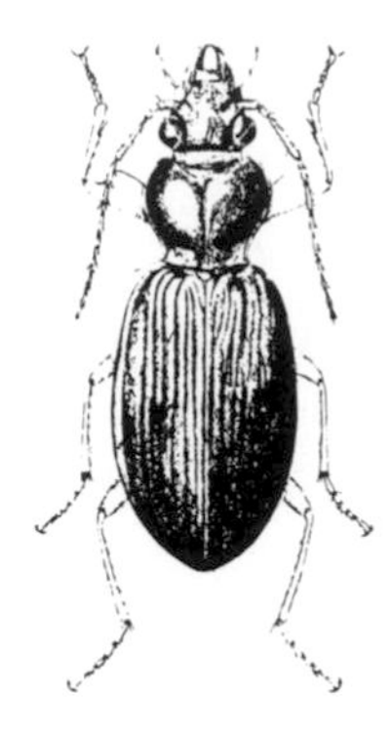

Figure 3.15. *Ground beetle (Coleoptera).*

backward between front legs (cicadas, leafhoppers, aphids, and allies) (fig. 3.16)........... HOMOPTERA

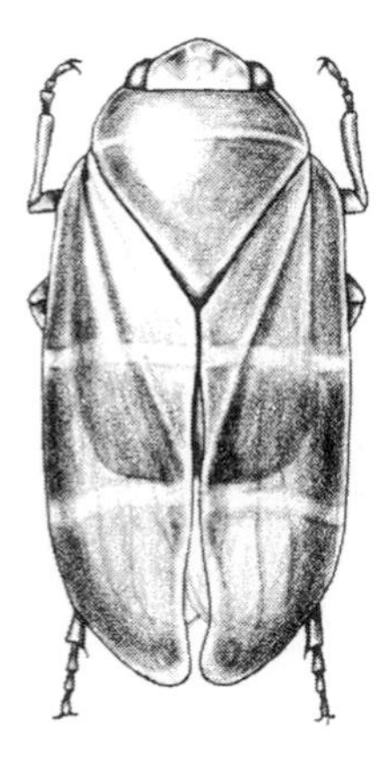

Figure 3.16.
Leafhopper (Homoptera).

17(16). Hind legs enlarged for jumping (grasshoppers, crickets, and katydids) (fig. 3.17)................. ORTHOPTERA
Hind legs not enlarged for jumping, or, if so (rarely in some groups) not long and projecting beyond abdomen 18

Figure 3.17. *Grasshopper (Orthoptera).*

18(17). Front legs enlarged, used for grasping prey (praying mantids) (fig. 3.18) .... DICTYOPTERA
Front legs not enlarged for grasping prey 19

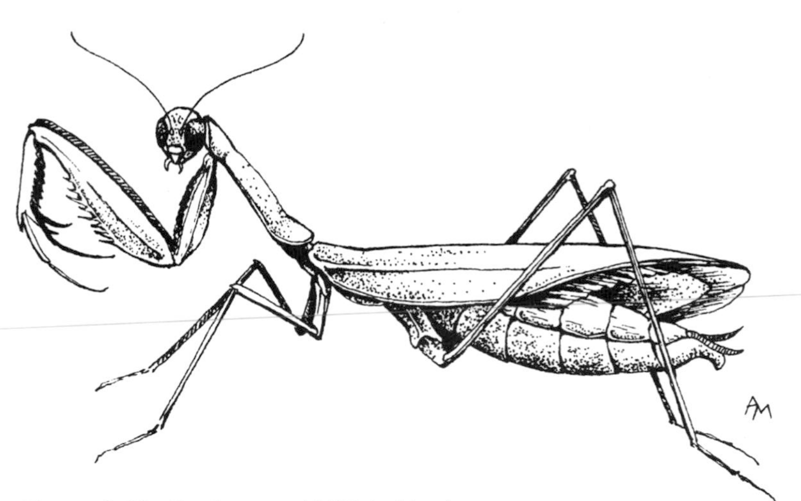

Figure 3.18. *Preying mantid (Dictyoptera).*

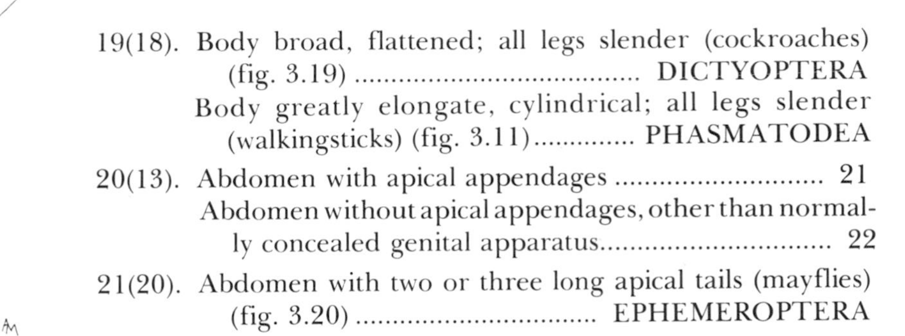

19(18). Body broad, flattened; all legs slender (cockroaches) (fig. 3.19) .... DICTYOPTERA
Body greatly elongate, cylindrical; all legs slender (walkingsticks) (fig. 3.11) .... PHASMATODEA

20(13). Abdomen with apical appendages 21
Abdomen without apical appendages, other than normally concealed genital apparatus................................ 22

21(20). Abdomen with two or three long apical tails (mayflies) (fig. 3.20) .... EPHEMEROPTERA

Figure 3.19. *Cockroach (Dictyoptera).*

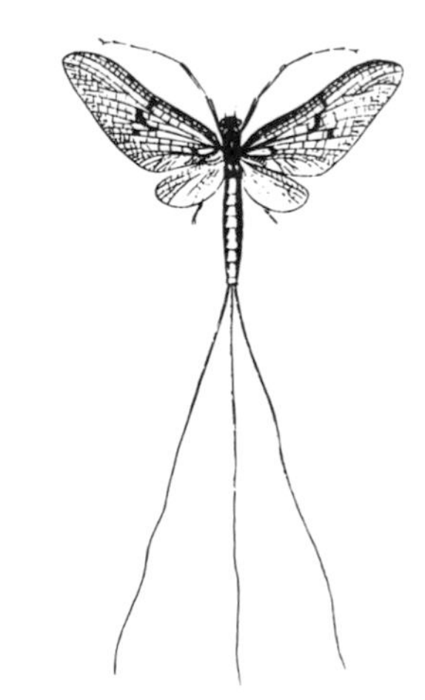

Figure 3.20. *Mayfly (Ephemeroptera).*

Abdomen with very short tails, or tails absent; body broad, more or less flattened; legs widely separate at base; body brown to black (stoneflies) (fig. 3.21) .... PLECOPTERA

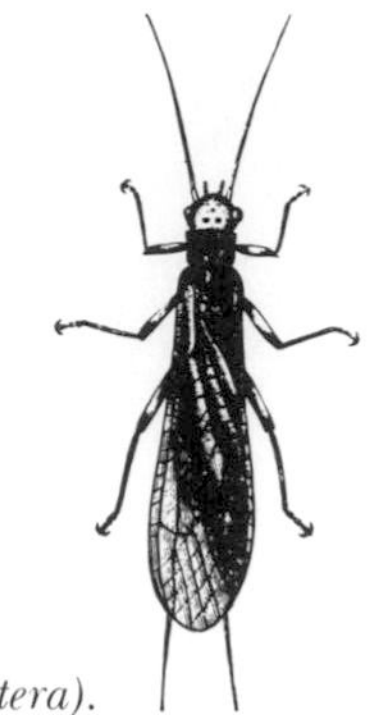

Figure 3.21. *Stonefly (Plecoptera).*

22(20). Wings with many short cross veins between the main veins; wings long and narrow 23
Wing with few cross veins 24

23(22). Antennae moderate in length, or long; wings held rooflike over and close to abdomen; chewing mouthparts (neuropterans) (fig. 3.22) .... NEUROPTERA
Antennae very short; wings held out at sides or closed well above abdomen; abdomen without tails, but with small exposed genital structures; sometimes body brightly colored; wings sometimes with spots (dragonflies and damselflies) (fig. 3.23) ODONATA

Figure 3.22. *Lacewing (Neuroptera).*

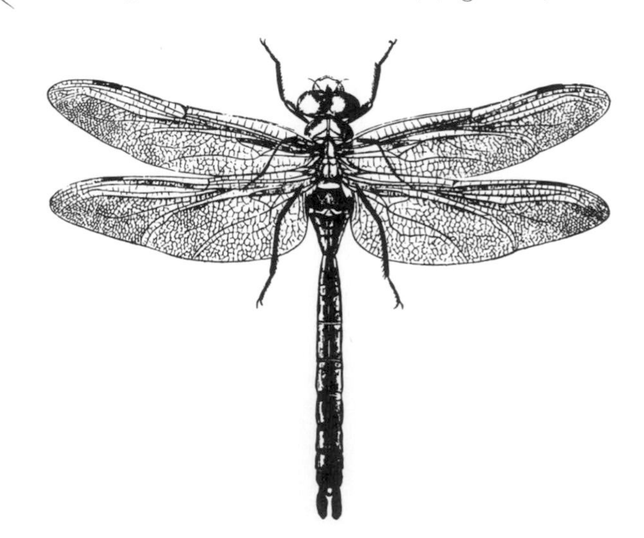

Figure 3.23. *Dragonfly (Odonata).*

24(22). Wings with scales or fine hairlike setae 25
Wings without scales or noticeable hairlike setae..... 26

25(24). Wings covered with fine hairlike setae, otherwise resemble moths; wings held tentlike over abdomen; antennae very long and slender; usually near bodies of water (caddisflies) (fig. 3.24)... TRICHOPTERA

Figure 3.24. *Caddisfly (Trichoptera).*

Wings covered with scales, rarely some with transparent areas, held tentlike over abdomen; body covered with scales; mouthparts, if functional, composed of a long, coiled tube (some very rare species with chewing mouthparts); (moths, skippers, and butterflies) (fig. 3.25)...LEPIDOPTERA

Figure 3.25. *Moth (Lepidoptera).*

26(24). Wings with long fringes of hairlike setae on each side; very small insects (thrips) (fig. 3.26) THYSANOPTERA
Wings without fringe of setae 27

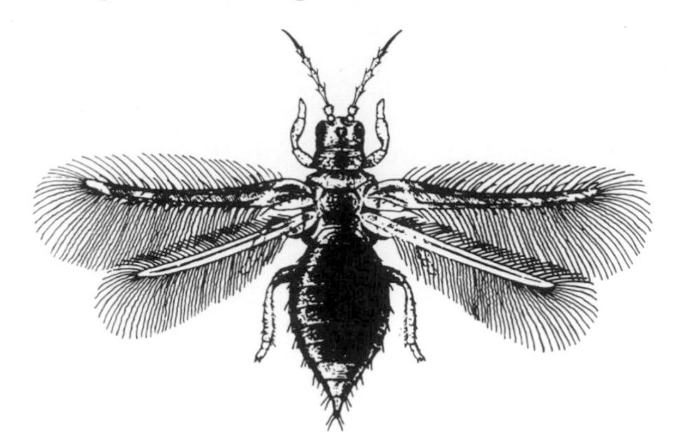

Figure 3.26. *Thrips (Thysanaptera).*

27(26). Hind wings shorter than front wings (ants) (fig. 3.27) .... HYMENOPTERA
Wings of equal length (swarming termites) ....ISOPTERA

28(1). Wingless.. 29
With two pairs of wings.. 35

29(28). With two tails or a pair of forceps at end of abdomen 30
Without abdominal tails or forceps.......................... 31

30(29). Tails long or forceps present; legs short; antennae stout (entotrophs) (fig. 3.28)................... ENTOTROPHI

Figure 3.27.
Bumble bee (Hymenoptera).

Figure 3.28.
Entotroph (Entotrophi).

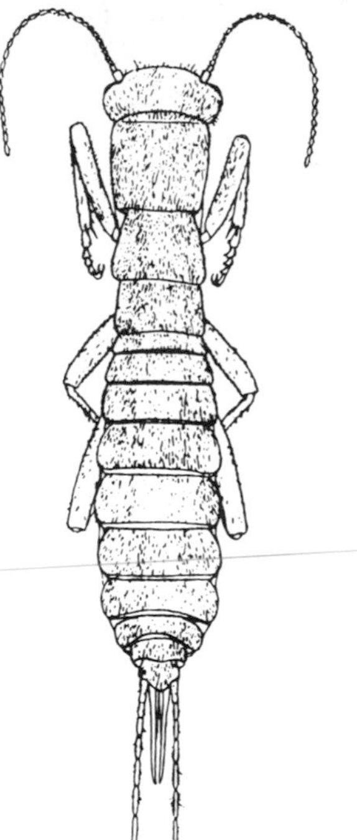

Figure 3.29. *Rockcrawler (Grylloblattodea).*

Tails short; legs long; head large; antenna long and slender (rockcrawlers) (fig. 3.29).....GRYLLOBLATTODEA

Tails very short, one-segmented; body compact (angel insects) .... ZORAPTERA

31(30). Apparently with only two pairs of legs, the front pair projecting forward and functioning as antenna; true antennae absent; very small (proturans) (fig. 3.30) .... PROTURA

With antennae.. 32

Figure 3.30. *Proturan (Protura).*

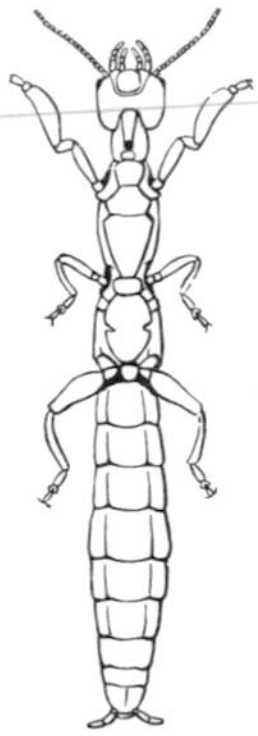

Figure 3.31. *Webspinner, female (Embiidina).*

32(31). Front tarsi enlarged, produce silk; live in silk galleries (webspinners) (fig. 3.31) .... EMBIIDINA

Front tarsi not enlarged, not producing silk 33

33(32). Ectoparasites of mammals; body flattened; somewhat resemble bedbugs, on sheep; some with or without a single pair of short wings, on bats (sheep keds and bat bugs) (fig. 3.32) .... DIPTERA

Free living 34

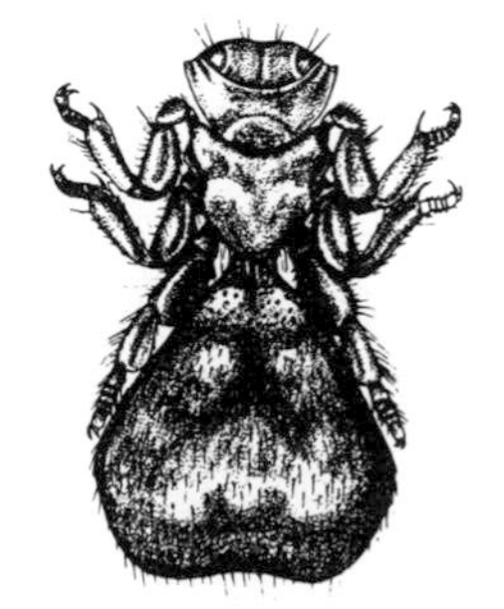

 Figure 3.32. *Sheep ked (Diptera).*

34(33). Long antennae present; head short; very small (barklice and booklice (fig. 3.33) .... PSOCOPTERA

Antennae moderately long; head elongated to form a snout (scorpion flies and allies) (fig. 3.34) .... MECOPTERA

35(28). Wings present in males only, long, narrow, capable of flexing forward; front legs with enlarged first tarsal segment; live in silk galleries (webspinners) (fig. 3.35)... EMBIIDINA

Wings not flexible, short; front tarsi not enlarged .. 36

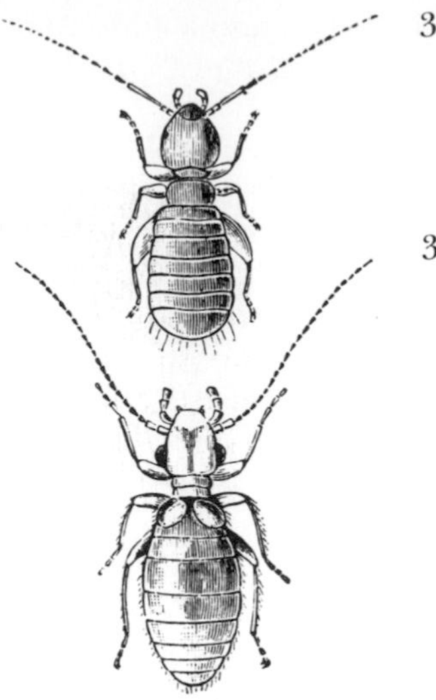

Figure 3.33. *Barklouse (Psocoptera).*

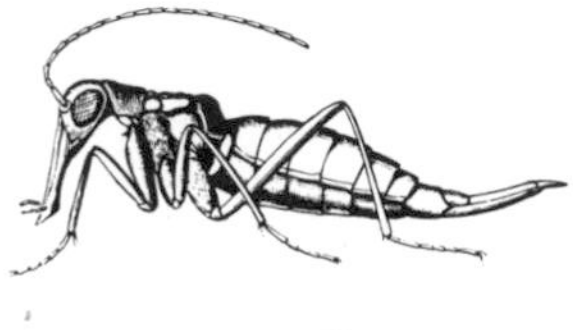

Figure 3.34. Boreus *sp. (Mecoptera).*

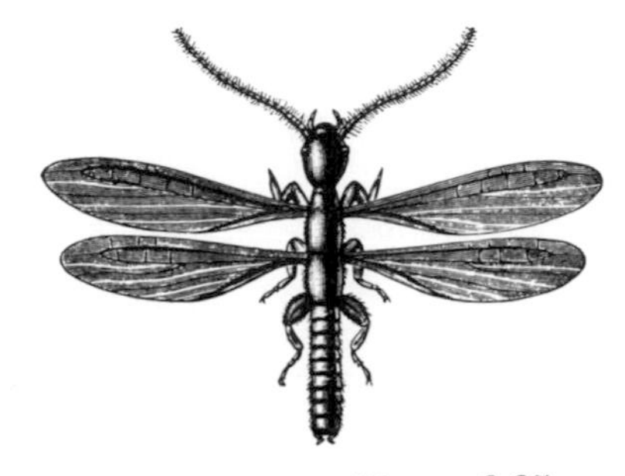

Figure 3.35. *Webspinner, male (Embiidina).*

36(35). Wing veins few (angle insects) (fig. 3.36) ZOROPTERA

Wing veins numerous; head elongate forming a snout; abdomen of some males with enlarged, scorpionlike apex; (scorpionflies and allies) (fig. 3.37) MECOPTERA

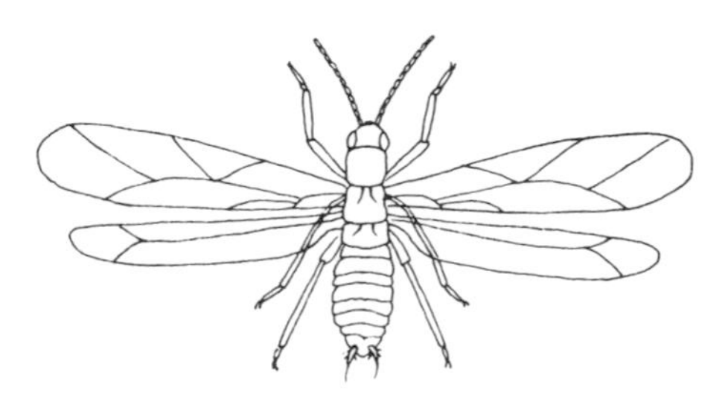

Figure 3.36. *Angel insect (Zoraptera).*

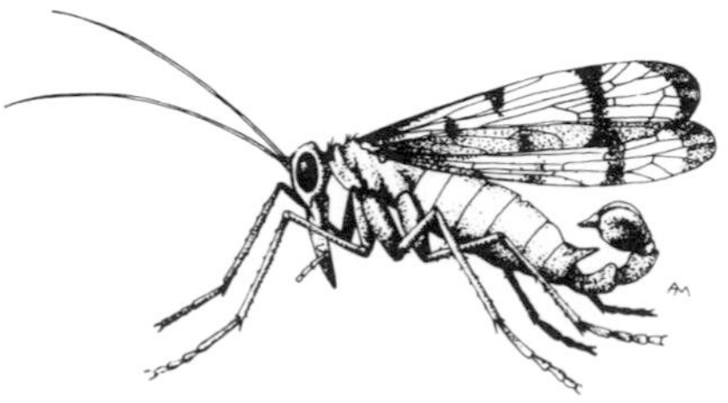

Figure 3.37. *Scorpionfly (Mecoptera).*

4

Family Trees

"A caterpillar 'tis an emblem of the Devil
in its crawling walk,
And bears his colors in its changing hues."

Martin Luther (1483–1546)

Imagine a boll weevil 1,000 years old. A few years ago an archeologist found a specimen in a cave in Mexico. The weevil, perfectly preserved in a cotton boll, identified by Rose Ella Warner, then at the U.S. Department of Agriculture, indicates that agricultural pests are not new to mankind. Indians certainly lost portions of their cotton crop to the ravages of these insects. The weevil gives us information about life in the past, and might be considered a fossil in one sense. As fossils go, however, it is hardly a fossil at all. Insect remains found in tombs among the buried dead, animal bones covered by landslides, and bodies frozen in ice are only a few hundred to several thousand years old and are not really fossils.

The nature of insect fossils

Some insects drop into tar pits after succumbing to the oil fumes, and, altered by the chemicals in the tar, remain perfectly preserved for many thousands of years. This fits the definition of a fossil—animal and plant remains *altered* and *preserved* from past *geological ages*.

Unfortunately, even though many mammal remains have been taken from tar pits, tar pit insects have been generally overlooked. Strangely, even though many kinds of insects are attracted to decaying animal and plant remains, they are rarely collected along with bones and other large, hard structures, for example, turtle shells. Undoubtedly, they are disregarded by vertebrate fossil collectors.

You certainly have seen insects stuck in the sticky material used on fly paper. Trees frequently exude a similar material known as gum. Tree gum, especially that of coniferous trees, traps insects now and did so in the far distant past—in fact, as far back as 40 million years BP (before present). Insects stuck in these tree secretions remain in much the same condition as if they were mounted on modern microscope slides (fig. 4.1). The preservative, Canada balsam, used by histologists, contains the same chemicals as amber, essentially fossil Canada balsam.

Amber insects, among the most perfectly preserved of all fossils, give entomologists valuable data on the age of insect species and their prehistoric distribution. Insects decay exactly the

Figure 4.1. *Fossil fly traped in amber from the Dominican Republic. (Courtesy of Robert E. Woodruff)*

same as other animal and plant bodies, except that their hard, sclerotized body covering resists digestive enzymes secreted by detritus bacteria and fungi. The internal contents of freshly killed insects will decay, unless preserved in alcohol or rapidly dried. Because of the hard exoskeleton, they change very little externally stored in collections, but their internal organs dry and distort.

When an insect dies, falls into water, mud, or is covered by a landslide, its body contents usually decay, but the exoskeleton remains, and, covered with silt or soil, with the passing of time becomes a fossil. Various changes in the earth's crust subject these relicts to great pressure, eventually producing a rock embedding a fossil insect. But fossils vary. Water may drip through the silt-turned rock. Chemical action changes insect exoskeletons by substitution of one skeletal molecule for one rock molecule until transformation into rock is completed. Fossils retain the form of the original insect, but pressure often flattens them, precluding the possibility of learning about the interior of these specimens.

Mummified insects do not always lose their soft tissue by decay. If internal features are preserved, as occurs under certain conditions, careful treatment will enable an entomologist to "dissect" these specimens to reveal the ancient internal structure. Dr. George O. Poinar, Jr., an entomologist at the University of California, Berkeley, examining a fly embedded in Baltic amber approximately forty million years old, noted perfectly preserved internal tissues, possible through rapid dehydration and chemical preservation in amber. The insect's tissues were so well preserved that cell nuclei and other, even smaller, intracellular structures were visible on photos taken with an electron microscope. The tissues took methylene blue stain, indicating original tissues were present and had not been replaced by mineral chemicals. So perfectly preserved was this specimen that DNA, the genetic coding material of living cells, could be extracted and studied, a rare and significant find.

Sometimes specimens are trapped in sediment before internal structures decay beyond recognition. Not long ago specimens found in concretions, stones which may be cracked open to reveal the fossil inside, were carefully studied at the Smithsonian Institution in Washington. By "dissecting" these specimens with a fine brush dipped into hydrochloric acid to dissolve away the calcium carbonate rock, and slowly working carefully with the rock in view under a dissecting microscope, the fine structures came into view. Even details of the nervous system were revealed.

Concretions are found in shale, rock composed of calcium carbonate or iron carbonate, sometimes in sandstone, and in coal. Deposits of rock-forming chemicals encrust the fossil specimen much the same as batter covers a shrimp, forming concentric layers around the fossil, until all trace of the original shape is obscured. Often concretions are discovered near a shale outcropping or unearthed during coal strip mining. Crack these open with a geologist's hammer. Most of them will contain a fossil plant or animal, and, sometimes, with luck, an insect.

How fossils are formed

Fossils are roughly classified according to the geological process occurring at the time of entrapment. Considerable variation takes place, resulting in a broad range of remains to study. Most are mere traces, but a few are perfectly preserved specimens, as we have seen previously. Let's review the principal kinds of fossils collected by considering how they are formed.

Freezing. Animals are often found in permanent ice, as in glaciers or polar ice caps. Few insects occur in ice, but in Montana large numbers of grasshoppers met unscheduled death during migration over the night mountains. The exact age of the grasshoppers is unknown. Entrapment could have started during the ice age as long ago as 10,000 years, but probably these particular grasshoppers are not more than about 4,000 years old. Layer after layer of perfectly preserved grasshoppers, exposed when summer sun melts the leading edge of these glaciers, may be seen by visiting Grasshopper Glacier, near the northeastern corner of Yellowstone National Park.

Drying. Preservation of soft parts through the action of hot, dry wind in desert regions often produces mummified vertebrate remains dating back thousands of years. The desiccated carcass does not mold or decay because detritus bacteria and fungi cannot live under extremely dry conditions. Sometimes remains are buried in sand and later exposed in a preserved state. Dried insect fossils have not been reported. However, pinned insects certainly are dried mummies, because they are preserved by desiccation—they only lack age to qualify as fossils.

Preservation of hard parts. Many animal parts do not decay because the detritus organisms lack enzymes to dissolve and digest those parts. Bones and shells are the best examples of this. The actual remains, barely changed from the living condition, are trapped in stone. Parts exposed to light, rain, and air pollutants will disintegrate, but once covered by soil without oxygen, they remain unchanged for millions of years. Under certain conditions, for example, in peat bogs, soft parts, pickled in tannic acid by soaking in bog water, are as well preserved as any study specimen used in biology classes. Many insects are known from such material, some reported by Dr. Philip S. Callahan, a U.S. Department of Agriculture entomologist, who describes insect fragments found in peat bogs at Dartmoor, England. Additional material, collected in peat approximately eight million years old from sites along the northern coast of North America, is listed by John V. Matthews, Jr., of the Geological Survey of Canada.

Petrification. A process of replacement by minerals of organic chemicals comprising an organism's body, petrification turns to stone the hard parts of a plant or animal. A similar process, termed permineralization, affects porous tissue. Most preserved prehistoric insects became fossils in this manner.

Carbonization. Great pressure reduces complex carbon compounds to simple, inert carbon. This is the state of insects found in amber and in stone. Their bodies are reduced to a caste of carbon. Many "perfect" insect specimens have been described from material of this kind.

Various fossil forms besides the preceding are known, particularly tracks, burrows, borings, and fossil excrement (coprolites), but these cannot be directly attributed to a particular species or even an insect.

This array of specimens, regardless of their form of preservation, gives entomologists some ideas about the way insects evolved. Remember, however, fossilization is rare. No one knows what percentage of the species living in the past has been fossilized.

Most insect fossils are wing prints (fig. 4.2), useful, but often incomplete. Much of our knowledge about the relationship of insect orders depends on deduction from a study of the characters of living insects instead of lines of descent being revealed to us by a series of fossil remains. Nevertheless, fossil remains of insects not only show us the age of various orders, which leads us to, and helps us construct, phylogenies (phylogenetic "trees" or charts) depicting the probable lines of evolution of the insect orders, but also tell us much about the relationships within orders. As can be seen on table 3.2, there are few entirely extinct insect orders.

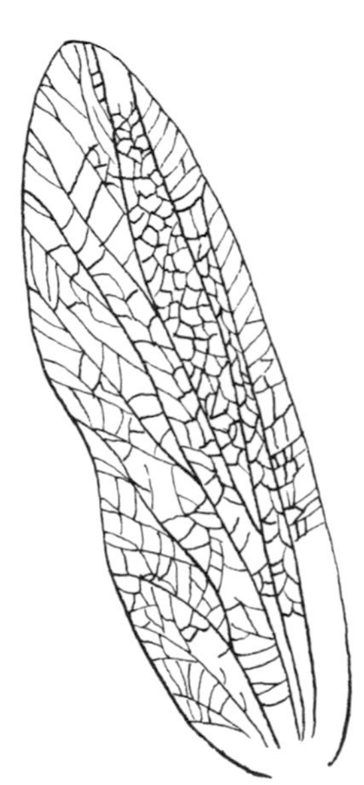

Figure 4.2. *The most ancient Carboniferous winged insect (redrawn from Laurentiaux, 1950).*

Earliest known insects

Examine the geological time scale (table 4.1). This was developed by geologists with the aid of biologists. Major physical changes of the earth's surface occurred at nearly regular intervals during the long period of the earth's existence. This chart

Table 4.1

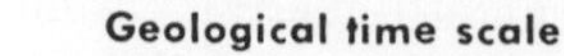
Geological time scale

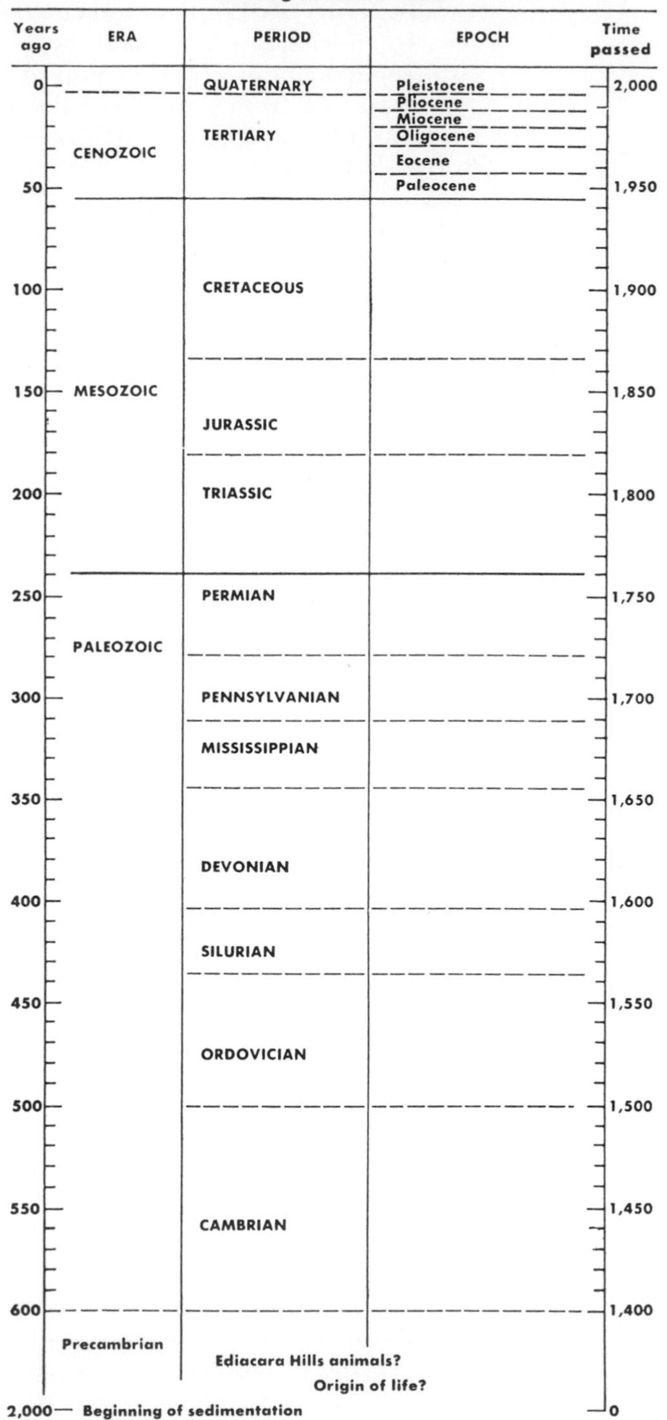
Years ago
ERA
PERIOD
EPOCH
Time passed
0
QUATERNARY
Pleistocene
Pliocene
Miocene
Oligocene
Eocene
Paleocene
2,000
TERTIARY
CENOZOIC
50
1,950
100
CRETACEOUS
1,900
150
MESOZOIC
1,850
JURASSIC
200
TRIASSIC
1,800
250
PERMIAN
1,750
PALEOZOIC
300
PENNSYLVANIAN
1,700
MISSISSIPPIAN
350
1,650
DEVONIAN
400
1,600
SILURIAN
450
1,550
ORDOVICIAN
500
1,500
550
1,450
CAMBRIAN
600
1,400
Precambrian
Ediacara Hills animals?
Origin of life?
2,000— Beginning of sedimentation
0

shows only a part of the time since the earth was formed, specifically, only the last two billion years. Exactly when life arose is not yet known, but there is evidence that it may be as long ago as 3.1 billion years.

Insects are comparatively recent inhabitants of the land, apparently first showing up during the Devonian as descendants of a land arthropod. They first appear in the fossil record some 350 to 400 million years ago, at least 100 million years after the emergence of plants and animals from the sea. The earliest known insect fossil appears to be the Devonian *Rhyniella praecursor* found in Scotland. It is assigned to the primitive wingless order Collembola (the springtails and allies).

The fossil record of other Hexapoda supports the existing classification. It tells us virtually nothing about the relationships of the various extant orders; that is, it is still impossible, with minor exceptions, to show how one order evolved from another. True, there are at least nine extinct orders (many more according to some specialists) of insects, but these have been placed in the existing classification according to the living order that they most closely resemble. In other words, we classify the fossils as extinct members of living orders. We do not separate the living orders according to evidence supplied by the fossils. Many fossils, however, are set aside within the order as separate suborders and sometimes families. Most of the fossils of the "higher" insects (see Division Neoptera on table 3.2) clearly fall within existing families, and sometimes even extant genera. Fossils of the Pleistocene epoch are often the same as living species.

Notwithstanding, all evidence pieced together by paleontologists makes a logical sequence showing evolutionary relationships of insect orders. This is to say that the features used to separate and group living orders fit exactly the evidence shown in fossil specimens.

Where did insects come from?

We might ask how insects evolved from their precursors. If we look for an insect prototype, we must find an animal that is segmented, one in which the body has lost its unity of structure and has become segmented by being divided into repetitive parts. Look at an earthworm and note the rings; these are what we call segments. The worm is the beginning of the prototype we are searching for, but segmentation is not enough. The insect prototype must have legs. Therefore, our next candidate is the centipede we see in garden soil or beneath a fallen log. These "hundred-legged worms" are just that—worms with legs—and probably indicate the next stage in the development of the ancestral insects: segmented legs as found in all members of the phylum Arthropoda to which the insects are assigned.

One significant feature common to all insects and related classes of the phylum is the presence of a coelom (true body cavity), which has been reduced to a small sack surrounding the gonads. The hemocoel has taken the place of the coelom. As a result, there is no vascular system, because the entire body cavity that remains is filled with body fluids. A single tube forming a dorsal heart is all that remains of the vascular tubes in insects. Open at each end, this tube pulsates to circulate the body fluids that comprise the "blood" or hemolymph of insects. This material flows freely between the sheets of muscle tissue carrying only food and metabolic wastes to and from the tissues.

This distinctive system requires entirely new respiratory structures. Gases, including oxygen and carbon dioxide, are carried into and out of the body by tubes called tracheae. The ports of entry, the spiracles, open along the sides of the body, and are usually regulated by muscle-controlled valves. All except the most primitive Hexapoda have this form of respiration (in a few higher insects tracheae are lost because of the small size of the body and their mode of living); hence, this feature should be present in any prototype we try to find (fig. 4.3).

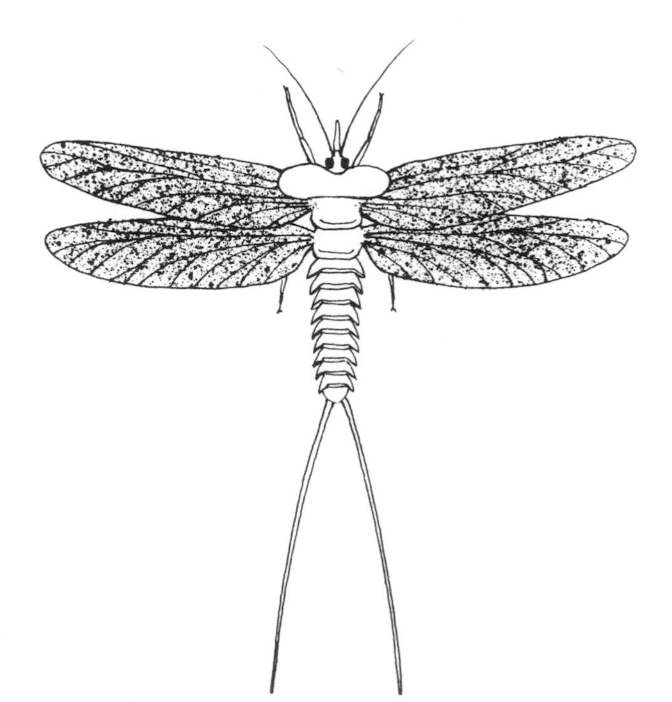

Figure 4.3. *Representative of an early extinct order, the Palaeodictyoptera, a cockroach relative (redrawn from Laurentiaux, 1952).*

We can't stop here, however, because all known insects have a modified body plan which divides the body into three major parts, or tagmata: head, thorax, and abdomen. Thus, several segments of the body were fused during their evolution into a portion containing the mouthparts and sense organs, including eyes (usually), and a single pair of antennae. The head, therefore, is a compact sensory organ equipped with food-handling parts—mandibles and other "tools" used for eating.

The second tagma is the thorax, the center of locomotion, always composed of three segments, almost always with three pairs of segmented legs, and with flying wings on thoracic segments two and three. But we cannot look for wings on our prototype, because they were not present in the early insects. Wings, however, make the Hexapoda the remarkable, dominant group of animals it is today.

To continue, the third tagma, the abdomen, is the center of food digestion, waste elimination, and reproduction. This portion probably had a pair of legs on each segment, but smaller than those of the thorax. Today these legs are absent in almost all hexapods. Small, unsegmented structures are located on the ventral side of the abdomen of the Thysanura (silverfish and allies) and a few other scattered groups, but there is some question whether these are atrophied true legs or really secondarily developed specialized organs.

Where then is the prototype? It exists only in the human mind and probably will never be discovered, unless, of course, in the future a time machine is developed that will allow us to view the images reflected into space, now traveling at 182,000 miles per second out toward a star a billion or more light years away. Once this is done, will the facts be as exciting as our present day speculations?

How many species of insects?

One of us has attempted to answer this question. A list of all species now known has not been compiled. Such a list would be massive. The catalog of the beetles of the world, many years out of date, occupies thirty-one volumes, and this merely lists the published names and their references. The rate of discovery of new species of insects continues at several hundred a year, year after year. The number of genera and species that live and reproduce in the U.S.A. and Canada has been counted. These and estimates of the number of described species in the world are, in round figures, about 90,000 species in North America north of Mexico, and about 750,000 species from all regions of the world. We do not know how many remain to be discovered—another million?—probably less. We may never know the exact number for two reasons. First, the natural habitats of the world are rapidly disappearing and with them the species. Second, many (in some areas, most) of them are undescribed. Further, no one knows how many of the species now described have become extinct since they were discovered and are known only from museum specimens.

This situation is complicated by there being so few systematists compared to the number of insect species. Much of the study material now on hand is not identified. For example, the Florida State Collection of Arthropods, Gainesville, has several thousand jars of specimens collected at light traps and stored in alcohol which represent literally millions of specimens awaiting scientific study. About a thousand systematists specialize on insects, and few of them are able to work full time at these studies. This means, at the least, 1,750 species per specialist and probably many more. A specialist can, at most, study and write a description of two species in a full day. This does not count the time it takes to sort specimens, collect specimens, write to borrow study material, study and identify the specimen, label and distribute specimens, record locality data, and make dissections. Several prominent systematists claim that it takes a lifetime to monograph one thousand species. Assuming this to be true, then it is easy to see that we are way understaffed.

How many individuals?

"Biomass" is a term used to describe the total weight at any given time of living matter on, in, and above the earth, including in water. This is a convenient measure of the number of individuals functioning in their ecosystems. The megaton estimates of this are beyond comprehension and mean little other than to those concerned with studies relating to oxygen available for respiration and carbon dioxide for photosynthesis. We need to know whether animal biomass (for example, the total weight of humans) is likely to require more oxygen for respiration than that being produced by trees in our forests or grass in our lawns. Everytime we mow grass along our superhighways, we are reducing the amount of oxygen available for burning gasoline and diesel oil by vehicles traveling along roadways, and, of course, the roads themselves take up space otherwise usable by plant life.

The easiest way to estimate biomass is to select an area, determine the weight of samples of vegetation (remembering that most of a tree is non-living), and so on. Then these figures can be multiplied by total area. Subtract the area used for roads, buildings, parking lots, and sports arenas, then compute the biomass. Most of this will be plant life. Understand, animal life is entirely dependent on plant life for all food and oxygen, either through direct feeding or as predators on plant feeders. In one sense, all animals are parasites or hyperparasites on plants. In addition, the metabolic rate of various kinds of animals and plants (these, too, use oxygen to keep alive) differ widely from group to group. For example, human biomass is approximately eight times bird biomass. However, bird metabolism is approximately eight times the human rate. This means, of course, birds and humans each consume about the same amount of food and oxygen. Paradoxically, we feed "our feathered friend" in the winter, while we let fellow humans starve to death!

Insect biomass varies, because some insects are extremely abundant and others are "rare." For example, ants, leafhoppers, or springtails seem to live everywhere. (Rare has now accurate meaning, but generally it is used to indicate that only a few individuals are known.) Again, metabolic rates of flying insects are likely to be much greater than that of a slow-moving beetle. Because insect biomass estimates are seldom attempted, we have little hard data on the subject. We surmise that there is no correlation between abundance of one kind of insect and number of species in a group. That is to say, a species-abundant group, beetles, for instance, do not actually have as great a biomass as other groups with far fewer species. Springtails, with about 6,000 species, may outweigh beetles with 290,000 species. Most of the insect biomass, and thus, the greatest number of individuals, probably is accounted for by springtails, ants, leafhoppers, grasshoppers, leaf-feeding moth caterpillars, termites, and maybe cockroaches. The accuracy of this statement has never been tested.

What is a species?

To understand evolution and speciation, you must know what a species is. Often figures on the number of species are confused with the number of individuals. Actually, it seems that most people believe only one species, "bug," is involved. Few realize that the great variation of size, shape, color, and habits of "bugs" indicates the array of species. They may have a vague concept of species of vertebrate animals, particularly mammals and birds, but even these may be confused in their minds. In thinking of the common dog, they believe that the great number of varieties or breeds each is a separate species. So do they think of each flowers and fruit varieties as a separate species; and, at the same time, they lump all mice into a single species, and so on. To correct these misconceptions, we must set forth the criteria defining the biological unit, species.

Species is defined as a group of organisms living in a definite geographical area, where this collection of individuals are capable of reproduction to produce fertile offspring (the breeding population), and to continue to do so for a significant period of time. All members of the breeding population physically resemble each other more closely than they resemble members of another species. Individual insects are usually found living in one or more isolated breeding populations, and one or more of these populations comprise the total living individuals of the species. Under favorable conditions these separate populations may communicate. For example, a population of beetles living in a moist canyon in Arizona is separated most of the year by the arid desert surrounding the canyon, but, during the short rainy season, individuals from one canyon may migrate to a nearby canyon and join individuals living in a neighboring canyon. In this manner, the gene pool (genes of the population) are kept to full strength. If this mixing of genes did not take place, some genes might disappear from one or another of the populations. This could weaken the population to the extent that it would be unable to survive adverse climatic conditions and die out.

Speciation and factors affecting speciation

Under certain circumstances, there may be strong gene selection that considerably alters a population. If these changes prevent the members of one population from crossing with members of another population to produce viable offspring, a new species is formed. This, primarily, is the mechanism of evolution. There are other ways evolution can take place, but they are variations of this selection process.

Various factors, environmental and genetic, affect a species, causing change. Gene change may take place at the time of gametogenesis (that is, during the formation of the sperm and egg). These changes generally cause the zygote to die and that ends it. Occasionally a "sport" is formed, a variation with some selective advantage. Since many genes are pleiotrophic, controlling both a physiological process and the formation of a particular structure, the "sport" or genetic variety will look and act differently. The progeny of these variants may outbreed the regular variety, and, through this gene shift, an entirely different species is formed. Genes always determine the final species, even when it results from environmental selection. Actually, genes must always have been present to permit change of appearance and habits, a phenomenon technically termed preadaptation.

Thus environmental and gene selection work together in the evolution of a population and a species.

Adaptation

The major reason for the many kinds of insects is because they are preadapted for living everywhere life can exist. Because special equipment is needed by every organism to live in its special habitat, adaptable life forms show great diversity in size, shape, and color. This is characteristic of insects more than almost any other group of species. Possibly this is because their basic body construction offers protection to internal organs needed to carry on their special physiology.

A review of the various orders of insects makes it easy to see which characteristics are most adaptable. Before doing so, we must understand a term frequently misused—diversity. Diversity and variation are two different terms. The former does not mean, as frequently thought, differences between species. Some people maintain that a family with a great number of species is, therefore, highly diverse, while actually a group with a great number of species may be much *less* diverse than a group with a few species. Diversity means a wide range of body types and systems in a taxon. For example, the large beetle family Staphylinidae is exceedingly diverse because, among other things, it shows every possible combination of tarsal segments. It has, therefore, a diverse tarsal formula, while a somewhat larger family, Curculionidae, shows almost complete uniformity of the tarsal segments. The term diversity is useful only when applied in a comparative sense. Accepting this use of the word requires us to restrict the term variation to differences between individuals comprising a breeding population. After doing this, we are ready to examine a few of the adaptive features of insects to see why they are so species abundant. The preadapted features have enabled insects to live in diverse habitats and to occupy diverse niches.

Why so many insect species?

Speciation tends to be a function of ecological habitats and niches. Some areas of the world have a wide range of habitats and others a narrow range. In the western mountainous areas of the United States, altitude determines much of this variation. A single mountain will have dozens of major habitats with each change of altitude and a comparable change of species. In contrast, a particular genus may have two or three species in the east, but be represented by two dozen or more species in the west. This is true for many kinds of organisms: birds, plants, and others, not just insects. Always insect species outnumber all others. Hence, we speculate that factors other than altitude affect the speciation process, and this is true.

The term *niche* is used to describe what a species does in its habitat. Knowledge about how a species lives in its habitat helps to explain speciation. Some species in an ecosystem may feed on fungus, some on decaying animal matter, all at a certain altitude, temperature, and relative humidity. See how these few factors enable many species to occupy overlapping geographical ranges. Specialties among closely related species cut out competition. At the same time, specialization is dangerous because it limits survival in ever-changing habitats. This, too, contributes to the speciation process.

To paraphrase the original question, "What factors enable so many species of insects to evolve?" We can answer by listing a few obvious ones:

Wings. Best developed in insects and birds, and probably better in most insects than birds, because insect wings of many species fold and do not get in the way when insects feed in cramped quarters. Insect wings are used to aid reproduction and to disperse eggs, and, sometimes, to escape predators. Adult insects, in general, do little feeding and, when they do, their wings are not in the way. Bird wings are also used for food-getting, as well as dispersal. They aid reproduction, but restrict their eating habits to a considerable extent—birds cannot bore into wood or hide under bark the way that many insects can. But much of this is speculation and without support of experimental data.

Exoskeleton. The hard, water impervious, sensory surface of insects permits many adaptations impossible for other animals. For example, they can close their spiracles and aestivate during adverse environmental conditions. Considerable protection from predators adds to the utility of this structure. Per square millimeter, it takes a larger animal to crush the insect body and extract the nutritious contents than it does to hold a bird, pull off its feathers, and shred its tissue. The exoskeleton also allows some insects to squeeze their bodies into cracks, crevices, under bark, or spaces beneath logs and stones. Specialized species bore into wood, living or dead. Beetles, in particular, have these characteristics, which presumably accounts in part at least for their great number of species.

Coldbloodedness. We usually think of coldbloodedness (poikilothermism) as a primitive condition and warmbloodedness (homoiothermy) as an advanced state, probably because humans are homoiothermy and refer to "coldblooded murder," obviously something degrading. Many examples among the insects give us a different view of this, however. Insects can go without food and water, surviving for long periods of time in cold, even freezing, weather by simply stopping, or at least slowing down all metabolic activity. Warmblooded creatures must "run their engine" twenty-four hours a day. Fortunately, our automobiles are "coldblooded," else think what our gasoline bill would be! Back in the days of the horse and buggy, the old plow horse had to be fed every day, work or no work. The best birds can do is to leave the area when it gets too cold—at a great expenditure of energy. In fact, one wonders sometimes why they come back. The advantages of coldbloodedness are even more apparent when one considers the insect fauna of the arctic regions. Insects there are extremely abundant during the few days of summer when the sun warms the air to above 15° C. One time we were collecting beetles in Death Valley National Monument in California on New Year's Day. There was snow in the mountains and the air temperature in the morning was at the freezing point. By noon the air had warmed up and there was considerable insect activity. Even mosquitoes came out and started biting. These are only a few of the advantages of poikilothermism for insect life.

Metamorphosis. The change from the earthbound feeding and growing stage (the nymphal or larval stage) to an airborne reproductive and dispersal stage through metamorphic life cycle changes offers insects the most possibilities for habitat selection. This Dr. Jekyll, Mr. Hyde existence probably is the greatest single factor accounting for the many insect species. It is specialization of specializations, a dual life, each phase adapted to its respective habitat. Thus, the larvae or nymphs have their food supply and the adults have theirs, if they eat at all. The immature stages concentrate on eating; the adults, on reproduction, each specialized for its particular habitat.

Senses. The preceding discusses only a few aspects of this tale of the success of insects. Much more could be said about each, and there is much yet to be learned. These are by no means the only factors, however, that account for insect abundance. Sense organs, in particular, offer much to think about. Sense organs are scattered over the body surface. Spines of many different shapes and sizes convey sensory receptions to the nervous system of the insect and to the central computer (brain or cephalic ganglia) where these are sorted and acted on. It is well known that all sensory information is received as electromagnetic waves. The waves that are received vary with the shape of the receptor, a subject that has become the life work of Dr. Philip S. Callahan, mentioned previously. Enter his laboratory, and you will see models of different sizes and shapes placed between devices emitting electromagnetic waves of many wave lengths. With delicate instruments, Callahan is able to determine just what each of these shapes means in insect reception. Suffice it to say, insects detect many different bits of useful information. One cannot conveniently refer to the five senses of insects, because this information is not broken down into the same "senses" as human senses. For example, several kinds of insects are tuned into smoke detection and will appear at a forest fire from miles away long before humans can smell smoke. This is a great advantage to many wood-boring insects, for example, because they generally infest fire weakened trees.

Each of these factors and certainly many more that we know little or nothing about combine in unique ways to bring forth the hundreds of thousands of insect species.

Population sampling

Population study by organized sampling challenges both amateur and professional entomologists. Insect collectors understandably collect in areas located near campgrounds, biological stations, and parks (see chap. 2). The data obtained is qualitative and usually does not give quantitative data needed to determine either abundance or distribution of species. Obviously, if the preceding description of speciation is factual, data must have been obtained by a study of populations. Much of our knowledge of populations is obtained in laboratories using fruit flies (*Drosophila* spp.), or from field studies of plant populations much more manageable and predictable than animals. Some species of insects do not respond to our lures (e.g., UV light traps), making it uncertain whether a species is present or not in a particular area.

Vertebrate biologists, especially ornithologists, have mapped the distribution of their species with great accuracy. Now certain groups of butterfly collectors are organizing, led by Dr. Paul Opler of the U.S. Fish and Wildlife Service, to survey and map butterfly populations in the same manner as is done for birds by the American Ornithologists Union members.

Several butterfly species have become extinct during this century (see chap. 10). Many amateurs, especially enthusiastic butterfly collectors, wish to protect rare species and learn more about their evolution. To do this, they must study populations.

Population studies may be made two ways: by field collecting and by culturing (see chaps. 2 and 18). Specimens are collected systematically by marking a grid on a map and then visiting the intersecting points to determine the presence or absence of a particular species. Determination of the distance between points on the grid will depend on the species to be studied. Ordinarily, start by mapping locality data taken from specimens in collections. Select the size of the grid from this mapping. In our experience, 100-mile interval grids are a good starting point. It is a simple matter to place this grid over the known distribution map of the species. Then field collecting helps to fill in the gaps.

A word of caution is necessary. Often the scientific literature gives the distribution of a species as "eastern United States." A study of the reported samples may show the specimens were taken in Indiana, New York, Maryland, and Georgia. These collections are indicated on a map as dots, and a line drawn around them indicates the area supposedly occupied by the species. A study of specimens in collections may fill in a few more localities, but there is no indication that the species really occupies the entire range between dots on maps. These species need to be studied by sampling the entire area. Doing so will usually reveal some interesting information about the nature of these species.

Once the species is located near one grid location, detailed studies may be made to determine the parameters of the population. How widely distributed is the population? You might be surprised. Common species may be concentrated in only a part of the few acres of a mountain meadow or between canyon walls. (Keep this in mind when searching for a species at one grid point; it may easily be overlooked, because you are searching a few yards from the breeding site.) When located, try some of the techniques used to determine population size (see chap. 10). If the species is abundant, crossbreeding experiments may be attempted using members of other populations. Particularly important is the study of morphological variation of several populations. A knowledge of statistics will help you to select ways to treat data obtained from these studies. The possibilities of experimentation are nearly unlimited in this great laboratory of the outdoors.

The second method of population study is by rearing specimens. Butterflies and moths often can be cultured in the laboratory, at home, or at "insect farms," large screened outdoor areas into which are placed pairs of the insect, released to breed as if in the wild, but kept in captivity for easy capture of specimens for study and even sale. Two important accomplishments are achieved from this. It helps preserve the wild populations by supplying fresh specimens for collections and living insects for release in the wild to increase the native population (obviously not pest species). Crossbreeding experimentation can be attempted with these specimens. Data from crossbreeding tests will be useful for phylogenetic speculations.

The basic taxonomy of more and more North American and European insects is being completed, meaning that fewer and fewer new species are described each year from these regions. Naturally, amateurs turn to more challenging studies. True, there are always rare species to be added to the collection, but there is more to insect study than filling boxes with specimens. Turn to the study of populations and get to know the insect species as living, active, and truly amazing animals!

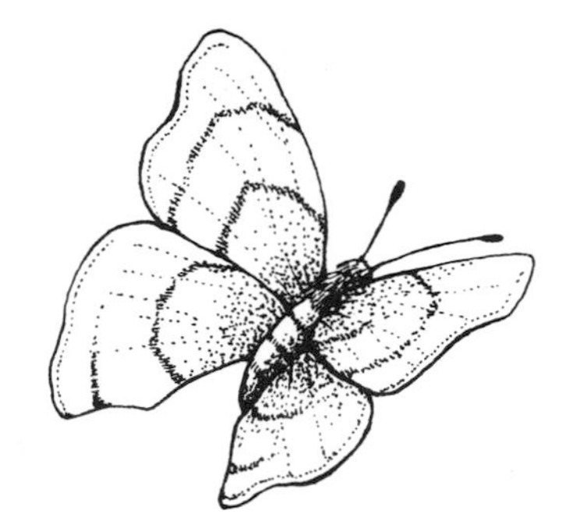

5

Insect Appetites

"Alice, 'for I can't understand it myself to begin with, and being so many different sizes in a day is very confusing.'
'It isn't,' said the caterpillar."

Lewis Carroll (1832–1898) *Alice's Adventures in Wonderland*

On a hot sunny day in July, a cabbage butterfly larva can eat as much as ten times its own weight in your cabbage patch, which makes Brer Rabbit look like a "good ol' boy." Two things contribute to the success of insects and the demise of the cabbage patch. First, most insects have a specialized feeding stage—we call this stage a nymph, naiad, or larva. Details of this are explained in the next chapter. The external covering of the larva is capable of expanding several times its original size, and then this covering is shed, and the new one formed can repeat this process. Second, mouthparts are highly specialized instruments designed for manipulation of food according to the needs of each species. Most insect research is related to the feeding stages of insects. Their role in an ecosystem pivots around feeding.

All insects feed in one or another life stage, often both the immature and adult stages, but, of course, never the egg or pupal stage. The adults of many species do not feed, and those that do feed, do so only to give them energy to fly, walk, copulate, or lay eggs. Therefore, most damage by pests is caused during the immature stage and this by the mouthparts. Occasionally plant tissue is injured by the ovipositor as eggs are laid.

The mouthparts are composed of modified segmented appendages, as is shown by a study of the various arthropod ancestors. Exactly how many segments were fused to form the insect head is unknown, but generally it is assumed that head segments four, five, and six bear the mouthparts as they are seen today. Actually, then, these complicated parts are highly specialized legs.

Types of mouthparts

Entomologists recognize six major kinds of mouthparts, each used for some specialized feeding process: chewing, used much the same as ours for feeding on plant and animal material; piercing-sucking, for pumping up blood or plant juices; cutting-sponging, used to make large wounds and then sponging up the juice instead of sucking it up; sponging, used to eat fluids; siphoning, for sucking up nectar; and chewing-lapping, a combination of the chewing type with mandibles to cut and a tube to suck. Many variations of each type can be found from species to species throughout the orders of insects. By far the most common type is the basic chewing type found in all the primitive orders, the beetles, wasps, and ants. The piercing-sucking type is characteristic of the bugs, and the siphoning types, characteristic of the Lepidoptera. The flies and the bees have many variations of the cutting-sponging and chewing-lapping types.

Probably all mouthparts are derived from the basic chewing type, the main feature of which is the mandible. Insects with mandibles are termed mandibulate insects. The common grasshopper and the house cricket have these structures and can be dissected for detailed study of the mandibles and the accessory parts, the labrum and labium (fig. 5.1). With chewing mouthparts, the insect is able to bite off part of a plant, feed on various kinds of animal matter, chew into wood, consume debris of various sorts and other kinds of solid material. The labrum forms an "upper lip" and is a more or less simple flap hinged to the front of the head in such a way as to partially or entirely conceal the mandibles beneath. The paired mandibles function much the same as jaws in vertebrates, but they work from side to side instead of up and down, as ours do. Various modifications of the

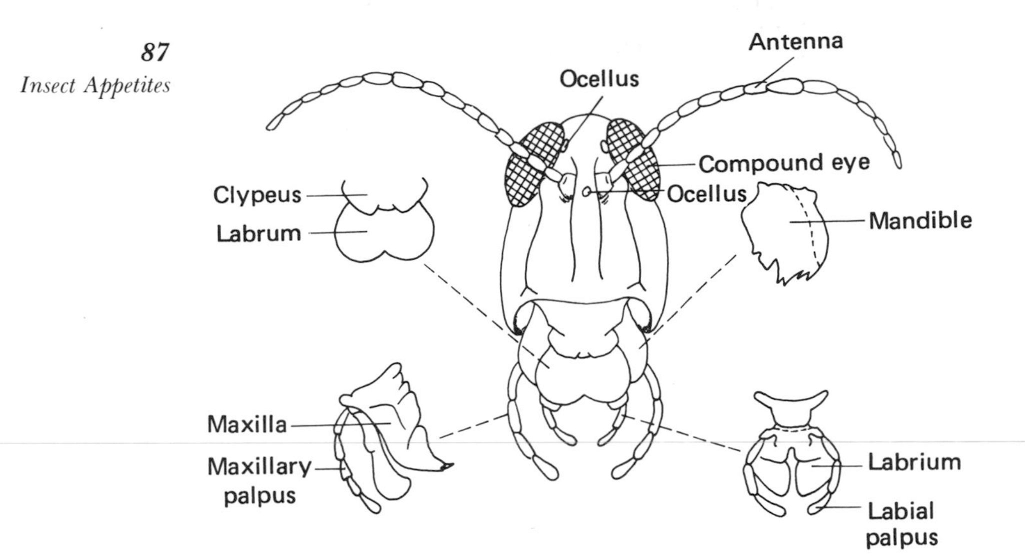

Figure 5.1. *Chewing mouthparts of a grasshopper.*

inner surface provide "teeth" or the molar area of the jaw. Powerful muscles anchored to the inner surface of the head capsule close the mandibles. An internal tentorium, a bar or strut inside the head between the eyes keeps the head capsule from collapsing when the jaw muscles are contracted. A pair of maxillae lies behind the mandibles. They are much the same as the mandibles, but not as stout and more complex by the addition of various substructures, particularly short palpi with sensory surfaces used, no doubt, to "taste" the food taken in and to inform the insect just where the food is, as it is crammed into the hungry creature. A cutting edge, the lacinia, helps mold the food bolus. Maxillae also are used as organs to clean antennae and other parts of debris. Behind the maxillae are the paired labia, fused to form the "lower" lip, the labium. This structure also bears a pair of palpi. The labial palpi are usually smaller than those of the maxilla. Inside the head capsule is a tonguelike flap termed the hypopharynx. In the simple mandibulates, this functions much the same as the human tongue, but it is greatly modified in the sucking insects.

Many insects have chewing mouthparts. Among the best known are the grasshoppers, katydids, earwigs, beetles, dragonflies, and termites. By using their mandibles to gather food, they cause serious damage to food and fiber crops, lumber trees, stored food, fiber and paper products, and even stored meats. Although differing in detail, the basic structure of these mouthparts is exceptionally uniform. However, in a few scattered species, one portion, the lacinia, may be greatly elongated to form a sucking tube used to take up nectar. Some larvae—for example, certain water beetles and lacewing larvae, both predacious—have sickle-shaped mandibles. Each mandible is hollowed out to form a tube through which the body juices of their prey are sucked. Other mandibular modifications include differences in the molar region adapted for such specialized feeding as eating pollen and other special foods.

Species that suck up fluids exclusively have various parts modified to form tubes. These new structures are variously termed "proboscis" or "haustellum," depending on the group and the author's preference. We prefer, for the sake of consistency, to use the term proboscis for any fluid-sucking type of mouthpart found in insects. Often the proboscis is formed by an elongation of the mandibles, maxillae, labium, and hypopharynx. The mandibles and maxillae are thin, bladelike structures used to cut into the skin of animals, the labium and hypopharynx form a tight tube, and the labium forms a case to hold these parts together while being inserted into the animal's blood vessel. The hypopharynx also contains an exceedingly fine tube through which saliva is pumped into the wound to prevent the coagulation of the blood while it is being sucked up. This is the structure of the proboscis of those pesky mosquitoes (fig. 5.2). When the mosquito feeds, she inserts the stylets into the skin of the victim. These are held in place by the labial sheath, which folds back out of the way during feeding. When she is finished with her blood meal, the stylets are retracted and returned neatly into the sleeve formed by the labium. Because of

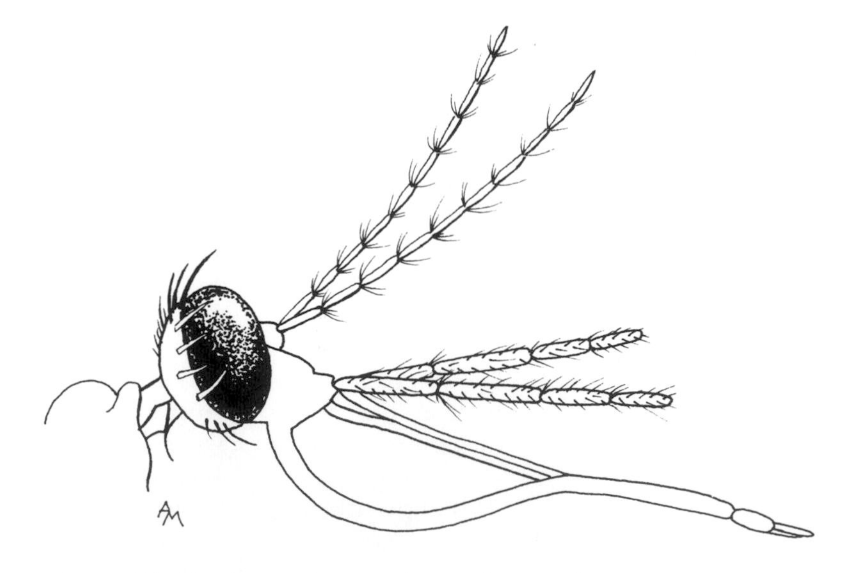

Figure 5.2. *Piercing-sucking mouthparts of a mosquito.*

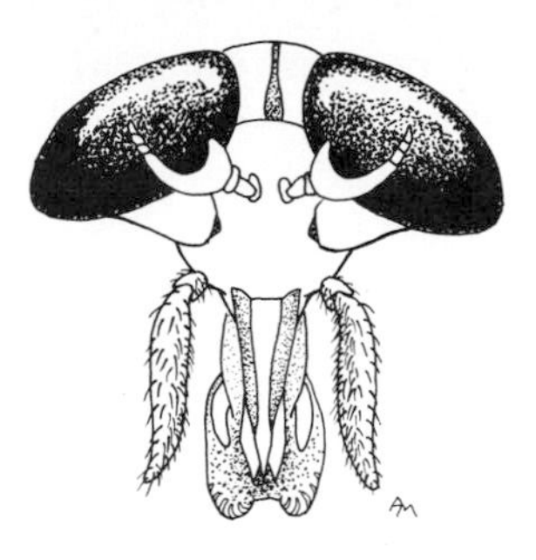

Figure 5.3. *Cutting-sponging mouthparts of a horse fly.*

the numbing effect of the anticoagulant, the mosquito feasts without the victim's being aware of its blood loss until the wound begins to itch due to the body's reaction to this foreign protein. Various modifications of this plan exist in blood-sucking fleas and lice. The mouthparts of plant juice suckers are similar, but, of course, they are not concerned with the coagulation of blood. No doubt, there are problems with certain types of plant juices, especially those with milky sap, and sugary solutions cannot be allowed to evaporate into sticky syrup. Plant-juice feeders include leafhoppers, aphids, and scale insects; all are economically important.

The cutting-sponging type of mouthpart, common among horse flies, deer flies (fig. 5.3), and stable flies, is not as specialized as that of the mosquito. When a horse fly lands on its victim, it punctures the skin with the mandibles and maxillae, which are modified into stylets. These are capable of a scissorlike motion, which enlarges the wound to permit the free flow of blood. When the blood reaches the surface of the wound, it is sucked up by the spongelike labium. The anticoagulant, a heparinlike substance, keeps the blood from clotting as it is absorbed by the fly. These insects perform this feat exceedingly rapidly, because it is painful, and they have only a short time to feed before the victim reacts. Multiple attacks on domestic animals, such as cat-

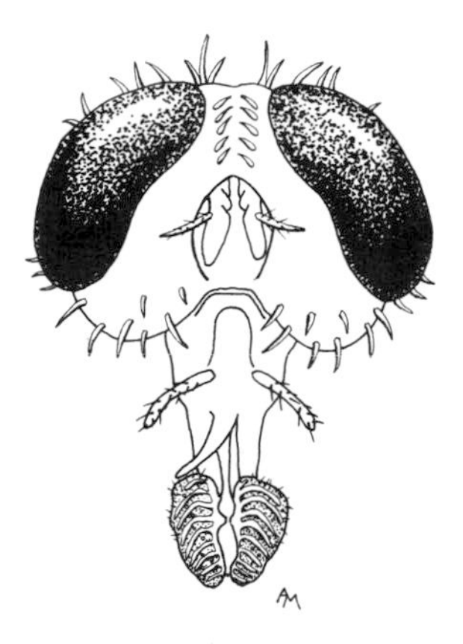

Figure 5.4. *Sponging mouthparts of a house fly.*

The common house fly lacks stylets (fig. 5.4); hence, its mouthparts are termed sponging, although its close relative, the stable fly, does bite. Both maxillae and mandibles are absent in the former, but the labium is modified into a structure called the former, but the labium is modified into a structure called the labellum, the "sponge" portion. This consists of an enlarged pad with tubes that open in such a way that when pressed to the surface of a substance such as sugar, they open. When a house fly lands on a surface, it "tastes" with special hairs on the labellum and also on the feet. If the fly finds the material to be possible food, it vomits some of the liquid from its stomach through the labellum. This dissolves, or suspends the food, to enable it to suck it up through the tubes of the labellum. This, of course, is an ideal way to transmit disease organisms from place to place. Imagine, if you will, a fly that first visits human feces in an open latrine, then decides to top off the meal with sugar from your sugar bowl for dessert. When you also eat from the same bowl, you run the risk of ingesting the pathogens left behind by the fly when it regurgitates the liquid from the fecal meal.

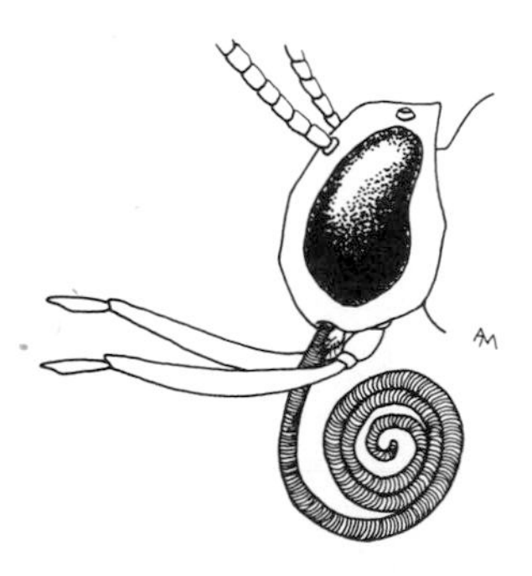

Figure 5.5. *Siphoning mouthparts of a butterfly.*

Siphoning mouthparts take up nectar and similar liquids through a tube formed from the lacinia only (fig. 5.5). The remaining parts of the mouthparts are greatly reduced or entirely absent. The tube or proboscis is used much the same as we

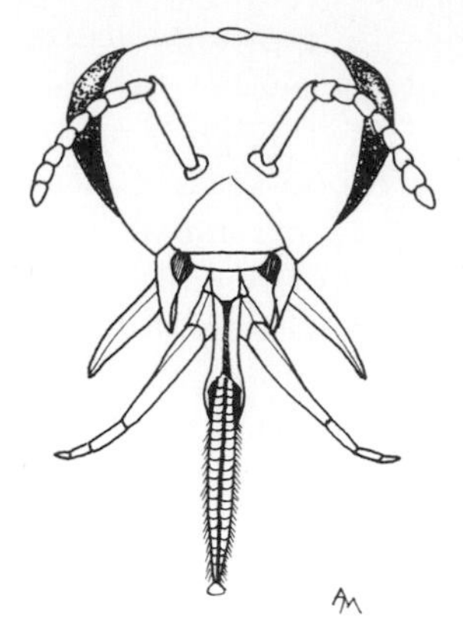

Figure 5.6. *Chewing-lapping mouthparts of a honey bee.*

use a soda straw. When not in use, the tube is coiled out of the way. It is uncoiled by increasing the pressure of the hemolymph in each lacinia that forms the tube. These mouthparts are characteristic of moths and butterflies. When a butterfly visits a flower, mud puddle, or sap flow, it uncoils the tube, inserts it into the fluid in a flower corolla, or elsewhere, sucks up the desired fluid, and then recoils the tube by relaxing the internal pressure.

Adult honey bees and other bees have their mouthparts modified for taking up both liquids and solids. This chewing-lapping type (fig. 5.6) uses mandibles for defense, molding wax in the combs, and other chores, including, by some species, cutting into flowers or cutting leaves. The maxillae and the labium are both elongate and modified for lapping up fluids.

Diets

Insect diets and nutritional requirements are as varied as the mouthparts. They require macronutrients in large amounts and micronutrients in trace amounts, much the same as any animal, including humans. In fact, it is because these requirements are similar to those of humans that we have so many pests. Those requiring different food from ours are not crop pests. As with all living organisms, water is essential in their metabolism. Water content regulation is a critical problem facing all terrestrial animals, including insects. The chitinous integument (outer covering or skin) contributes to water control, along with special excretory organs, the Malpighian tubules. By using ureic acid (a solid) instead of urea (a liquid), the use of water in the excretory process is kept at a minimum. These facts must be considered as we try to understand the food and feeding habits of insects.

Types of food

Insects feed on all kinds of organic matter. Many feed on living plants, no species of which, living or dead, seems to be immune. Animal tissue is attacked by insects, but that of large animals much less than that of fellow insect species or animals smaller than the insect predators. Sea animals and plants are entirely free (except that seals have lice) of insect predation, but, if their dead bodies reach shore, they will be fed on by carrion- and other detritus-feeding insects. Rather than attempt a discussion of the chemistry of insect food requirements, it is more interesting to approach the subject by classifying the feeding preferences of insects. Mouthparts just described are modified to permit feeding by these processes. Often other parts, particularly the front legs, are modified to assist the mouthparts in feeding. The classification of feeding is shown in table 5.1.

Table 5.1. *Feeding habits of insects.*

*Classification**	*Food Type*	*Examples*
Phytophagous	Feed on plant tissue	Aphids, treehoppers
Carnivorous	Feed on animals	Mantids, assassin bugs
Mycetophagous	Feed on fungus	Certain beetles and flies
Saprophagous	Feed on decaying animal	Rove and carrion beetles and certain flies
Coprophagous	Feed on feces	Dung beetles
Haematophagous	Feed on blood	Female mosquitoes
Entomophagous	Predatory on insects	Carabid beetles
Parasites	Internally feed on living insects	Ichneumon wasps

Monophagous: Organisms that feed only on a single species
Oligophagous: Organisms that feed only a few, usually closely related species
Polyphagous: Organisms that feed on many and various species

*NOTE: Other terms for feeding types are included in table 2.2, chapter 2.

Many insects have restricted diets, feeding only on a single species of plant or animal. The sunflower beetle, *Zygogramma exclamationis*, feeds as larva and adult only on the sunflower. This beetle is, therefore, monophagous, by restricting its diet to a single plant. *Zygogramma suturalis*, a related species, is oligophagous, because it will feed on a few plants, for example, goldenrod and ragweed, or other combinations. However, another species in the same group, the Colorado potato beetle, *Leptinotarsa decemlineata* feeds, despite its common name, on a wide variety of plants in the nightshade family: for example, potato, tomato, and eggplant. This species, and others with similar feeding habits, is termed polyphagous.

Digestion

Usually the digestive tract of insects is complete; that is, it extends from the mouth opening through the body to the anus, located at the end of the abdomen (fig. 5.7). Aphids and a few other sap-feeding insects have an incomplete digestive tract specially modified for this kind of food processing. They have a filtering mechanism that removes water from the fluid and enables them to process large volumes of this dilute sugar solution. Similar modifications occur among blood-feeding insects, but their digestive tracts are complete. Other less drastic modifications include "gizzards" or teeth in the proventriculus to help break up food particles for easier digestion.

The gut of the insect has three basic parts: the foregut, used for the preparation of food for digestion; the midgut,

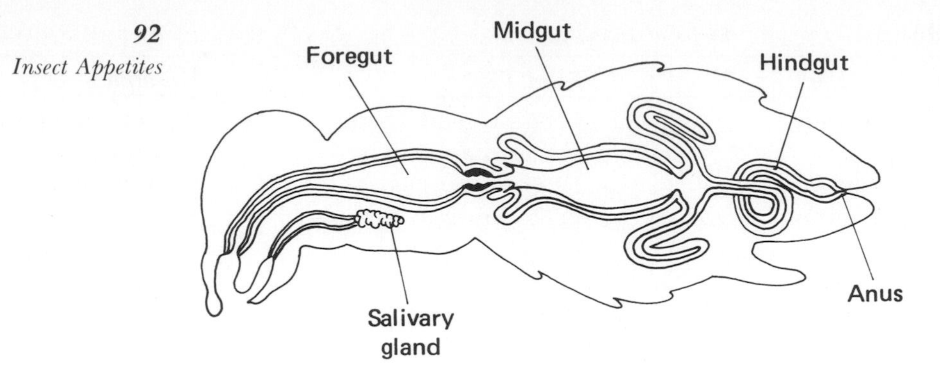

Figure 5.7. *Digestive tract of a representative insect.*

where the actual digestion takes place; and the hindgut, where excess water is absorbed and usually dry feces are formed. The digestive enzymes vary according to the food to be digested. Wood-feeding insects, such as termites and certain cockroaches, maintain a colony of protozoans, or bacteria, in the gut. These secrete the enzymes necessary to change the high molecular weight carbohydrate, cellulose, to simple sugars usable by both the insects and their friendly stomach parasites.

Nutrients

Insects, as do humans, require certain macronutrients and micronutrients to maintain life processes. Carbohydrates for energy, amino acids for protein formation, and sterols for hormone synthesis are the macronutrients required in various proportions by humans and most insects. Those insects that feed exclusively on one of these macronutrients are capable of using the basic elements, plus certain minerals contained in the fluids they absorb, to manufacture the other two macronutrients. Nitrogen is required by all living organisms. Thus, sap-feeding insects must be able to use inorganic nitrogen compounds dissolved in the water contained in sap to make the proteins that they need for tissues and growth. This same fluid also contains the needed micronutrients. They are calcium, copper, iron, potassium, zinc, and sodium. The vitamins required by insects differ from those needed by humans, and they are able to obtain these either from their food sources or by synthesis during their metabolic processes.

The amount of food required by insects is directly related to the activity of the stage they are in. As previously pointed out, the immature stages usually eat large amounts of food, most of which is stored as fat to be used by the active adult for the energy needed for flying and for producing eggs. Foods high in cellulose, usually indigestible by insects, as it is with other animals, must be eaten in much larger quantities than highly concentrated foods, such as blood. The same is true of sap as food. It must be "boiled down" (concentrated), much the same as we boil maple tree sap to form maple syrup and maple sugar. Animal tissues, living or dead, are high in food value and, therefore, a little bit goes a long way toward providing the nutrients an insect needs. This is sometimes almost on a one-to-one basis. For example, the wasp parasite that emerges from the completely eaten caterpillar host may be almost the same size as the host.

Excretion

Insects use a special organ, the Malpighian tubules (fig. 5.7), mentioned above, as excretory organs. These structures function in somewhat the same manner as the advanced kidney of the higher vertebrates. In insects, blind, thin-walled tubes, usually clustered around the junction of the mid- and hindgut, extend into the hemocoel, the insect's body cavity. Waste material is absorbed through the walls of the tubules and dumped into the alimentary tract, not separately expelled as in the mammals. Excess water is reabsorbed from the hindgut.

Feeding adaptations

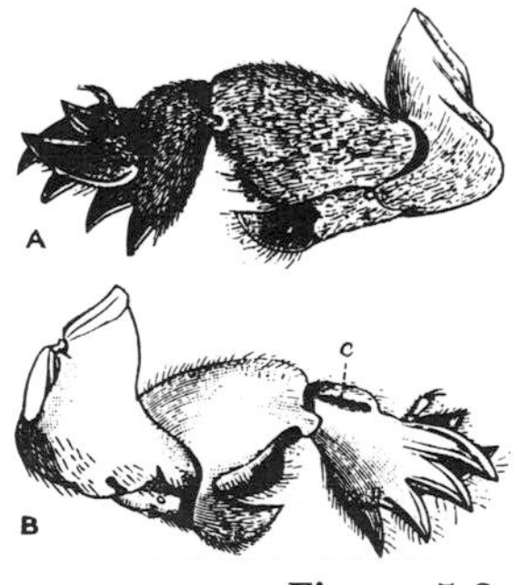

Figure 5.8.
Shovellike forelegs of a mole cricket: A, outer aspect; B, inner aspect; C, ear slit.

A wide variety of feeding structures, besides the modifications of the mouthparts, aid insects in food procurement. The raptorial forelegs of the mantids, the shovellike forelegs of mole crickets (fig. 5.8), and the hinged mouthparts of dragonfly naiads are some examples. The forelegs of the praying mantids are fitted with spines (fig. 5.9) for holding prey. Once the legs strike out and close around the victim, the spines impale it, and there it is stuck while the mantid chews off its head and, eventually, consumes the entire body, except for the indigestible wings, which float down to the ground, the only reminder of the hapless prey. The mole cricket's legs are mindful of a shovel and spade and are used to dig through the soil until roots are uncovered and eaten. The mole cricket's forelimbs greatly resemble those of its mammal namesake, the mole that feeds on larger roots, tubers, and rhizomes. Both species are the bane of gardeners. Dragonfly naiads feed on any aquatic creature small enough for

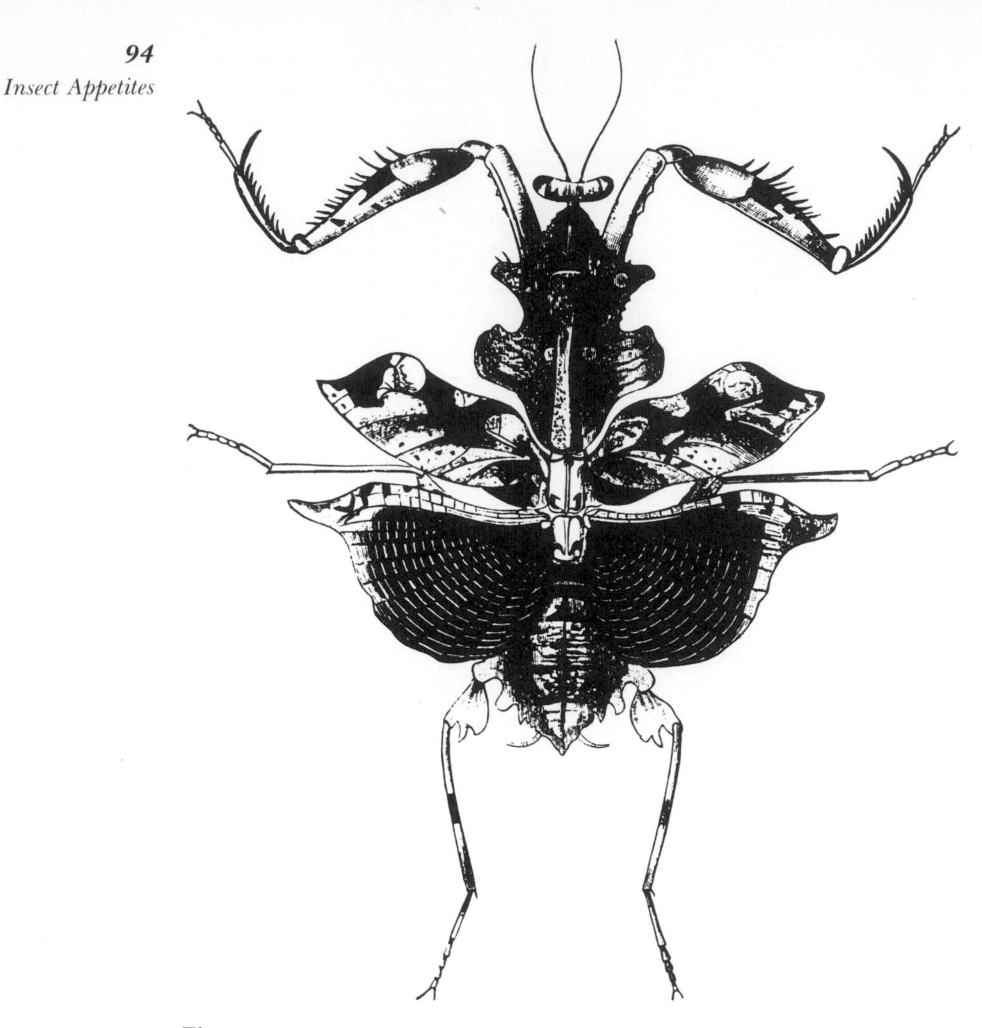

Figure 5.9. *A preying mantid from Borneo showing spines on front legs used to grasp prey (after Westwood).*

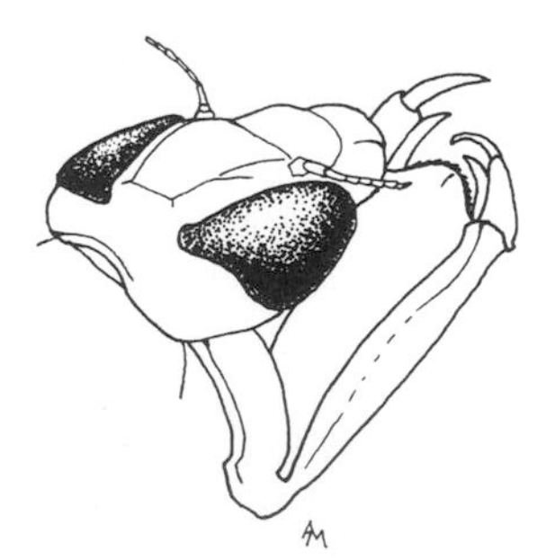

Figure 5.10. *Protrusible mouthparts of a dragonfly naiad.*

them to capture. Their hinged labium (fig. 5.10) has two clawlike lobes. When not in use, these are folded under the body, but when prey comes within reach, they strike out and capture it.

Traps constructed by some insects supplement their morphological adaptations for good gathering. The larvae of antlions, known as doodlebugs in some regions, hide at the bottom of funnel-shaped sand traps. The mouthparts are upturned to expose the mandibles, the only parts seen as they lie in wait for the next victim to fall into this "pitfall" trap (fig. 5.11).

Some caddisfly larvae construct silken seines stretched across a stone to trap water creatures. Certain mantids resemble flowers that attract bees, only to meet spines and jaws instead of nectar and pollen. On and on goes the list of clever, or so it

Figure 5.11. *Doodlebug in its funnel-shaped trap.*

would seem, ploys to lure food. Perhaps the most bizarre is that of certain female fireflies. They use their light flashes to trap males of another species, and then eat them (chap. 8).

Although neither trap nor mouthpart adaptation, the production of galls is still another way to enhance the food supply. Certain wasps, flies, and aphids lay eggs in plant tissue causing the development of a gall, abnormal plant growth caused by the injury of egg laying. The resulting stimulation of cell division and growth is akin to cancerous growth found in animals, but usually this does not kill the plant. It does, however, disfigure the plants used as ornamentals and cuts down on crop production. The egg hatches and the larvae or nymphs feed on the tissue produced until they reach maturity. After the adult leaves, this cavity may become the home of other insects. Sometimes parasites invade the gall, attack the original owner, and take over in its place.

Feeding captive insects

Details about rearing insects in captivity are given in chapter 18, but since success in those endeavors depends on proper feeding, it is not out of place here to preview some of those requirements by applying the information given in this chapter. To successfully feed insects in captivity is not an easy task, especially if you hope to raise considerable numbers. Phytophagous insects are the easiest—once you learn their favorite plant, keep plenty of it handy, and you will have healthy insects. Predacious and haematophagous species are exceedingly difficult to rear. They are fussy about what they eat and especially about the conditions under which they feed. The temperature and humidity must be just right, and so on. In summary, it may be stated that it is absolutely necessary that you know how the insect lives in the wild before it can be reared in captivity. Study it in the field, then duplicate those conditions in captivity.

6

Life Cycles

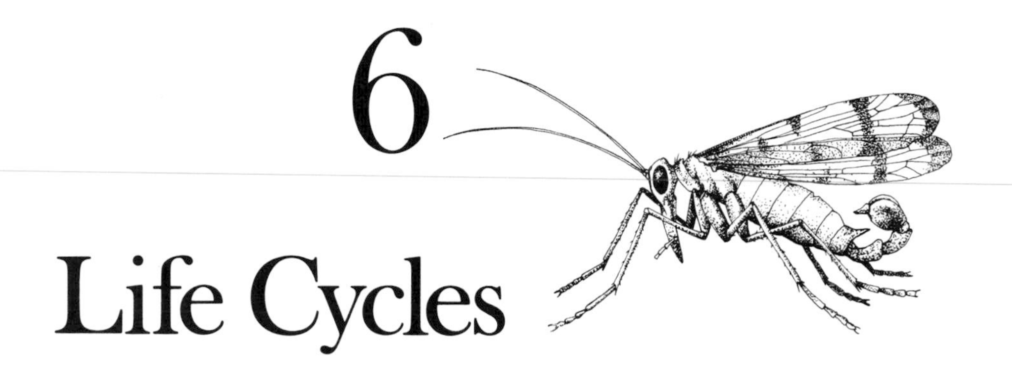

" 'Well, perhaps you haven't found it confusing yet,' said Alice; 'but when you have to turn into a chrysalis—you will some day know—and then after that into a butterfly, I should think you'll feel it a little queer, won't you?' "

Lewis Carroll, (1832–1898 *Alice's Adventures in Wonderland*

Each order of insects has some distinctive feature in its life cycle. Unlike mammals, which grow larger gradually, most insects pass through striking developmental stages, a process termed metamorphosis. This refers to all changes in shape and function from egg to adult. Metamorphosis is not unique to the insect world. For example, the well-known amphibians, typified by the frog, have three distinct stages: they start life as an egg, change to an aquatic tadpole, and develop into a different-looking semiterrestrial adult.

Insects also have three distinct types of growth patterns referred to by entomologists as metamorphosis. The first kind is

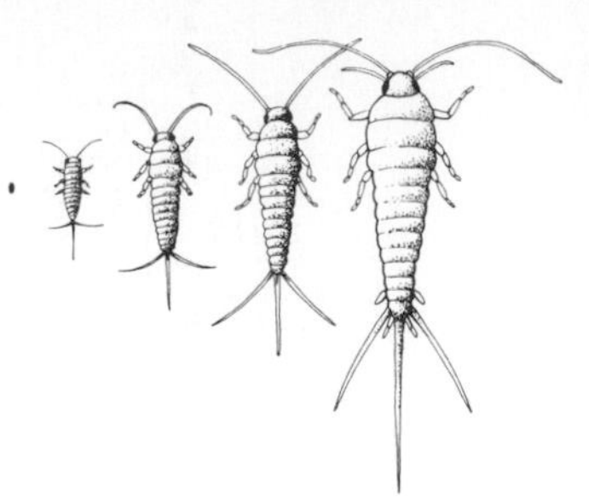

Figure 6.1. *Ametabolous metamorphosis of a silverfish.*

ametabolous metamorphosis; that is, they lack metamorphosis. Growth is gradual without obvious changes in shape (fig. 6.1), but with change in size only. Five orders of insects, all wingless, are ametabolous. The common household silverfish is an example. They represent less than 10 percent of the insects of North America, about 800 of the 87,100 species in the United States and Canada.

Seventeen orders of insects have incomplete metamorphosis. These insects develop gradually after hatching from eggs by a series of skin sheddings or molts into the adult stage (fig. 6.2). After hatching from the egg, the feeding stage is termed a nymph or naiad. A nymph resembles the adult, except that it lacks functional wings, having only wing pads and an undeveloped reproductive system. Insects with this type of change are referred to as having paurometabolous metamorphosis. The common grasshopper is an example.

Naiads, the nymphs of three aquatic orders, Ephemeroptera (Mayflies), Odonata (Dragonflies and Damselflies), and Plecoptera (Stoneflies), differ from their adults by the presence of gills and other adaptive structures and shapes that fit them for aquatic life. This type of metamorphosis is termed hemimetabolous metamorphosis (fig. 6.3).

Many insects have several molts before they reach the adult stage. The stage of an insect between successive molts is called the instar. The first instar is the stage after hatching and before the first molt, and so on.

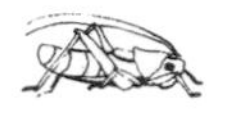

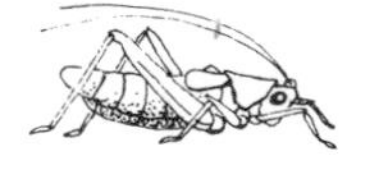

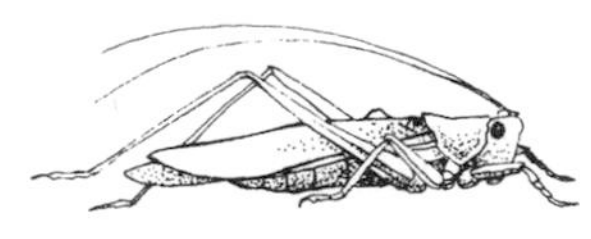

Figure 6.2. *Paurometabolous metamorphosis of a grasshopper.*

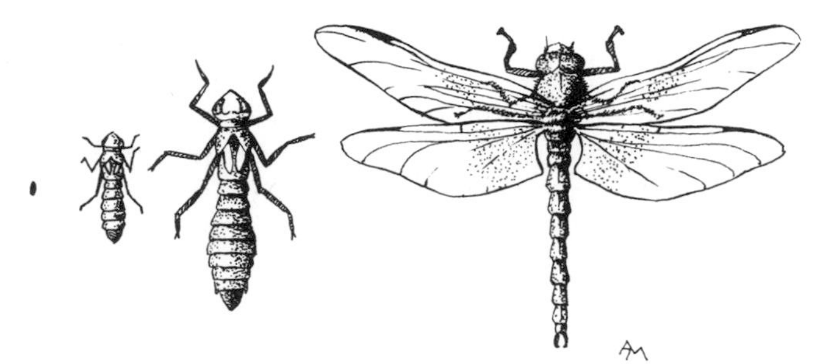

Figure 6.3. *Hemimetabolous metamorphosis of a dragonfly.*

Holometabolous, or complete, metamorphosis is the most common and advanced type. Although this occurs in the species of only eight orders, this represents about 71,680 of the 87,100 species found in North America north of Mexico, and this same proportion holds true throughout the world. These species all pass through the egg stage, develop as larvae, and transform in the pupa stage to the sexually mature adult (fig. 6.4). The larvae and adults are so dissimilar that it is impossible to tell the one is merely a stage of the other.

Metamorphosis is regulated by hormones, actually a balance of three, produced by three separate glands: the juvenile hormone, which keeps the insect in the larval stage; the brain hormone, which stimulates other glands in the insect's body to produce ecdysone to stop growth; and the hormone responsible for the actual transformation from larva to pupa and pupa to adult. Varying the amount of one in proportion to the others controls the entire, complex, growth process.

The famous insect morphologist, Robert E. Snodgrass, asked in his booklet, *The Caterpillar and the Butterfly*, ". . . how or why did the young moth or butterfly ever become such a thing as a caterpillar, a creature so different in every way from its parents?" Snodgrass answers this question by pointing out facts about the two opposing hormones. But there must be factors to control the sequencing of the hormones, and these are genes. It seems that there are two sets of genes in the developing egg—one set operates during the larval period, while the other set is suppressed. Then, when the larva has attained its maximum growth, the other set takes over and controls the molting to transform to the adult stage. The "ugly duckling" turns into the beautiful "swan"—its inner soul was beautiful all along. As Alexander Pope (1688–1744) puts it, "first grubs obscene, then wriggling worms, then painted butterflies."

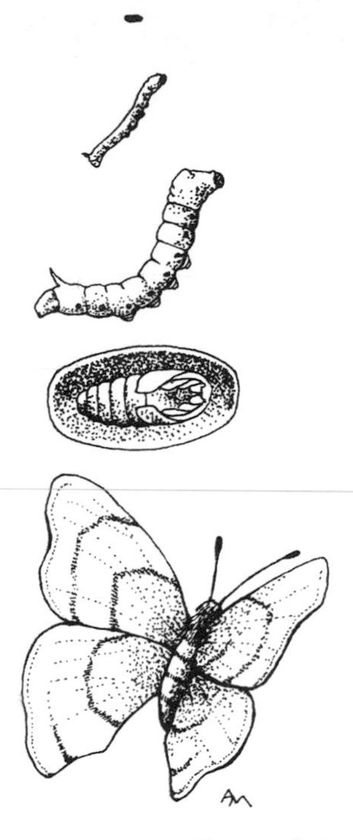

Figure 6.4. *Holometabolous metamorphosis of a moth.*

Eggs

Insect eggs occur in many shapes (fig. 6.5) and sizes and are laid in as many different places. Usually the developing insect within is protected by a shell, but not always. The adult female usually places her eggs in protected places away from predators, parasites, and the direct rays of the sun. Some eggs, for example, those of the Colorado potato beetle, are laid openly on leaves. Their bright color attracts many "egg knappers," predators on this rich source of protein, but this does not seem to cut down on the population of this abundant pest. Grasshoppers lay their eggs in the soil in the fall where they remain throughout the winter and hatch into nymphs the following spring. Mosquito eggs are dropped into the water, or, sometimes, into depressions in the woods that will fill up with water in the spring, while others make floating egg rafts. In these insects, water is home for the larvae. The female, who in no way is adapted for aquatic living, instinctively places her eggs in the habitat of her offspring. Human body lice deposit their eggs in the seams of clothing, while head louse eggs, known as nits, are attached to hairs. The eggs are kept warm by the host's body heat, permitting them to hatch shortly after laid. Nits are fastened to hair or

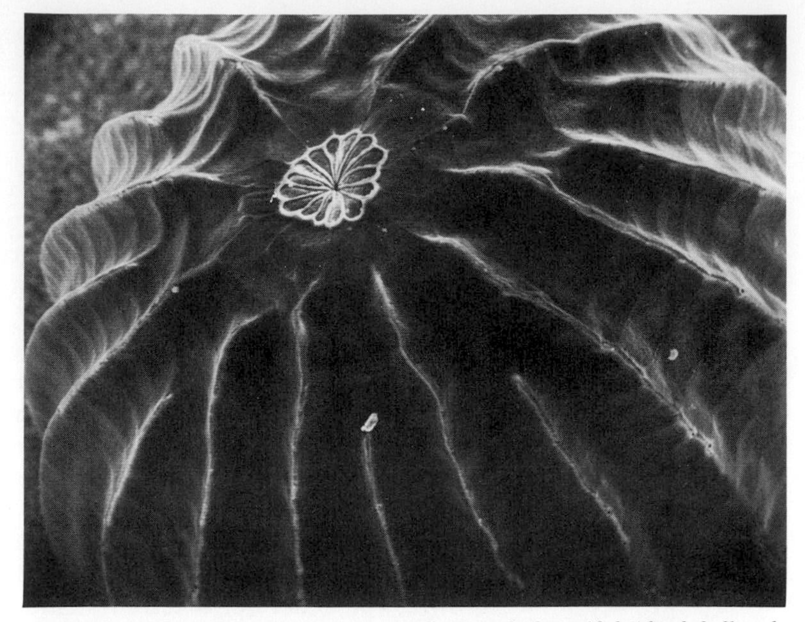

Figure 6.5. *Egg of the corn earworm moth. Note the beautiful ridged shell and the chrysanthemum-like pattern around the micropyle. This pattern is different in different species of noctuid moths (800X). (Courtesy of Dr. P. Callahan, U.S.D.A.)*

clothing by a gluelike material that prevents them from being knocked or brushed off (fig. 6.6). Some insects protect their eggs by enclosing them in protective secretions. Cockroaches surround their eggs by this material to form a capsule called the ootheca (fig. 6.7), which may contain as many as twelve developing cockroaches. These are scattered around the house or buildings in which they live, including the bottom of paper sacks into which your groceries are packed. This assures a new supply of roaches, even after the exterminator has visited. Mantids'

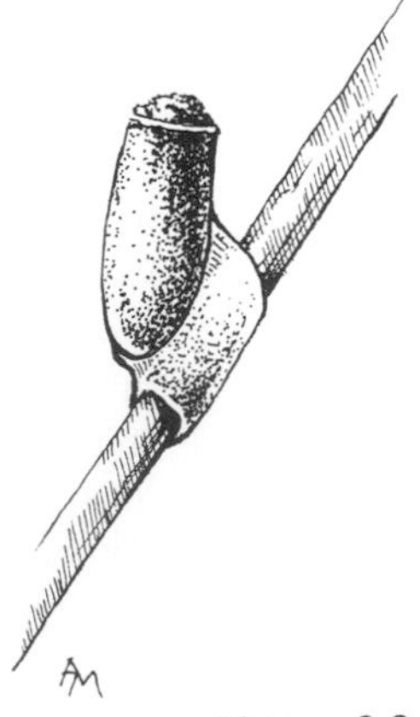

Figure 6.6.
Egg of a louse, a nit.

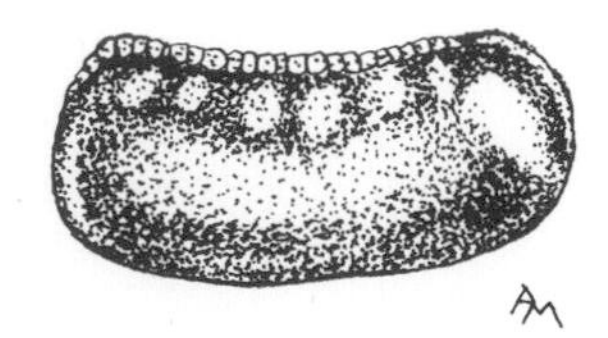

Figure 6.7.
Egg case of a cockroach, an ootheca.

oothecae are brownish capsules containing as many as 250 developing nymphs. The egg cases are glued to woody plants, or sides of buildings, where they overwinter. Scale insects lay their eggs under the scale covering of the female, where they remain protected until they hatch. The female gypsy moth deposits her eggs in a mass of body hairs attached to the sides of trees, buildings, or travel trailers. Most insects lay many eggs, which helps to counteract their high mortality rate and continue the species.

Nymphs, naiads, and larvae

Nymphs are the immature stage of insects with incomplete, or paurometabolous, metamorphosis. They closely resemble their parents, but they are smaller and lack wings. Instead, wing pads develop and will eventually become the functional wings. Their active feeding, as with larvae, is responsible for considerable crop damage, particularly nymphs of grasshoppers, plant bugs, and aphids. The number of nymphal instars varies from species to species. The milkweed bug, for example, has five nymphal instars. After the last nymphal molt, the wings become functional, and the reproductive organs mature. Thus the adult stage is reached without a pupal stage, although considerable change takes place between the last nymphal instar and the adult.

Naiads are equipped with gills to absorb oxygen dissolved in water, their habitat before they become adults. Because of this, and other adaptive features, they differ considerably from the appearance of the adult (fig. 6.8). Naiads live in freshwater ponds, lakes, streams, and rivers, where they may take as long as three years to develop. The food of some is algae and other plant life, while others are predators, feeding on other insects and other small aquatic invertebrates. Stonefly naiads are found under stones in water; hence, their common name. Some are plant feeders; others are predacious. Although difficult to rear in an aquarium, they are interesting subjects and demonstrate a variety of adaptive features.

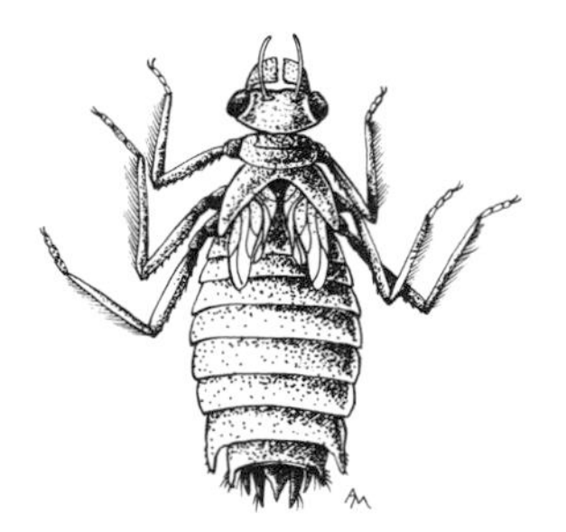

Figure 6.8. *Naiad of a dragonfly.*

Larvae, characteristic of complete, or holometabolous, metamorphosis, develop within the egg. After the egg is fully developed, that is, capable of living an independent life, it leaves the egg and feeds. This is usually the principal feeding stage. Most are more or less wormlike, but variously shaped and usually white or brownish, but many are dark or colorful. The number of molts is fixed for the species, but this is influenced by the type of food it eats. Leaf feeders, for example, have four or five molts and then transform into the pupal stage. Other situations may require more molts, and, in certain species, for example, the larvae of the tsetse fly, there may be only a single molt.

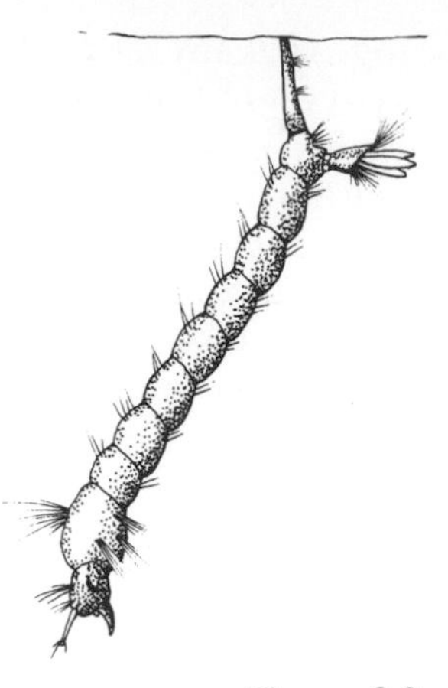

Figure 6.9. *Larva of a mosquito, a wiggler.*

Once the larva reaches its genetically determined maximum size, it ceases to feed. It may wander around looking for a suitable place to remain during the pupal stage. Some shorten, become less active or entirely inactive. The larvae of most moths attach themselves to a twig or other suitable anchor, spin a cocoon, and change to the pupa inside.

Many common names are applied to the various kinds of larvae. Grubs and wireworms are two common names used to designate the larvae of various Coleoptera, the beetles. Caterpillars, cutworms, earworms, armyworms, and hornworms are only a few of the names used for the larval stage of Lepidoptera, the butterflies, moths, and skippers. Maggots, wigglers (because they wiggle in the water) (fig. 6.9), bots, warbles, and screwworms are terms for the larval stage of Diptera, the flies. Hellgrammites are the aquatic larvae of the dobsonflies, and doodlebugs, the larvae of antlions (both Neuroptera). These names have been designated as a means of identifying the culprit causing damage to a particular crop. Because of the amount of food consumed by larvae, it is this stage that is of greatest economic importance. For example, adult moths cause no damage to crops, but their larvae cause millions of dollars of damage each year to the world's forest and field crops.

Prepupae and pupae

Pupae do not feed and are usually inactive. As the stage between larvae and adults of holometabolous insects, their "job" is to remake the larvae into adults. This stage should not be considered a resting stage, as it is sometimes called, for, while a pupa, all larval characteristics are converted to adult characteristics. A tremendous amount of biochemical and histological activity takes place in this stage. Some pupae are covered with a silken cocoon, characteristic of moths, secreted and formed by the larva before it transforms. Butterflies form a chrysalis, a naked case secreted by the larva (fig. 6.10). Other species form a puparium, a structure made from the last larval skin, characteristic of flies. Many pupae are "naked," that is, without any protective cover. They are usually white or pale yellow and resemble mummified adults, typical of ants and bees.

Figure 6.10. *Pupa of a moth.*

The pupal stage is vulnerable to desiccation and predation. As are eggs, pupae are often hidden, protected from severe weather and hidden from predators. Protective coloration and careful concealment assist pupal survival. Pupae often overwinter and emerge the following spring. The chrysalises of butterflies and cocoons of moths are collected in the early spring and soon thereafter adults will appear. To observe this, gather some cocoons or chrysalises and place them in a dry fish tank or other large container with a layer of soil in the bottom to keep the chamber humid. If these are collected too early, their disapause may not be broken, and the adults will not emerge. Put them outdoors or in a refrigerator for two or three weeks at 7° C. If this fails after a week or so, repeat. Place sticks against the side of the jar to serve as a landing place for the winged adults that emerge. Do not be surprised if flies or wasps emerge instead of moths and butterflies. If the larva was parasitized, this will happen. Keep these specimens; they may be new host records.

We mentioned in the preceding discussion of the larvae that the last instar may shorten and change somewhat even before the pupa is formed. This is normal for holometabolous insects. However, some paurometabolous insects have a prepupal stage. This is also an inactive stage resembling the pupa of the holometabolous species. It differs, however, because it is a stage between a nymph and an adult instead of a larva and an adult. This occurs in thrips (Thysanoptera) and scale insects (Homoptera). No doubt it is a primitive form of complete metamorphosis.

Subimago and imago

"Imago" is another term for the adult stage. The main focus of the imago is reproduction. Although many adults feed, and sometimes cause considerable damage and may be of great economic importance, their main function is to mate (usually, although some insects are without males) and lay eggs. Some adults, such as those of mayflies (Ephemeroptera), do not feed. They mate, deposit their eggs in the water, and die. They are short-lived. Most other insects live longer, feed on various materials to obtain energy for mating, for egg manufacture, and for flying or walking to find mates. They may live two or three months, occasionally for a year, but rarely longer.

Rarely there is a "subimago" stage, a winged, sexually immature stage. This occurs in the mayfly after it emerges with wings from the last naiad instar, probably as a way of getting the adult from water to the safety of land. The subimago flies to a nearby bush or tree, and a final molt takes place immediately by shedding the skin of the subimago to reveal the fully formed, sexually active adult beneath.

Most common names of insects apply to the adult stage. The gypsy moth, boll weevil, mydas fly, drone fly, tiger swallowtail, and eyed click beetle are all names used for adults. Therefore, we refer to the gypsy moth caterpillar, the boll weevil grub, and so on, when we are talking about the immature stages of these common species.

Fecundity

Most of us think of the adult stage as the end or goal of these life-cycle changes, but, as the chapter title implies, it isn't. Perhaps because we think our adult life is our reason for being, we ascribe this same meaning to the insect's life. A more biological way of thinking about this is that these life-cycle stages are merely what takes place to produce more eggs, for isn't the egg the most important thing for the continuation of the species? If something happens to an egg between the time it is laid and the time more eggs are produced, that segment of the species population disappears forever.

Granted this, then an important feature of every species is its fecundity, or fruitfulness of offspring. Obviously the more eggs laid, greater is the chance of survival of resulting offspring, and better the chance the species will continue its evolutionary course. Thus, we refer to some species as having a high fecundity, meaning that many eggs are laid and many young produced. We think of the species as being more successful, an attribute characteristic of cockroaches, corn earworms, and house flies! We should point out that most species of insects can't make this claim of success. Thus we find many "rare" species, and pride ourselves on having added these to our collections. Therefore, whether a species is to be considered rare or common depends, partly at least, on its rate of fecundity. We say partly, because a species may be rare in a collection, because we do not know where to find it. Using the terms "rare" and "common" must always be qualified by a consideration of the species' fecundity.

Diapause and dormancy

Strange as it may seem, most of the life of temperate region insects, and many tropical species, is spent without movement and with the rate of metabolism great reduced. Three principle terms apply to these periods, depending on the reason for this period of inactivity. Dormancy is a resting, or quiescent, period with reduced protoplasmic activity due to high carbon dioxide concentration within the tissues. Diapause is a spontaneous period of dormancy, which may occur at any time during the development cycle of the insect. During diapause, there is no growth, no metamorphosis, and the metabolic rate is reduced. Quiescence is another term for diapause, and it refers to arrest or temporary cessation of development because of unfavorable environmental conditions. When dormancy occurs during the winter, it is often called hibernation, but, if during the summer, it is aestivation.

Diapause is controlled by temperature and photoperiod. For example, as the days grow shorter in the fall, the pupae of hornworms enter diapause instead of completing development and permitting the adult to emerge. Obviously, if the latter happened, the adult would die without mating and laying eggs. Diapause lasts for the entire winter and is broken only after a fixed period of low temperature. This means survival for insects living in areas with low-temperature winters. This phenomenon is under gene control and will differ in northern and southern populations of the same species.

Dormancy may occur in any stage. It is least common during the adult stage, normal in larvae or nymphs during the summer as a result of dry or cool weather condition, and characteristic of the embryo in eggs. For example, silkworm moth eggs laid in autumn will not develop until spring. They must be exposed to 0° C for several months. However, eggs laid during the summer will develop and hatch into active larvae.

Factors controlling diapause also determine the number of generations each year. Species with only a single generation each year are termed univoltine. This is characteristic of northern species, and it is possible because of diapause regulation, which spreads out the stages throughout the year. The same species, living in a warm, southern climate, may be bivoltine, with two generations each year, the first completing its life cycle rapidly in the spring and early summer, but the second brood entering diapause during some stage and overwintering in this condition. Even more generations may occur in rapidly breeding species, such as fruit flies and aphids. Even in tropical regions, where temperatures are moderate throughout the year, species will have a diapause period. This may be due to the alternation of wet and dry seasons, or other factors such as the flowering of host plants.

The most important inducement of diapause is the photoperiod. Critical day length induces pupation much the same as it induces flowering among plants. Temperature changes obviously induce or end diapause, or at least speed up or slow down development. Low humidity slows development of eggs. Completely desiccated eggs will survive long periods of drought and, when wet, will quickly develop into larvae. Mosquito eggs laid on dry land may develop within a matter of two or three hours once submerged in water, but if not, they may survive dried out for three or four years and sometimes even longer. Obviously, this is an excellent way to survive in severe desert conditions. Dormancy and diapause are peculiar to each species, since they are genetic adaptations.

Diapause in jewel wasps

The jewel wasp, *Nasonia vitripennis* (Walker), is a parasitic wasp easily reared in the classroom and will demonstrate some of the principles of diapause. This insect does not sting; hence, it is safe to study without extra precautions. The female wasp oviposits in the hard puparium of flies. Their larvae feed and eventually destroy the developing pupa. In fact, they are an effective means of controlling fly populations.

During the larval period, dormancy sets in, and the parasite stops growing. Diapause can be prevented or induced by regulating the temperature of the culture. Light also is a factor. Thus, if the culture is kept in the dark at a temperature between 26°C–30°C, diapause is prevented. However, if the temperature is lowered to about 6° C for three or more days, diapause is induced. Diapause can be broken only by keeping diapausing larvae for three months or longer at 2–5°C and then transferring the larvae to room temperature. Jewel wasps infect house fly or blow fly larvae, and materials for this experiment are available from biological supply companies.

7 Doing Their Thing

"Go to the ant, thou sluggard; consider her ways, and be wise."

Proverbs 6:6

Sluggards and ants are supposed to have contrasting ways of life, but, in a sense, laziness and stereotyped behavior are the same. The lazy man refuses to think. The active ant cannot think. Both have what we term "frozen brains" and cannot change their ways. To understand these implications, we must examine the factors that control the behavior of insects.

Behavior depends on the evolution of the nervous system to a certain extent, but not completely. Single-celled protozoans, *Vorticella* sp., for example, react to simple stimuli in a variety of ways, showing that response is an innate feature of protoplasm. Obviously, the more complex the receptor organs and the more highly evolved the nervous system, the more complex the behavior of the organism. Behavior is generally reserved as a feature of animals, since plants have no nervous system. At the same time, it cannot be denied that plants react to external stimuli, as witness the closing of the Venus's Fly Trap.

The study of insect behavior has excited generations of entomologists. Insect behavior, or ethology, combines factors involving the morphology, physiology, and life history of insects. Some ethologists specialize on the anatomical components of behavior, others on the chemistry and physiology of the subject; all relevant data are used for a comprehensive and cohesive understanding of behavior.

Ethology has become of increasing interest to biologists in recent years, especially since worldwide attention was given to the field by the awarding of the Nobel Prize in Physiology and Medicine in 1974 to three pioneers in the field: Konrad Lorenz, Niko Tinbergen, and Karl von Frisch. Von Frisch is noted for pioneer work on communication among honey bees, and the other two scientists for various experimental studies, as well as field observations on the behavior of a wide variety of animals.

Two common pitfalls are encountered when studying insect behavior. First, anthropomorphism, the attributing of human characteristics to insects, distorts the observations and clouds the scientific results. Extreme examples of this are the proposition that crickets sing because they are happy or ants tend to their young because they love them. Teleological explanations of natural events ascribing purpose to fulfill a goal is the other common mistake in behavioral studies; for example, the notions that termites build nests and eat wood to provide for their young or that lightning beetles produce light to enable them to see in the dark. All statements made about observations must be tested with controls before conclusions may be drawn, if one is to be scientific. Keep in mind that "Occam's razor" says that the simplest explanation is the most likely.

A variety of terms have been used to describe insect behavior; the most generally used is the term "instinct," meaning that all behavior is the result of inherited action patterns that cannot be changed or stopped. Recent writers have tried to avoid using this word, because it is vague, without a precise definition. Instead, they prefer to use reflex and taxis. A reflex is a simple, inborn response to a stimulus, dependent on and involving the nervous system. It is an automatic response, which usually can be predicted. Most insects fly when a shadow falls across the insect's eyes, a response that always seems to take place under such circumstances. A taxis is movement in a particular direction in response to an external stimulus. This may involve several reflexes and is either positive or negative. There is an intensity threshold; that is, the stimulus must reach a particular level before a reaction is initiated. We might reserve the use of the word instinct to describe the inborn pattern of involuntary responses combining reflexes and taxes.

Learning, implying the modification of behavior through

past experience, probably does not exist in insects, or at least it is difficult to show experimentally. It is true that behavior changes may be brought about through continued stimulation, but all known cases are simply a change in reflex action due to fatigue, starvation, or similar effects, as will be shown in some of the following examples. Finally, reasoning, the formation of judgments from given facts without previous experience, is not demonstrable in any but the mammals. Some actions may appear to be examples of reasoning, such as those described later for the sexton beetle. But, as we will see, such is not the case.

With these terms in mind, then, it is possible to define insect behavior as all the activities and reactions of the whole organism in relation to its environment and the nature (or genetics) of the insect, that is, its ability to behave. Most, if not all, behavior is directly related to physiological needs, as shown in table 7.1.

Table 7.1. *Biological drive and behavior.*

Motivation	*Mechanism*	*Result*
Food getting	Aggression	Competition with other species
Survival	Protective devices and action	Irritation of other species
Reproduction	Courtship and mating	Promotion of the species

Orientation behavior

All organisms are constantly bombarded with a full range of stimuli from the environment. They react either positively or negatively to light, humidity, temperature changes, chemicals, and other organisms. Part of this reaction is orientation. They face the stimulus, turn their backs to it, or simply ignore it, which in itself, is a passive reaction. We already know that some insects avoid light, others are attracted by it; most insects avoid drying and seek moisture; all insects tolerate a particular temperature range, avoid caustic chemicals, and are attracted to those things that "smell" like food or a mate. It is certainly a basic and remarkable feature of all animals from the simplest to the most advanced to be able to recognize their own species. This means that they categorize every individual animal as one of four kinds: 1) a member of the same species; 2) a member of a different species that may be food; 3) a member of a different species that is not food, but may be dangerous; and 4) a member of a different species that may be safely ignored. All decisions in this respect are made as reflex responses to intricate combinations of stimuli. The process is not simple, nor is it perfect, but it is, nevertheless, characteristic of all animals.

Directive orientation is usually by trial and error (kinesis), due to unequal stimulation of sense organs. We know which direction to go in when we are called, because of differences in the sound reacting on one ear, as compared to the other. The same is true of insects, and this is, of course, the advantage of being bilaterally symmetrical and thus having paired sense organs. This is much more efficient than that exhibited by the lower, nonbilaterally symmetrical animals.

Chain behavior (instinct)

Insect response is more complicated than simply a reaction to isolated stimulation. Each species, through the long process of selection, has inherited distinctive patterns of response following a particular sequence of events. Many authors refer to this as instinct, but we prefer the more descriptive term, chain behavior. Thus, we find that when there is a series of seemingly complex reactions, it is because each is stimulated by the performance of a previous action. External stimulation initiates the changeover from one main behavior action to the next. The completion of one action acts as a releaser of the next. It is this chain that permits seemingly complex patterns by allowing for these changes through otherwise stereotyped behavior.

The famous French naturalist, Henri Fabre (1823–1915), describes many simple examples of chain behavior. The observations he made and experiments he performed are the basis of a lifetime of writing in a series of books that still retain their charm, though they fail the test of modern experimental science.

Fabre observed a marching cadre of caterpillars descending from a tree in search of another to replenish their food supply. He noted that the lead caterpillar must find the tree, the others merely followed. Holding a ceramic garden pot in front of the lead caterpillar, Fabre succeeded in getting a line of caterpillars to completely encircle the pot. Once the lead caterpillar reached the last caterpillar in the circle, it too became a follower, leaving the chain without a leader. This removed all responsibility to find a new food tree and resulted in the demise of all the caterpillars from fatigue and starvation. We see the same reaction in a herd of sheep, stampeding cattle, or certain political movements among humans.

Fabre performed similar experiments with ants, discovered that they laid down scent trails (not knowing about pheromones described further in this chapter), and got them to run in circles. Once he grasped the meaning of this simple mechanism, he made additional tests on more complex behavior patterns. For instance, Fabre watched potter wasps of various species and noted that one species followed a certain pattern in building its nest, others, another. The basic nest building sequence consists of bringing small balls of mud to the building site and adding bit by bit to form a potshaped nest. Once the nest reaches the size characteristic for the species, a new action is triggered. If the nest is partially destroyed during the building process, it is repaired; if damaged after it is completed, no repairs are made in some species; in others, repairs are made even after the next action, nest stocking, has started, unless the damage is extreme. Stocking, in some species, consists of filling the nest with stung caterpillars on which the female lays her eggs. Some species fill the nest to the top; others seem to be able to count (probably because of the limited number of eggs she carries) and put in a more or less fixed number. Once stocking begins, a man-made hole in the side or bottom of the nest is generally ignored. Stocking will go on either until the specified number has been added, even though they fall out through the hole, or the nest is capped. Capping is triggered by a full nest. Those species that cannot "count" eventually weary, and the chain is broken. The others complete the cap and go away leaving behind an empty nest.

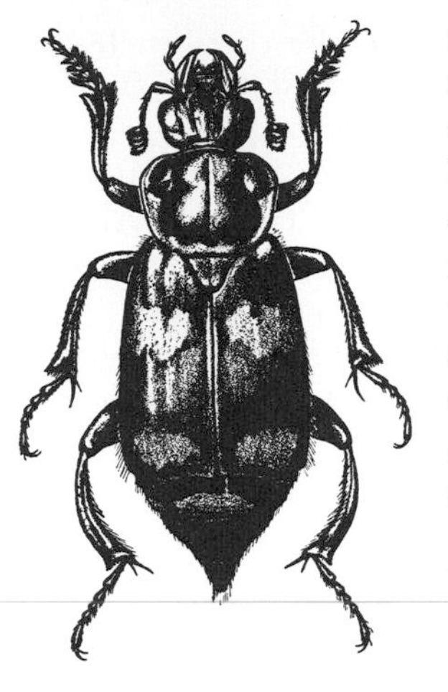

Figure 7.1. *A sexton beetle,* Nicrophorus sayi.

One of the most profound experiments is that of testing the sexton beetle (*Nicrophorus* spp.) (fig. 7.1). These beetles are attracted to freshly killed small animals, such as snakes, toads, mice, even fish. To protect these choice bits of food from other necrophilous animals, several pairs of beetles dig out under their prize and drop it into a grave (fig. 7.2) (thus their common name; their Latin name, incidentally, means "bearer of the dead"). To confuse the beetles, a mouse was tied between two sticks pushed into the soil. The string held the mouse in the air above the ground, which, of course, prevented the mouse from being buried. By all rules of chain behavior, digging should cease and the interruption discourage further digging by the beetles. Such was not the case. Instead, the beetles climbed up

 Figure 7.2. *Sexton beetles burying a dead mouse.*

onto the sticks and chewed the string in two, allowing the mouse to drop to the ground. Then burying proceeded normally.

Unless you apply the rules set forth at the beginning of this chapter, you might erroneously conclude that this action is an example of reasoning by these clever beetles. Not at all! Almost always the sexton beetles meet rootlets as they dig, and these must be chewed in two and removed before the untethered animal can be buried. What is the string to the beetle but a rootlet? They simply searched for and found the offending obstacle preventing burial and went on with their normal chain reaction.

Insects have a central nervous system, which lies beneath the digestive tract (the opposite position from ours). The ganglia of the head end are enlarged to form what we like to call a brain, but this structure is not used in exactly the same way as ours. Most incoming messages are dealt with by the ganglia serving each segment of the insect body, and the reflex action that results is not through any consideration of the action by the brain. It is for this reason that decapitated insects continue most of their activities as if nothing much had happened, at least for awhile. Decapitated male praying mantids will continue to copulate with the female, and decapitated female silk moths will continue to lay eggs, regardless of the loss of that seemingly necessary part. These maimed individuals will walk or run, although with slightly different gaits, but they seldom fly, because they lack the necessary organs for orientation. The brain controls muscle tone and sensitivity, as well as body functions, such as metamorphosis.

Insects lack "consciousness." They merely differentiate between groups of stimuli and select accordingly. Once they are "on course," they are hard to sway. Beware of the hornet bent on stinging. Only death seems to discourage this compulsive action.

Insect communication

Communication between individuals occurs when one provides visual, auditory, tactile, or chemical signals that influence the activity of another individual above and beyond its reaction to environmental stimuli. This happens between members of the same species, as honey bees in a hive, or between different species, as between aphids and ants. Signals may stimulate mating or induce males to defend their territories. Assembling or dispersing among social insects, and cooperative behavior of castes, are other results of these signals. Communication between different species is of three types: that benefiting the signalers; that benefiting the receiver, and that resulting in mutual benefit. Communication may become more complex by involving several types of signals. Signals may be tactile, chemical, visual, or audible. Light and sound signals are treated in the next chapter.

Tactile and visual communication. Communication by touch between insects is possible only after other signals bring the insects together. Rarely do insects depend on touch as their sole means of communication. If it were otherwise, insects would have to run blindly into each other before any type of interrelationship could take place. They do, of course, examine many surfaces by touch to determine the nature of food or to find a suitable place to lay eggs.

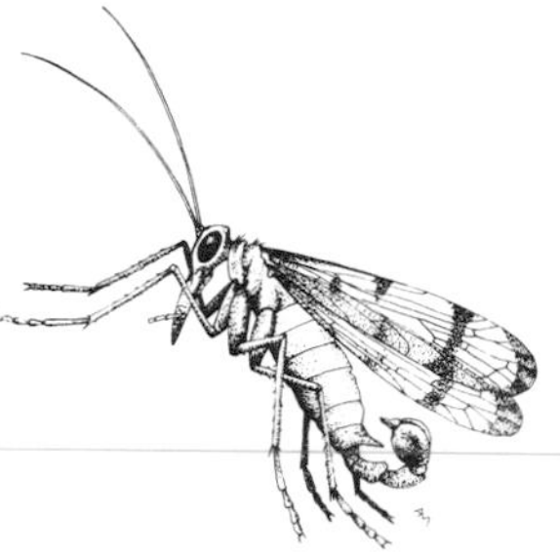

Males of the common fruit flies (*Drosophila* spp.) are first attracted to females by recognizing their shape, but final recognition depends on the male's tapping the female with is front legs, after which they successfully mate.

Figure 7.3.
A scorpionfly (Mecoptera).

Scorpionflies (Mecoptera), so-called because the males of some have an upturned abdomen (fig. 7.3), resemble scorpions, at least to us. Unlike scorpions, these insects do not sting. Exactly why this resemblance is uncertain, unless, of course, it is a way for the female to first recognize the male. Certain male scorpionflies (*Panorpa* spp.) court females by offering a gift of food, which is received before copulation takes place. The male appeases the female with droplets of saliva, which harden and serve as a snack for the female during copulation. Males of other species bring seeds or even petals to "calm" the female. This appeasement behavior seems to guard against exciting an aggressive response in the female, who might possibly attack the male.

Chemical communication. A group of chemicals known as pheromones, chemical messengers secreted from the body of an insect and released into the air, elicit responses of various types. Pheromones are produced by insects and many other animals as well, including some mammals. Sometimes these are called ectohormones, because their effect externally is much like the effect of hormones internally. Pheromones, produced in special glands, are released at specific times by the organism, and, although their chemical composition is not species specific, they usually affect only members of the same species that secrete them. Many of these complex compounds have been chemically identified, but more remain unknown, except for their specific function.

Insect pheromones are grouped according to the type of behavior or activity they coordinate. Sex, alarm, trailmaking, aggregation, and caste-regulating pheromones are discussed below. The typical pheromone is a chemical or mixture of chemicals released by one individual to induce a response, often several, by the opposite sex of the same species. An allomone, on the other hand, is a chemical released by one species, which induces a positive response by another species. This is aimed at another species for defense. Another, a kairomone, is a chemical released by one species to elicit a favorable response from another species, usually as a host attractant. We are concerned here only with the typical pheromones, though you should know that insect communication through chemicals is a complex subject.

Sex pheromones. These, probably the best known, are produced by either males or females and attract members of the opposite sex. In this manner, they act as a congregating mechanism; that is, they bring scattered individuals together. Pheromone production is controlled by endocrine glands. It has been studied in many insect pests where it is used as a means of sexual communication, but, of course, it is not limited to pest species. Certainly this occurs in many other species, if not all. The active chemicals in the sex attractants of many economically important insects have been isolated and identified (table 7.2). Sometimes they are used in pest management, and some have been prepared to use to disrupt mating behavior patterns, either to lure males to their death in traps, or to confuse the sexes by releasing large quantities of the material and saturating the atmosphere.

The Gypsy Moth, *Lymantria dispar*, causes considerably damage to deciduous and evergreen trees in the northeastern United States. This member of the tussock moth family, Lymantriidae, was imported into Massachusetts from Europe in 1869 as a source of silk. Unfortunately, it escaped and became established in the wild. Now the caterpillars damage millions of acres of forest and shade trees throughout Northeastern North America and are spreading rapidly elsewhere. The female, a weak flier, usually only flutters about from place to place. The male, however, is a strong flier. Females produce a pheromone that the

Table 7.2. *Synthetic compounds developed from isolated sex pheromones.*

Synthetic Compound	*Scientific Name of Insect*	*Common Name*
Grandlure	*Anthonomus grandis*	Boll weevil
Disparlure	*Porthetria dispar*	Gypsy moth
Riblure	*Argyrotaenia velutinana*	Redbanded leafroller
Hexalure	*Pectinophora gossypiella*	Pink bollworm
Bombykol	*Bombyx mori*	Silkworm
Trimedlure	*Ceratitis capitata*	Mediterranean fruit fly

males can detect in minute quantities with their antennae. They home in on the females, attracted from considerable distances, even a mile away. Without this sex pheromone, it would be only by chance that the male and female would find each other, since she rests near the base of a tree, while he flies high in the forest canopy. Entomologists are able to isolate this sex pheromone and use it in a trap (fig. 7.4) to attract the males. Once destroyed, mating is prevented. This usually augments other means of control in operation at the same time. The use of the Gypsy Moth sex pheromone is not effective enough to cause the moth to become extinct (no known method of control has ever achieved this), but it does reduce the populations to acceptable levels in some areas.

Females of the American cockroach, *Periplaneta americana*, attract males by pheromones. After mating, and once the female produces an ootheca (egg case), the pheromone is no longer produced.

Another interesting pheromone occurs in the conenose bug (also known as the kissing bug), *Rhodnius prolixus*. This bug feeds on human blood and is the vector of American trypanosomiasis, or Chagas disease. Before females mate, they must acquire a blood meal, necessary for maturation of the female sex organs and to provide protein for eggs. Once the blood meal is acquired by the bug, she produces a sex pheromone that attracts the male. This is perfectly timed, since she is now ready to bear young. Obviously pheromone production is controlled by the insect's endocrine system.

Alarm pheromones. These warn individuals that danger is nearby. This is most common in social insects and others that are gregarious. The effect of alarm pheromones is not restricted just

Figure 7.4. *Gypsy moth sex-lure trap.*

to the species that produces the substance, but may excite related species, as well. This pheromone must be produced in large amounts to cause a response and is only effective for short distances and for a short period of time. Ants commonly secrete these alarm substances when disturbed. Termite soldiers, especially nasutes, rush to the source of the danger, attracted by the alarm pheromone. These termites use a special chemical spray to destroy, or drive back, predators (chap. 17). Honey bees produce an alarm pheromone as they sting a victim, which alerts other bees to nearby danger.

Trailmarking pheromones. These are common among social insects, especially termites and ants. They are laid down as the workers forage. Once they locate a food source, they return to the nest to communicate this information to other workers. Trails are marked either with a continuous line of pheromone scent or by intermittent deposits. This substance is short lived, unless the source of food is extensive, in which case, as more and more workers pass along the path, more of the pheromone is deposited.

Both ants and termite workers have special glands for the secretion of trailmarking pheromones. The next time you see a column of ants moving to and from a source of food, you may wish to try an experiment. Place an additional source of food, say cookie crumbs, about two to three feet from the opening of the nest. Next, place a piece of paper between the food and the nest opening. Soon you will see ants heading toward the food; note that these few ants are the scouts in search of food. Once the ants (or, sometimes, a single ant) find the food, after passing over the paper, they will return the same way to the nest. When these ants have returned to the nest, remove the paper. Shortly a column of workers will be on its way toward the food; when they reach the spot where the paper was placed, they will become disoriented, because the trailmarking pheromone was removed with the paper. It will take them some time to find the food again. This simple test illustrates the use of the trailmarking pheromone and how it enhances the food-finding ability of a nest of ants. Similar observations have been made using tropical termites.

Aggregation pheromones. These are produced by one or both sexes to attract members of the species to food or to the mating area. They have been extensively studied in the bark beetles (Scolytidae). When bark beetles attack a tree, an aggregation pheromone is produced in the hindgut of the beetle, combined with the feces, and dropped. This frass attracts other bark beetles. As more beetles bore into the tree, more aggregation pheromone is deposited with the frass from the borings. After a certain period of time, the pheromone is no longer produced, but the number of beetles at the site is large enough to establish a colony in the tree.

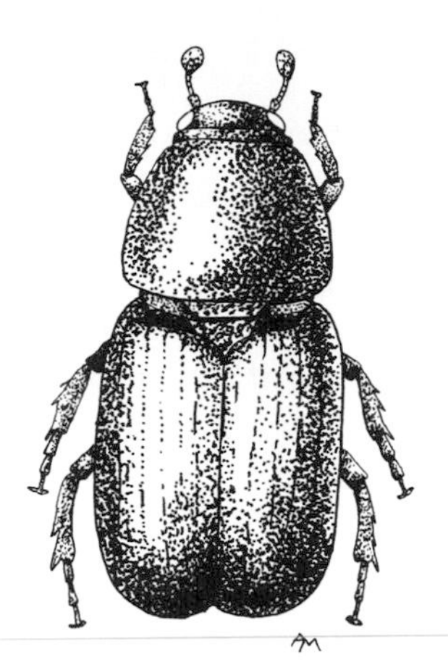

Figure 7.5. *A bark beetle (Coleoptera: Scolytidae).*

Bark beetles do serious damage to trees by boring into live wood and by transmitting plant diseases. These small beetles, 3–5 mm in length (fig. 7.5), usually black, brown, or reddish brown, feed on the inner bark of numerous species of trees, including elm, hickory, walnut, pecan, and conifers. Probably the most notorious of all bark beetles is the European bark beetle, *Scolytus multistriatus*, a species that was accidentally imported into the United States bringing with it the fungus that causes Dutch elm disease. Many American elm trees, *Ulmus americana*, a shade tree in many areas, have been killed by the fungus. The adult bores into the trees and deposits its eggs, and the fungus that will eventually grow, in the vascular system of the tree. This become clogged. The fungus spores are transported by the adult in special pockets on the beetle, adapted for this purpose. This symbiotic relationship helps the fungus by spreading it from tree to tree and the beetles by killing the tree into which they bore.

Other bark beetles, species of *Ips* and *Dentroctonus*, are similar pests. Coniferous trees often resist attacks of bark beetles by drowning the attacking insects in sap, but this resistance is overcome when enough bark beetles attack a tree and the fungus they carry plugs up phloem, which prevents the flow of sap. The beetles lay eggs as they bore under the bark. When the eggs hatch, the larvae branch out in all directions as they feed, leaving an interesting, engraved pattern on the wood (fig. 7.6). Some are

Figure 7.6. *Burrows of engravings under the bark of dead wood made by the adults and larvae of a bark beetle. (Courtesy of Dr. Stephen L. Wood)*

utterly spectacular and may be seen on firewood logs after the bark falls off.

Bark beetles attract other insects to the trees. The pheromone apparently has some components that signal other insects as well. Some are predators of the bark beetles, and others also attack the tree. A wide variety of beetles and other insects may be found under the loose bark of the dying trees, ranging from wood borers to predators, eggs, larvae, pupae, and adults, all fine additions to any insect collection. Although most of these are small, they are extremely interesting, because of the variety of adaptions they exhibit. Probably most insect species have aggregation pheromones, but these remain unknown to entomologists.

Caste-regulating pheromones. Pheromones precisely regulate caste members in insect societies. Termites, ants, social wasps, and bees, each with castes, maintain their colonies by pheromone communication. Caste regulating pheromones limit the number of reproductives, the queens and kings, the workers, and the soldiers. Caste regulation is necessary to keep order and direction in the colony. If caste-regulating pheromones were not produced, there might be disproportionate numbers of reproductives or soldiers, for example, in termite nests. Workers could not provide for the needs of the colony and would soon die.

Insect defensive reactions. Insects employ an array of defensive and offensive reactions to protect themselves and their offspring. As with most behavioral reactions, these depend on both morphological and physiological adaptations, and some are most effective. Defenses involve behavior, structure, chemicals, and coloration, which includes protective and warning colors and mimicry.

Behavioral defense. Probably the most commonly observed behavioral defense in insects is their reaction when approached. Generally, they will fly away, scurry under a leaf, stone, or anything they can squeeze into, and, if they are equipped with jumping hind legs, away they go, quicker than the eye can follow. Because of their small size and their ability to move quickly, they are able to employ these tactics to avoid any danger, suspected or real.

Grasshoppers, katydids, and crickets simply jump out of the way of potential danger; flies, bees, and wasps fly away. Reflex dropping is a common form of protection, especially in leaf, long-horned, and snout beetles. Most species of these insects will fall to the ground when disturbed. Other species of beetles and many other insects as well, will feign death when touched. They will lie still, and, without any movement, many of their predators do not recognize them as anything but a small stone or a seed. This works best, of course, with those predators that will eat only live prey.

Structures used for defense. Numerous morphological adaptations have developed among insects as protective devices to prevent predation. Spines, spurs, and extremely hard cuticle are a few adaptations used to ward off attacks from other insects, spiders, and, sometimes, even bird predation.

The first line of defense for any insect is its cuticle, which forms the nonliving exoskeleton, a covering of the true integument lying beneath. Not only is this structure of great importance in preventing the loss of vital internal fluids of the insect, but also for protection from the action of chemicals, invasion of bacteria, and warding off of predators. Some insects have heavily sclerotized exoskeletons tough enough to require a jeweler's drill to penetrate their surface—no insect pin is strong enough to push through it for mounting. This is true of many weevils, buprestids, and darkling beetles, especially species that live in deserts. Mouthparts play a major role in defense. The sharp mandibles of some ants, dragonflies, tiger beetles, and termite soldiers, among others, are used to ward off potential predators. Even piercing-sucking mouthparts of the ambush bug, water bugs (aquatic Hemiptera), and assassin bugs have proven to be excellent defensive weapons, and most field biologists are aware of the painful bites of certain horse flies and the stable fly. Some insects, for example, the common lubber grasshopper used in biology classes and praying mantids, are fitted with numerous spines and spurs. Predators find these feeding deterrents and often avoid such insects. Attempts to eat these spiny creatures often result in the frog, lizard, or snake predator's vomiting up the insect before it is affected by digestive juices. We have seen large toads capture and swallow large horned scarab beetles, only to have the beetle walk out of the toad's mouth a short time later. The cerci, modified into a pair of pincers, of the common earwig, serve as a defensive tool when the insect is under attack. It can also give the unsuspecting collector a good pinch!

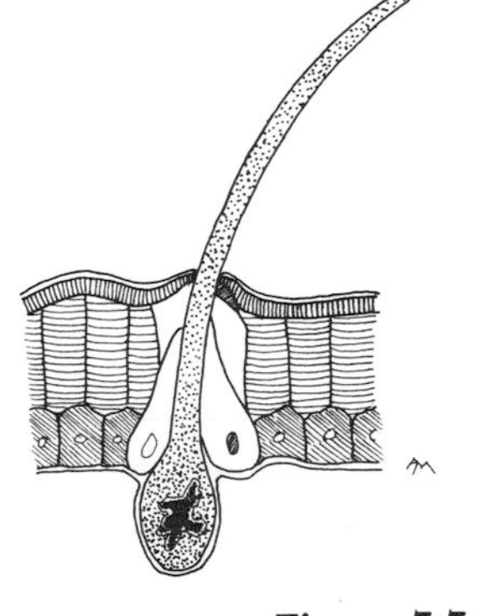

Figure 7.7.
Urticating seta of a moth.

One of the most spectacular structural defenses is the urticating setae of many caterpillars, such as those of the io moth, *Automeris io*, saddleback caterpillars, *Sibine stimulea*, and the flannel moths, *Megalopyge* spp. Their irritating setae (fig. 7.7) may be hollow tubes filled with fluids that cause a burning or urticating sensation when applied to the skin. These materials are known as vesicants. Some of these so-called "hairs" may not be hollow, but rather are similar to "angel hair" used on Christmas trees as

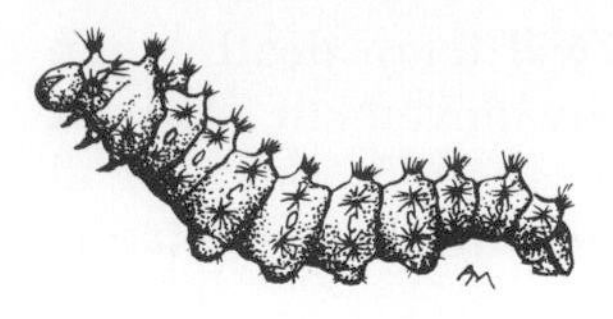

Figure 7.8.
A caterpillar showing groups of urticating setae.

ornaments, and can cause skin rash and a great amount of discomfort, if taken into the respiratory tract. Caterpillars of this type (fig. 7.8) should never be handled because of this.

Chemical defense. Many kinds of chemicals produced by insects are used as defensive secretions. Earlier in this chapter, we discussed pheromones that are primarily used as chemical signals to members of the same species. Chemical defense involves odoriferous or repugnant substances used against predators. Many of these are terpenes, caustic organic chemicals capable of causing severe injury to other insects and other small predatory animals. These substances are produced in special glands in the epithelial layer of the integument. Muscle contraction discharges the material stored in a small sack close to the gland.

Chemical defense secretions are produced by some species of insects in most orders. Some walkingsticks (Phasmatodea), already well protected by their cryptic shape and color, have glands in the thorax capable of emitting these materials. Many cockroaches secrete foul-smelling substances characteristic of their species. The Madeira cockroach, *Leucophaea maderae*, for example, is capable of giving off a particularly offensive odor in addition to its normal cockroach smell. Stink bugs (Hemiptera: Pentatomidae) (fig. 7.9) earned their common name because they produce a disagreeable odor, which seems to ward off predators. If you pick one up, you will smell a heavy, unpleasant odor on your fingers and hands. Although this substance does not blister the skin as do blister beetles and certain caterpillars, you will want to wash your hands after being covered by their secretions.

Three interesting beetles, the bombardier beetles, *Brachinus* spp. (Carabidae), the blister beetles (Meloidae), and the false blister beetles (Oedemeridae), use caustic fluids for their protection. Bombardier beetles eject a fluid that reacts to air. When secreted from the anus, it produces a puff of smoke and explodes with a popping sound. Some entomologists refer to this fantastic defensive action as a "cannonshot," as it certainly gets one's attention and wards off its attempted predators. Although this material is nearly harmless, it burns the skin, causing a numbing of the part, and slightly stains clothing and skin.

Figure 7.9.
A stink bug (Hemiptera: Pentatomidae).

Blister beetles produce cantharidin, a vesicant that causes skin blisters. When handled, they "bleed" cantharidin at the knees, or the fluid leaks out if the beetle is squashed. This substance is extremely poisonous. A few cases of death among horses from eating these beetles in hay have been reported. The bodies of these insects, because of this substance, are so poisonous that even a few individuals in baled hay, the usual way an animal inadvertently eats them, are enough to kill horses and cows.

Figure 7.10.
Larva of a swallowtail butterfly with osmeteria everted.

Swallowtail butterfly larvae have an interesting structure, a pouch-like sack, the osmeterium, located behind the head. This is responsible for the production of defensive fluids. Odoriferous chemicals from plant materials eaten by the caterpillars accumulate in these sacks. When the larvae are disturbed, their pouches evert, showing a brightly colored interior, and, when this is done, the odor is released (fig. 7.10). Not only does this material repel the potential predator, but the bright color of the osmeterium startles it as well.

Social insects produce their share of defensive secretions. Probably the most notable are those used by the nasutes (soldiers) (fig. 9.5) of certain termites. The front of the head is modified to form a nozzle-like structure capable of spraying a sticky fluid at enemies. Ants produce a variety of chemicals for defense, discharged from the mouth or anus, and often in combination with stings. Some Dolichoderine (soil inhabiting) ants produce foul-smelling fluids, which can be ejected from the anus aimed at enemies. Most ants produce formic acid, hence, their family name, Formicidae. Modified ovipositors of Hymenoptera form a stinger, which provides an effective offensive and defensive organ known and avoided by almost all terrestrial animals. (Further information on stinging is given in chaps. 9 and 11.)

Protective coloration. Protective color patterns provide still another method of escape from the insects' predators. This includes camouflage or cryptic coloration, flash patterns, warning coloration, and mimicry.

Cryptic coloration. This camouflage seems to be the most effective method of protection, judging from the number of species of insects that are so colored. Most insects either blend into their natural background or resemble a natural object by disruptive color arrangement similar to that used by military personnel to camouflage their equipment and themselves. Some insects resemble pebbles, twigs, leaves, bark, and even bird droppings. Such camouflage is characteristic of diurnal, more or less sedentary, insects living in the open.

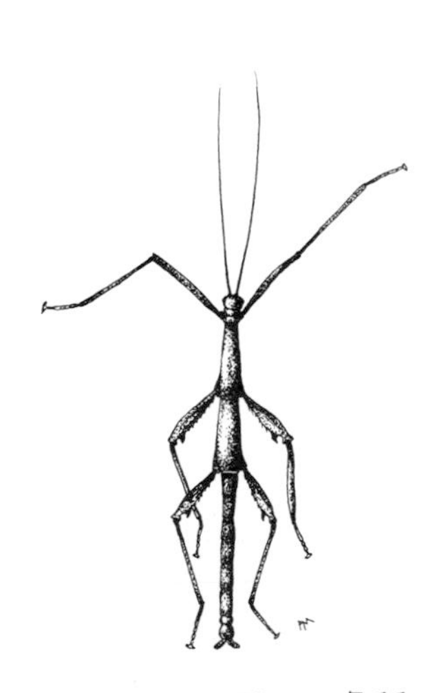

Figure 7.11.
A walkingstick (Phasmatodea: Heteronemiidae).

The green of katydids (also called long-horned grasshoppers) matches vegetation, the veins of their wings even mimicking veins of leaves. Certain treehoppers, particularly the thorn mimic treehopper, *Campylenchia latipes*, resemble the sharp thorns of plants. Sometimes these insects drop to the ground and become a menace to barefooted tourists and bathers. The superior camouflage of walkingsticks, possible because of their close resemblance to twigs, sticks, and small branches in their environment, assists them in avoiding predators (fig. 7.11).

Several families of Lepidoptera (moths, butterflies, and skippers), including Sphingidae and Ithomiidae, have species with clear wings; that is, the membrane between the veins is transparent. When some of these species are at rest on a tree or even on the ground, their transparent wings help them blend into their background and they seem invisible, especially to vertebrate predators. Many examples of this type of camouflage can be seen in the tropics. One outstanding example is the dead-leaf butterfly, *Kallima inachus*. This butterfly resembles a dead leaf when it lights on a twig; its position with wings folded above the abdomen completes the mimicry (fig. 7.12).

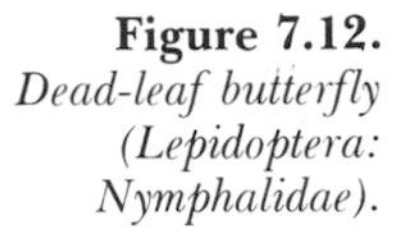

Figure 7.12. *Dead-leaf butterfly (Lepidoptera: Nymphalidae).*

Some predators rely on camouflage to hide them from their prey. Tropical mantids, ambush bugs, and antlions all blend into their background as they wait for their food to go by.

Flash patterns (revealing coloration). Flash coloration is used to confuse would-be predators. Bright color is exposed while in flight, but hidden when at rest. Grasshoppers are often well protected by this type of cryptic coloration. Most are the same color as the vegetation they feed on; others are the same color as the soil or sand on which they live. When a predator (including an insect collector) attempts to capture of flying grasshopper (one marked with brilliant orange or blue on the hind wings, exposed while in flight), which then drops to the ground and folds its wings, the grasshopper is hidden, because the eye of the predator is looking for the bright color. The grasshopper is later able to fly to freedom. Some butterflies have eyespots on their hind wings. Similar markings are also found on certain larvae and on the common click beetles, *Alaus oculatus*, the eastern eyed click beetle. These eyespots are thought to detract and confuse a predator into thinking that the insect is much larger than it is. Barbas, the little elephant in the children's storybooks, used the same ploy to ward off the rhinoceroses by painting large eyes on their rears and presenting these breeches to the invading rhinoceroses, scaring them away.

Warning coloration (Aposematic coloration). Many insects seem to flaunt their presence; they almost seem to say to predators, "Go ahead, eat me!" Many of these are brilliant orange or red, which seems to be a definite warning to amphibians, reptiles, and birds, because these insects are either poisonous to eat, or mimic a poisonous species. Some displaying these warning colors feed on milkweed. These plants contain cardiac glycosides, substances similar to digitalis. The Monarch Butterfly, *Danaus plexippus*, the Milkweed Bug, *Oncopeltus fasciatus*, the Red Milkweed Beetle, *Tetraopes tetraophthalmus*, and the Swamp Milkweed Leaf Beetle, *Labidomera clivicollos*, all feed on milkweeds, species of *Asclepias*. All are orange as adults and all have the milkweed toxins in their tissues. These have no ill effects on the milkweed-feeding insects, although they poison vertebrate predators.

Some interesting experimental studies were conducted by

Dr. Lincoln Brower, formerly of Amherst College, to demonstrate this poison and its effects on the Monarch Butterfly. These are marked with a deep orange with black wing veins. Brower reared one strain of Monarchs on cabbage and one strain on its normal host plant, milkweed. First, he fed a group of blue jays the cabbage-fed butterflies, which the birds ate with no ill effects. Then they were offered the milkweed-fed butterflies, which made them ill and induced vomiting. After this the birds refused all Monarchs on sight, apparently due to their earlier bad experience. It appears that blue jays need only the one experience with poisonous Monarchs to teach them to refuse orange and black insects.

Distastefulness is not the only factor involved in the conditioning of vertebrate predators. The stingers of bees and wasps are well known to birds, reptiles, and amphibians. Toads fed bees that have their stingers removed have no adverse effects. When offered a bee with a stinger, they will feed, but vomit up the bee when they are stung. From then on, they avoid any insect with the distinctive black and yellow color pattern. All are mimics.

Mimicry. Common in the insect world, mimicry is the resemblance of one species, the mimic, to the model, a species avoided by predators because of some protective feature. Although gaining protection from previous predator conditioning, the mimic is usually less abundant than the model, for, if the opposite were true, the selection pressure (survival rate) would be greater and protection of the mimic greatly decreased.

The two types of mimicry are Batesian mimicry, named after the English naturalist, Henry W. Bates, and Muellerian mimicry, named after Fritz Mueller. Batesian mimicry involves a palatable mimic and an unpalatable model. The mimic is protected by its resemblance to the model, because the model is more abundant than the mimic. The classic example of this type of mimicry is that of the Monarch Butterfly (the model), already mentioned, and the Viceroy Butterfly, *Limenitis archippus* (the mimic). Both species have similar color patterns, which apparently confuse birds. They avoid both species. However, an inexperienced bird will attack both species until it captures and starts to eat a Monarch. Once this happens, it will then avoid both species and probably be wary of other orange and black insects.

Muellerian mimicry combines unpalatable models and mimics. This type of mimicry benefits both. More species with close resemblance provide more protection. Hence, the orange and black color pattern is widespread among insect species.

Other types of mimicry are known besides those using color as warnings. Certain flies mimic bees; some bugs mimic ants; grasshoppers mimic wasps; and so on. Some strange mimics, for example, the sphingid moth larvae that mimic small tree snakes, or caterpillars that resemble bird droppings must remain for your future investigation.

Mating and oviposition

Courtship patterns sometimes are complex and unique for each species.The first need of an adult is to locate a mate. Specific signals must be used to assure that the pair are of the same species. In a sense, they must know the password! Several stimuli are used, depending on the species, including sight, hearing, touch (the most common), and chemicals (pheromones already discussed). Mating behavior, then, may be divided roughly into four stages, each varying with the species concerned. Some courtship patterns are elaborate, others have hardly discernible stages. The insects must locate and recognize a mate. This is followed by a courtship ritual. Then copulation takes place, followed by postcopulatory behavior. This is to assure that the female is inseminated with the sperm that will produce viable eggs. It should be noted that insects do not always recognize their own species with precision. Some congregating mechanisms attract more than one species. We have observed pseudocopulation between members of different families of beetles, probably simply because they were both marked with orange and black. We do not know how much wasted energy is spent in this direction among the various insect species, but, we presume, because of the abundance of males, it does not threaten the existence of the species or even a population. Unfortunately, few observers have reported on the rituals of courtship in the lives of insects, and, for that reason, we do not know why pseudocopulation takes place.

Location and recognition of a mate is usually accomplished by pheromones, attraction to food sources, less commonly by sound, and rarely by other means, such as light flashes, color flashing, and so on. Each of these methods serves respective species. Rarely is there a combination of two or more. Each species has its stereotyped pattern developed along the course of its evolution. It seems that closely related species with similar feeding habits have evolved the more elaborate mate-locating and courtship patterns. This is an important isolating mechanism for many species.

Courtship routines precede copulation. If the specific ritual is not followed exactly, copulation is not permitted. Also, the female must be in the proper physiological state before she recognizes these advances. This seems to be necessary to enable the female, particularly, to recognize her mate. Courtship patterns often include some type of stroking by the male, using either his antennae or legs. Others flutter wings, perform some dance pattern, and so on. Careful observations of mating pairs of insects will reveal almost as many courtship details as there are species.

Nearly all insects are internally fertilized by the insertion of the male intromittent organ into the female genital tract for the deposition of a sperm package, the spermatophore. This is not universally true, because the males of Collembola (springtails), which represent a distinct group of tiny, wingless, usually soil-inhabiting insects, produce spermatophores, but do not copulate. Instead they place thesesperm droplets in a circle around the female, who is then attracted to the male by his courtship dance. The fenced in female then grasps one droplet with her genital opening and, in this manner, is fertilized. Male bed bugs actually pierce the female's body with their penis and deposit sperm in the body cavity. The released sperm must migrate through her body to find and fertilize her eggs.

Copulation may take place on the wing, as may be seen by watching flying damselflies, dragonflies (Odonata), or Florida's famous "lovebug," a fly that swarms along highways and annoys motorists by smearing eggs on their car and truck windshields. Others, such as male New Zealand mosquitoes of the genus *Opifex*, copulate with the female as she emerges from the pupal skin floating on the surface of the saline water in which she spent her larval life.

The way the sexes are attached during copulation varies from group to group. Male beetles usually mount the back of the female, but not always. Most moths copulate end to end, the wings obviously prohibiting back mounting. And so on. All ways are a part of the complexities necessary to maintain animal life in a terrestrial habitat, as opposed to that of life in the ocean, where sex cells may be cast into the water and chemical compatibility determines their fate.

Postcopulatory behavior of the male is usually one of complete indifference, but not always. Especially among predatory species, the male must protect himself from being eaten. Male preying mantids usually are not successful in escaping the jaws of the female. We suppose this is a part of the so-called "economy of nature." Why not let an otherwise useless tasty morsel go to waste when it can be used to help provide the protein needed for egg production? This "attitude" is general among insects. Witness the stung and dying drones cast from the beehive as winter approaches. Why feed these useless creatures all winter? They have already served their purpose! Sometimes, it has been discovered, the male will bring a nuptial gift, a morsel of food, or a sweet bit of honey, and present this to the female. He will stroke her with his antennae or palpi, certainly not as a sign of endearment, but to divert her attention from him to prevent her savage attack.

The method of oviposition is as complex and involved as the courtship patterns. Oviposition pertains only to the actual laying of eggs, something shared by the male who may fly along with the female. Obviously, eggs are items of food for many other insects; therefore, the female does her best to place them where they will be protected. Since most insects do not tend their eggs, but, instead, leave them to their own fate, the female must select the proper place to lay the eggs. This results in an egg-laying pattern that is unique for each species. To attempt to describe all variations will take more than a single book—it would be a description of all insect life. In general, we may conclude, insects lay their eggs near, on, or in the food of the immature stage. But this is not always so. Some insects seem to have retained ancient egg-laying habits. Now, however, the newly hatched offspring of some species are required to hunt for their food some distance away from the egg-laying site. This sometimes involves travel by one means or another over a considerable distance. For example, cicada females lay their eggs high in the tops of trees under the bark of twigs. But their nymphs feed on the rootlets of the tree deep in the ground. This is a long trip for the tiny nymphs. The first stage larvae (triunglins) of meloid beetles must climb up on a blade of grass to await transportation to the host they will parasitize. The dangers involving the newly hatched are many; the loss of young lives, great. This is offset only by the fecundity of the species; that is, enough eggs must be laid to ensure that at least one pair survives to again lay eggs, as we explained in chapter 6.

Insect-made structures

Many different types of insects make nests and hiding places. These are constructed by both solitary and social insects. The nests of termites, ants, wasps, and bees are the most obvious. Some solitary insects construct hiding places to conceal themselves from possible predators, for protection from environmental hazards, especially as a place to overwinter. These nests may be as simple as rolled leaves or as complex as a plant gall (chap. 12) or the paper nests of wasps.

Leafrollers and leaf miners. As many as 50 percent of all insect species are phytophagous, and, therefore, depend entirely on plants for their food. Some of these use this same plant host as a site of protection for themselves and their young.

In the eastern United States, the leaf-rolling grasshopper, *Camptonotus carolinensis*, spends its days in a shelter formed by a leaf rolled up and tied with silk that is spun from the insect's mouth. By night it feeds on aphids. At least eight species of weevils, *Attelabus* spp., roll leaves of trees and deposit their eggs in the roll. The larvae feed on the inner portion of the leaf and pupate in the rolled leaf or on the ground after the leaf drops. Probably the most familiar group of leaf-rolling insects are the pamphiliid wasps (Hymenoptera). Some of their larvae resemble lepidoterous larvae. They live in silken nests formed by tying several leaves together. Solitary species of the same group form a nest by rolling a single leaf.

Leaf miners (larvae of Lepidoptera, Diptera, and Coleoptera) feed on the leaf tissue between the upper and lower surface of the leaf. They often cause odd-shaped leaves and leaf discoloration. Extensive mines kill the leaf. Each species of miner digs a distinctive mine, forming a recognizable pattern on the leaf (fig. 7.13). Often it is possible to identify the species more easily by the mine than by a study of the adult. The larvae of the locus leaf miner, *Odontota dorsalis* (Coleoptera, Chrysomelidae, Hispinae), can be a serious pest of black locust trees. Small moths of the family Opostegidae are also leaf miners. These are just a

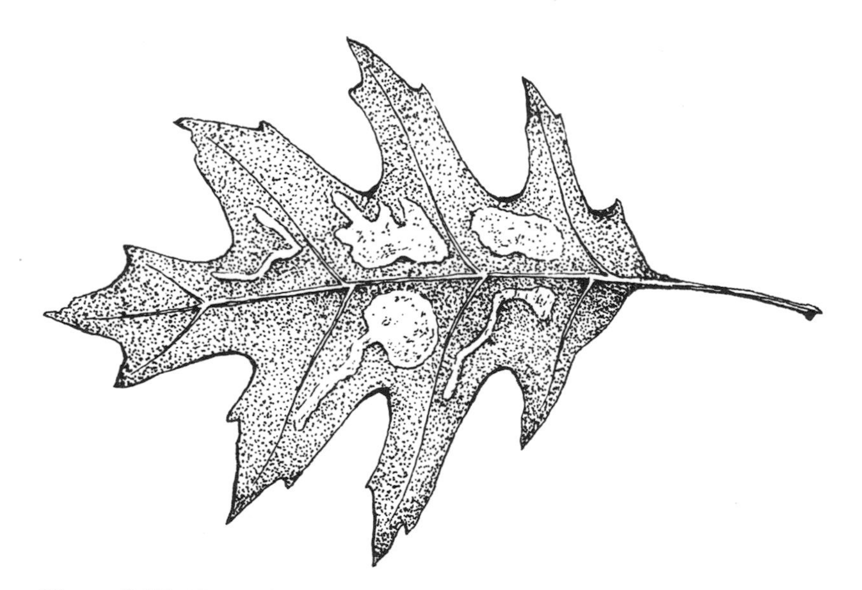

Figure 7.13. *Types of mines made by a variety of insects.*

Figure 7.14. *Caddisfly case (Trichoptera).*

few examples of many hundreds of species, sometimes whole families, that use the tender internal tissue of leaves for food, and, of course, do considerable damage to the plant by eating this photosynthetic tissue.

Protective cases. The eggs, immature stages, and adults of many insects are protected in partial or completely formed cases. Cockroaches and mantids form ootheca to protect their eggs. The nymphs of spittle bugs live in a frothy secretion, the spittle, which they secrete. Sometimes wild grass is so infested with these insects that merely walking through such a field will wet your garments.

Most larvae of the order Trichoptera, the caddisflies, form protective cases as they feed. These carefully designed cases have intrigued naturalists for at least four hundred years, being mentioned by Walton in his book on angling. The female caddisfly deposits her eggs in fresh water, and the larvae construct portable cases in which they live. They add to these cases as they grow. Some construct cases of twigs (fig. 7.14); others, of sand grains or small pebbles; and still others, of leaves. One species makes a case of sand that so closely resembles the spiral case of a snail that it is difficult to distinguish between the two.

Larvae of some chrysomelid beetles place their fecal excrement on their backs, where it forms a defensive shield and serves to camouflage the insect as well. Scale insects live under a protective scale, a waxy secretion that covers the body of the female (more on this in chap. 12). Bagworm moth larvae (Psychidae) and the adult female (fig. 7.15) construct a case made of the needles of various species of conifers. The larvae feed in this case and move it around from place to place on the host plant. Pupation takes place in the case. Only the male emerges; the female remains in the case. Copulation takes place through a hole in the bottom, and the eggs are laid inside. The adult female dies after egg laying. The eggs overwinter; then in the spring the larvae emerge, each constructs a new case, and the cycle continues. As the larvae grow, the bag is enlarged to accommodate their increased size. Bagworms are common on ornamental junipers. If you see these in the winter, open the bags and note that about half of them are empty. These bags contained males. The other half will have clusters of small yellow eggs in the bottom, laid by the now-dead female. You can make this an interesting winter collecting project by collecting and opening the bags. Keep a record of the contents. Some will contain small pupae, not of the bagworm, but of one of its parasites. The most frequently encountered parasite is a small ichneumon wasp, *Ictoplectis conquisitor*. Record your results and speculate on the success of this species in your area.

Figure 7.15. *Bagworm case (Lepidoptera). (Courtesy of Florida Department of Agriculture)*

8

Physics, Sex, and Communication

"I sit among the leaves here, when evening's zephyrs sigh,
And those that listen to my voice I love to mystify;
I never tell them all I know, altho' I'm often bid,
I laugh at curiosity, and chirrup, 'Katy did' "

W. S. Blatchley (1850–1940)

"Glories, like glow-worms,
afar off shine bright,
But look'd too near
have neither heat nor light."

John Webster (1580–1625) *Duchess of Malfi*

Much of the biological world is dark and silent—blind, deaf, and dumb. Plants, fungi, amoebae, worms, corals, all live without eyes and ears, at most with chemicals reacting to light, and bodies feeling sound—none see and hear.

The evolution of light-receptor organs, correlated with increased body activity, came about, no doubt, through selection, because of this survival advantage. More acute senses increase the chance of survival, which, in turn, makes possible a reduced ratio between the number of individuals needed to mature to the reproductive stage, and, therefore, the number of survivors. For example, it takes 10,000 eggs to assure that a single pair of blind springtails survives to produce another 10,000 eggs. The keen eyesight of a tiger beetle improves the chances of one pair's reaching adulthood from only one hundred eggs. Sight also aids prey recognition, and, perhaps equally important, mate recognition, an obvious aid to reproduction. Thus, we see a relationship between the development of sight acuity and fecundity. Some insects evolved better sight than others, probably because of an intricate combination of circumstances in the gene-selection process, which, of course, is characteristic of every species. We tell you this to point out the extremely rare conditions that must have existed when insects developed special sound-producing and light-producing structures.

Sound and light are strange physical phenomena. Both are by-products of other physical happenings. Sound does not exist as sound, unless it can be heard. Essentially it is only a change in atmospheric pressure caused by some other physical force. The same is true of light. It is the result of the physical conversion of energy, usually incidental to the production of heat. Light, in one sense, is a luxury, because it can affect behavior only if it can be seen. True, it is enjoyed by the majority of species on this earth, but only because of a quirk of nature that made the insects so numerous.

In this context, then, those organisms that signal by sound or light as a part of their behavior patterns, must be rare indeed. Both of these signals are produced by insects. We must continue to consider communication among the insects by this special study of those that sing, and, rarest of all, those that actually signal by sending out light to those that can see—it would be wasted effort if there were none to see—these are the fiddlers and the flickerers. We are learning that a considerable number of species of insects produce sound by rubbing two surfaces together and amplifying the sound that is produced. In contrast, few insects produce light. Those that use these methods of communication do so for the three basics of life—sex, food getting, and (rarely) protection.

Sounds in the world of insects

The loud, almost "ear-splitting" racket produced by the hundreds of newly emerged, sex-starved cicadas (called locusts in some places) cannot be missed by anyone near these hordes. The quiet "fiddling" of the cricket or the gentle songs of the katydid are sounds that not only entice members of their own species, but also amuse and inspire the bards and lovers of our own species. Alas, we are so busy making sounds ourselves—rarely music, we must say—that almost none of us have ever heard the many songs of insects, for indeed, many species have songs that rival those of birds, if we would only but listen.

Insect sound, as all others, is measured in decibels, a function of intensity and distance. This is expressed as a volume adjustment on radios or television sets. Insects may be grouped according to the intensity of the sound they produce. Some are extremely loud, that is, they are high-intensity sounds; others are soft and can barely be heard, the low-intensity sounds.

One time during a short visit with the famous sociobiologist, Professor E. O. Wilson of Harvard University, we were told about how startled he was one night. He was staying on a Florida Key enjoying that setting as he wrote one of his many books. An eerie sound awakened him one night. He described it as something akin to the sound TV programmers use to represent the language of Martian invaders—it wasn't a buzz, nor a chirrup, but perhaps a cross between the two, and certainly attention getting during the silence of a Florida night. He sat up in bed, and the sound stopped, but lying down again, it started. Eventually he discovered, not a tiny man from Mars, but a long-horned beetle trapped in his pillow case. When the pressure from the weight of Wilson's head squeezed the beetle, it protested with its species-distinct stridulation, a sound strange to those who first hear it.

Thus we find that there are insects producing high-intensity sound, heard by everyone, unless drowned out by a portable radio worn as we jog along, and the low-intensity, private sound produced by many well-hidden insects.

Some of the sound produced by insects is extremely high-frequency stuff (fig. 8.1). If we can hear it, we consider it to be of "normal" frequency. We are sure, you know, that the low-frequency sound produced by the bass violin or by drums is a frequency near the lowest point detectable by our ears. Lower frequencies are more felt than heard. The converse, sound in the high octaves, the voice of the high soprano opera star, for example, are high frequency, near the highest our ears can detect. But sound does not stop after humans cease to hear it, as the frequency increases. We can hear, when we are young, between about 50 Hertz (cycles per second) to 15, or perhaps as high as 18, kilohertz, the range of most hi-fi sets. To get higher, you pay more money, but probably you can't hear it anyway. As you grow older, you can hear less and less of the high frequencies and can easily get along with cheaper hi-fi sets.

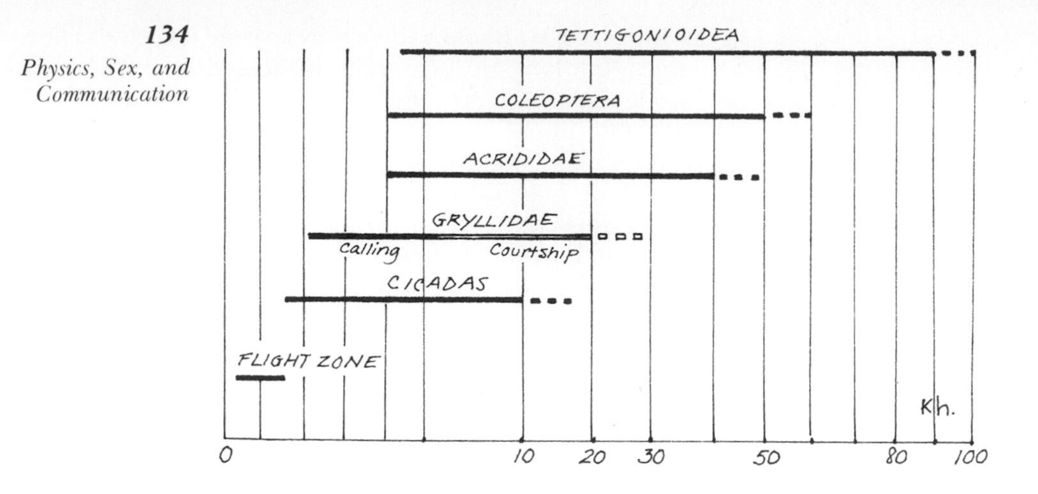

Figure 8.1. *Range of sounds produced by various insects (scale in Kilohertz).*

A lot more physics is involved in this than we care to go into here. Let's explain, however, that many insects, if they can hear at all, hear sounds at a much higher frequency than we can hear, some as high as 100 kilohertz. Expensive tape recorders can record these frequencies. Armed with good sound-detecting and recording equipment, we have learned much about insect communication with sound. We can tape at a high speed and slow down the tape on playback. This produces the same effect for insect sound as it does to play a 78 rpm record on a 33 rpm turntable. When we do this with sound we can't hear, the frequency of the sound is lowered to a frequency we can hear, and, of course, the tempo is slowed down. Once we asked an ornithologist at the famous Cornell University Sapsucker Woods Laboratory (where the sounds of birds, frogs, and toads are taped) which he thought were the most beautiful sounds; those of birds or those of insects. Obviously, this was a catch question designed by a professor. The ornithologist replied, as we expected he would: "Why, birds, of course!" Naturally he was thinking of the somewhat dull musical chirps of a katydid (fig. 8.2), as compared to the flutelike sound of a wood thrush. We fooled him and slowed down the songs of some beetles of the family Cleridae, forcing him to admit that the beetle songs rivaled those of the birds.

From this you must conclude that there are hidden sounds all about you. This is not a silent world, even to the generally mute rabbit. His big ears probably hear many of the secrets of insect talk, probably some of the things he told Alice about his "Wonderland."

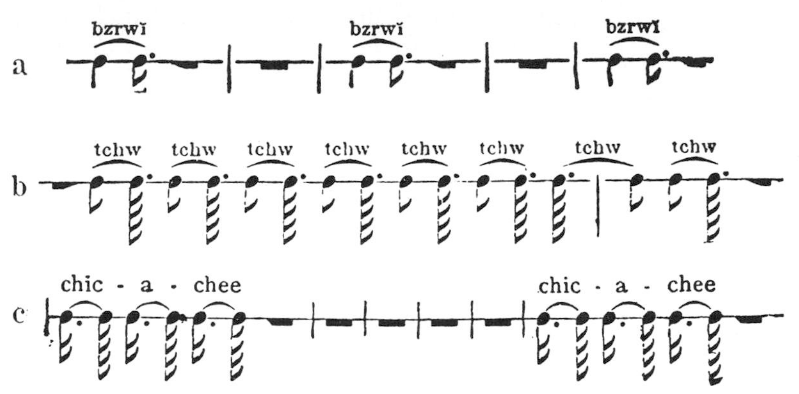

Figure 8.2. *Musical notes of a katydid. a, notes by day; b, notes by night; c. notes of another species.*

What insects produce sound?

We do not yet know all the kinds of insects that use sound for communication, mainly because too few of us are interested. To find out is costly, both in money for equipment and in time to make this kind of study. The authors know sound production occurs in the insect orders shown in table 8.1. This means, of course, that some species in each of these orders produce sound used for communication.

Table 8.1. *Insect orders known to have sound-producing species.*

Order	*Type of Sound Produced*
Odonata	Stridulation
Orthoptera	Stridulation; expulsion of liquid; wing clicking
Dictyoptera	Expelling liquid or air
Isoptera	Tapping or snapping
Hemiptera	Wing vibration
Homoptera	Stridulation
Psocoptera	Drumming
Coleoptera	Stridulation; wing buzzing; tapping; expulsion of air and gas; drumming; snapping; rasping
Lepidoptera	Wing clicking; expulsion of liquid; wing buzzing
Hymenoptera	Stridulation; wing buzzing; tapping; rasping
Diptera	Buzzing

We have not distinguished between sounds produced by nymphs, larvae, pupae, and adults. Each of these stages is known to produce sound in some groups.

How sound is produced

Loud or high-intensity sound is produced by stridulation, rubbing two specialized surfaces together (fig. 8.3). One is called the file, a part of the insect's body on which ridges, ribs, spines, even teeth, have developed. The second part is the scraper, usually a sharp edge, tooth, or a row of small teeth. Usually it is the scraper that moves across the file, much the same as the bow is drawn across the strings of a violin. Obviously it is possible to move the violin across the bow, or to move them both at one time, and such is possible with insect parts so modified. Usually high-intensity sound production also involves an amplifier, a hollow across which is stretched a membrane. On the other hand, low-intensity sound may be produced in the same manner, but without an amplifier, or it may be tapping, rasping, or other kinds of sound. Other forms of sound are produced by vibration of membranes, expulsion of a fluid or gas, tapping or thumping, and vibration of wings. Some sounds are incidental. Chewing, for example, is sound production, but probably not for communication, even though it may be specific for each species. It is possible that these sounds have meaning, but there is little evidence to support this. Are the chewing sounds of larvae or adults and larvae noticed by a group of feeding individuals? The cessation of chewing may serve as a warning signal, but again, the results of our experiments made some years ago to determine this were not positive. Sound from wing vibrations is usually incidental, but sometimes signals are deliberately produced, generally as a warning signal.

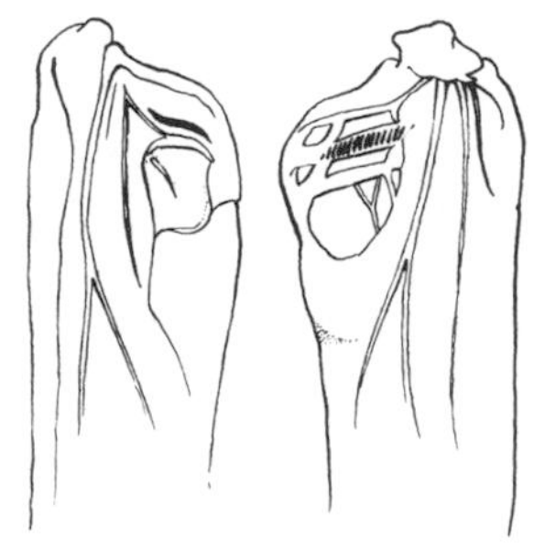

Figure 8.3.
Stridulatory mechanism of a grasshopper, the two wing covers separated. a, the stridulatory file; b, rudimentary file, the edge of left wing scrape across file on right wing.

How sounds are heard

It would be useless for insects to produce sound, if it could not be heard. The simplest form of hearing organ is a vibrating hair set in resonance with insect-produced vibrations. This is probably the most widely used ear in insects. However, katydids and crickets have ears on their front tibia (fig. 8.4), and grasshoppers have large typania on the thorax. This consists of a membrane stretched across a cavity, much as skin over a drum. It works in the same way our ear works. The membrane vibrates with the sound and a connecting nerve sends an electric wave to the nearest segmental ganglion.

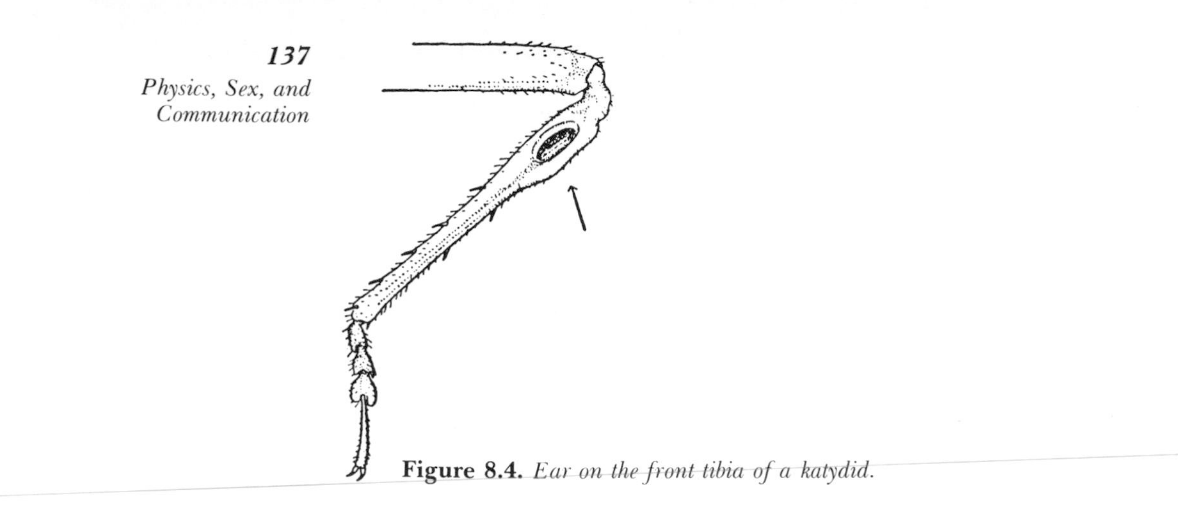

Figure 8.4. *Ear on the front tibia of a katydid.*

How the sounds are used

Sex. The songs of insects are the same as some songs of humans, primarily "love" calls. Of course, insects do not love, even if they court. But they do have problems finding mates. We pointed out some of these in chapter 1. A congregating mechanism is needed, and, again, much the same as in humans, music is the attractant among those species that produce sound. One or both sexes produce the music. Remember that it may be difficult for insects to recognize their own species, unless there is a specific signal of some type. Most species have barriers to help prevent crossbreeding. These are termed isolating mechanisms, as mentioned previously, and they are of various kinds. Physical and, especially, physiological barriers are usual. Genetic barriers are certainly effective, but, between closely related species, it is possible to successfully crossbreed without producing viable offspring. To do so is certainly wasted effort from the standpoint of the propagation of the species involved. This probably happens between newly separated species, or what may be called sibling species. Many of these sibling species have taken advantage of the occupation of distinct habitats or distinct niches, a consequence of population pressures and a part of the selective process. Assuming this theory to be true, then, among the singers, each has its own song. Each sings in its own way, "that's our song, darling." The music brings them together; other courtship procedures consummate the marriage.

One group of crickets, members of the genus *Oecanthus*, it was discovered by University of Florida's Professor Thomas Walker, are so similar morphologically, that it is almost impossible to separate the species. Yet once their songs are recorded and analyzed, pronounced differences in the frequency, notes, and intensity easily separate the species.

We have explained that low-intensity sound is not heard for more than a few inches, at most several feet from the sound producers. What happens then in really sparse populations? Professor Vivienne Harris at the University of Arkansas has shown that certain stink bugs combine pheromone secretion with sound production. Pheromones travel considerable distances, lure males to the immediate vicinity of the females, but, since these chemicals are not specific, the species depend on sound to locate females of their own species. Furthermore, Professor Harris learned that the females have three songs. One is a calling song to bring them close together. Once near, they change to a premating song, and, finally, a song that sounds like the clicking of castanets. Males also have a vocabulary of "words," some to answer the female; another, a calling song; and a third, probably a territorial or rivalry duet, as termed by the researcher.

Food. We know that sound plays a part in the dance of bees. Their buzzing sounds help to attract workers to pay attention to their dance signals to learn the location of food. It is possible, but not yet known, that sound could attract potential food; that is, it could lure prey to the predator.

Protection. Man, and other dangerous animals, have learned to avoid the buzzing of bees. Knowing, as we do, that bees sting, we don't have to test the individual bee by letting it sting us. Other bees, some flies, and even a few beetles also buzz, but, of course, they can't sting. Neither flies nor beetles have stingers, but buzzing, plus a characteristic shape, color, and flight pattern, all mimicking insects that can sting, scare off any potential predator or other animal that might intentionally or unintentionally harm the buzzing insect. Some adults, for example, adult bess beetles (Passalidae) signal to their larvae to warn them of danger, or to locate wandering "children." Certain ants tend the pupae of butterflies. The latter stridulate to help the ants find and move them to a safe place.

Insect language

The complicated dance of the bees, so well documented by von Frisch, has led researchers to look for other "language" signals among insects. No doubt a vocabulary does exist, but not a learned one. No coherent thought process permits an exchange of any communication signals other than those triggered by chain reactions. However, some unpublished work by Dr. Eileen R. Van Tassel of Michigan State University demonstrates that tiny water beetles of the genus *Berosus* send out faint stridulatory signals, which vary according to the message they impart. Courting pairs have a sex call, which differs from a warning signal that is given whenever disturbed. Premating calls are emitted by both males and females, but the exact meaning of these calls has yet to be determined. No doubt many other insects have vocabularies even larger. Of course, recording these sounds under laboratory conditions may give results that would differ in the field. What effect captivity has on the complex composition of the sound produced is unknown. It is known, however, that variations in temperature, humidity, and altitude will affect the quality of the signals, including changes in frequency, intensity, and beat. This is shown by the "thermometer" cricket (Snowy Tree Cricket, *Oecanthus nivens*) discovered in 1897 by Professor A. C. Dolbear. It is possible to determine the ambient temperature by counting the chirrups of this insect using Dolbear's Law, which states: count the number of chirrups in fifteen seconds and add forty. The sum is the temperature in degrees Fahrenheit. The law doesn't account for changes in altitude, and certainly it is an approximation.

Much more could be said about sound signals in insect groups than the brief introduction given here. Although books have been written on the physics of insect sound, the behavior of those producing and receiving sound, and what happens as a result of the signals, the subject is far from exhausted. Many insect sounds have never been recorded, and much of that that has remains to be analyzed. Unfortunately, the equipment to make these studies is expensive, and those with the time to use it are few. This exciting field of study is open to those with the proper background in physics and behavior biology.

Light production

Light production is a rare biological phenomenon restricted to only a few widely separate groups of organisms, most of them marine, some terrestrial, almost none in fresh water. The obvious question, then, is why? Do these plants and animals have anything in common? No vascular plant produces light. Among the other plants, only some species of fungi and bacteria have acquired the means for carrying on this process. Since neither of these plant groups have light receptor organs, it is difficult to determine the real function of the light and its selective advantage. Some evidence exists to indicate that this light attracts insects that then act as vectors to carry spores from place to place.

Several phyla of animals produce light as shown in table 8.2, but we concern ourselves here only with the insects, members of the phylum Arthropoda. Still, even among the insects,

Table 8.2. *Animal phyla with light-producing species.*

I.	Protozoa: Single-celled animals
II.	Coelenterata: Jellyfish and allies
III.	Ctenophora: Sea combs
IV.	Nemertinea: Proboscis worms
V.	Annelida: Terrestrial and sea worms
VI.	Mollusca: Snails, shells, and allies
VII.	Arthropoda: Insects, crustaceans, and allies
VIII.	Hemichordata: Sea squirts and balanoglossids
IX.	Chordata: Deep-sea fish

only a relatively few species are known to produce light; the most noticed are certain groups of beetles and some flies. The beetles that produce light are all larvae and most adults of Lampyridae (fig. 8.5), larvae of Telegeusidae, Homalisidae, species of the Elateridae genus *Pyrophorus* (fig. 8.6), and the larvae of Phengodidae (fig. 8.7). The larvae of a few species of the fly family Mycetophilidae also produce light. The latter are a tourist attraction in New Zealand, where they are so abundant that they light up a large room in a cave. Similar species live in the United States.

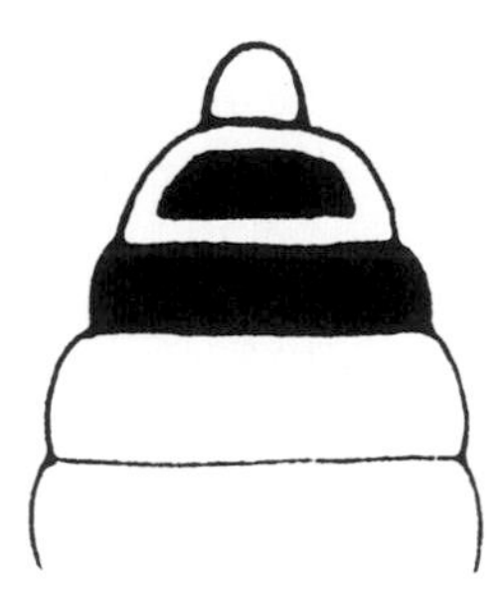

Figure 8.5. *Lampyrid beetle showing abdominal light organ.*

Figure 8.6. Pyrophorus *sp. showing thoracic light organs.*

Figure 8.7. *Larva of a phengodid beetle showing lateral light organs.*

How is light produced?

Two substances, luciferin and luciferase, are produced in specialized cells and brought together in light-producing organs. Luciferin is a natural protein whose oxidation is made possible by contact with luciferase, an enzyme, in the presence of ATP (adenosine triphosphate), the universal energy currency of the cell. The reaction, now well known through the biochemical research of many workers, is related to the energy-releasing process that takes place in muscles. Here, instead of motion, light is produced.

The light-producing organs (fig. 8.8) in fireflies (Coleoptera, Lampyridae) are best known. The organ in common fireflies is visible externally, located ventrically on the apical portion of the abdomen, or is inside on some. It consists internally of two layers of cells. A ventral layer produces light, while a dorsal, internal layer provides a reflecting surface. Because oxygen is necessary in almost all reactions releasing energy, we find, as we might expect, a rich supply of tracheal endcells where the oxygen comes in contact with the light-generating cells. Nerves also reach the organ and control the flashing, it seems, by controlling

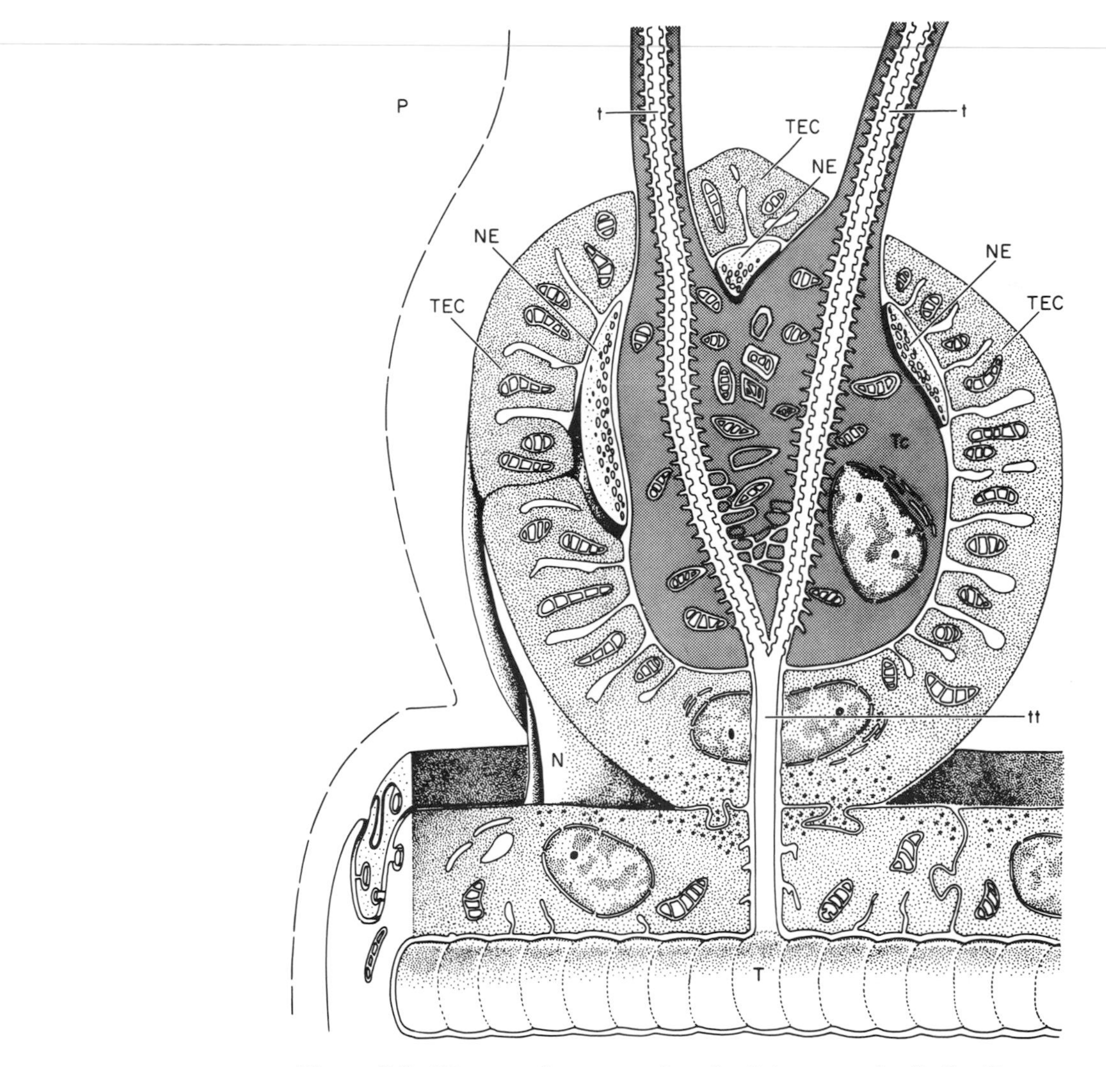

Figure 8.8. *Diagram of a cross section of a light organ of a firefly. H, hypodermis; O, hypodermal gland opening; P, photocyte layer; R, reflective layer (redrawn from Barber and Dilly).*

the amount of oxygen that is available for the light-producing reaction. You might wonder how this light can be produced in living cells without generating so much heat that the cells are destroyed. After all, look how hot an electric light bulb gets. Actually, the light produced is nearly 100 percent efficient; that is, it produces from 87 to 92 percent light; the remaining 8–13 percent is heat. Compared to our electric light bulbs, the firefly light is ten times as efficient. How this is done still remains a question. If we could duplicate the process economically, we would be able to greatly improve the current method of producing light. The light itself falls between 450 and 670 mL, a yellowish green color.

What do fireflies do with light?

Larvae of all known Lampyridae produce light in two tiny lanterns on their tail ends, even the larvae of few adults, who may have secondarily lost the light-producing ability. These larvae are mostly predatory, feeding on snails and earthworms, or they are general feeders on fruit and other plant material. They may produce light to attract fellow larvae to help them to their food.

The pupae are bioluminescent, able to turn on the lights of the transforming larvae when stimulated, although in this vulnerable stage they do not wish to attract attention. The newly emerged adults, after hardening and testing their new life style, are ready to use their new light organs.

Most of us are aware that fireflies flash their lights intermittently, but it took an amateur, Frank McDermott, a chemist by profession, in 1911, and H. S. Barber of the U.S. Department of Agriculture in the 1930s, to study several species and to note the differences in flash patterns. These seem to be more or less specific, but the species are close morphologically, and few people believed Barber when he published details of this in the 1940s. Although this had been published previously, Barber was the first to recognize that what were once thought to be a single species were actually a complex of species, several of which had never before been described and named. Several more recent students, notably Dr. James E. Lloyd, of the University of Florida, have worked out the details of these flash patterns and their taxonomy for a considerable number of species. He has travelled extensively in various parts of the world and finds the same phenomenon in all regions. All told, he has, to date, spent the equivalent of six years, at night, recording flash patterns.

In general, it may be stated that males flash to answer

female flashes. A female of the same species respond with the secret flash password sequence that identifies her as a member of the same species and a potential mate. Obviously, groping around in the dark to find a mate is risky business in a world of predators. Therefore, this coding system saves time, energy, and prevents, to some extent, the risk of being eaten—unless, of course, some predators catch on.

The code of flashes is not perfect. Timing is important, and to some extent, so is color, at least as an additional distinction. Since adults of some species emerge at a different time from others, the number of flash patterns used at any one season equals the number of species on the wing at any one time. Regardless, the male must arouse the correct response in the female for the process to continue. When the right mate is found, copulation takes place and the life cycle continues. Therefore, light flashing is an isolation mechanism for these species, the same as sound is for species of Orthoptera and others.

As is to be expected whenever a subject is intently pursued, it is learned that what appears to be simple is actually extremely complicated. Armed with sophisticated equipment, Dr. Lloyd, a professional research entomologist, spends many nights each year analyzing the flash patterns of fireflies. The equipment he uses may be as simple as a penlight used to imitate flash patterns and demonstrate to students the way a female may lure a male to her wedding bed. But when serious analysis is undertaken, he uses a flash generator that can duplicate more complicated patterns. He also converts flash patterns into sound and records the sound on tape. In the laboratory, he reconverts the sound into flashes, plots the results as graphs, and feeds these data into a computer for final analysis.

These studies show some surprising results. For some species, the key to "turning on a female" is the length and pulse of the flash; that is, flash length and flash intervals are key elements in the advertising patterns of some species. Answering delay is important in species separation signals. Males discriminate and are attracted to females that respond only after the correct time delay. Many other details have been recorded by Dr. Lloyd.

Among the strange sights encountered by Lloyd was his observation of the *femme fatale* of the insect world—females that lure the males of other species by mimicking the proper response of the true females of the species (fig. 8.9). Once nearby, the dastardly female recycles these males, turning them into

Figure 8.9. *"Femme fatale," a predacious female firefly.*

eggs of her own species, by eating these fickle, easily distracted males.

And there is still much to be learned about the ways of these insects.

9 Life in the Colonies

"These craggy regions, these chaotic wilds,
Does that benignity pervade, that warms
The mole contented with her darksome walk
In the cold ground; and to the emmet gives
Her foresight, and intelligence that makes
The tiny creatures strong by social league;
Supports the generations, multiplies
Their tribes, till we behold a spacious plain
Or grassy bottom, all with little hills—
Their labour, covered, as a lake with waves;
Thousands of cities, in the desert place
Built up of life, and food, and means of life!"

Wordsworth, *The Excursion*

The social relationship in insects shows various degrees of development (table 9.1). Truly social, or eusocial, insects include all the termites (order Isoptera) and many members of the order Hymenoptera, particularly, all of the ants (Formicidae) and some of the wasps and bees. We call these insects social only

Table 9.1. *Classification of insect social behavior (Michener, 1974).*

Type of Behavior	***Definition***	***Example***
Solitary	Insects do not care for young, no division of labor, no assistance to parents.	Most insects
Subsocial	Adults care for their own immature offspring.	Earwigs, bess beetles, some dung and carrion beetles
Communal	Adults of the same generation use the same nest, but do not cooperate in the care of young.	Some Halicitid bees
Quasi-social	Adults of the same generation use the same nest and cooperate in the care of the young.	Some carpenter bees
Semisocial	A worker caste of the same generation as the reproductive caste cares for the young.	Some Halictid bees
Eusocial	Overlap of generation, so that the offspring belonging to the worker caste assist their parents in the care of the young.	All termites, all ants, some wasps, and some bees

because they pool their work during their lives. Having no material possessions, except perhaps a store of food and the nest they have constructed, their only wages are the food they eat, or are fed, and the survival of their species. Remember that working in this manner (or manor!) is not their decision. They are born to serfdom as workers, "loyal" subjects of a queen. Even their kings, dullards though they may be, rank with the lowly, soon to be discarded when their service is done. Yet, with our eyes, we see only the wonders of these highly organized extracellular systems, ruled, as we have seen, by chemical drugs over which they have no control.

Eusocial insects are distinguished from all other insects by the cooperation of the members of a single family. Normally, there is only one queen, and, hence, one mother, making all members of a nest or hive siblings, even including the incestuous fathers of these polygamous families. The family division of labor includes fertilization of eggs (or not) by the relatively few males of the family. The caring of the young of the queen, the eggs, nymphs or larvae, and pupae is usually done by a sterile caste, the workers. The gathering and storing of food and the defense of the nest are subdivisions of work. This division of labor is the result of specialized castes formed genetically, as well as chemically. Although there is often an overlapping of generations, this is not the same as generations of other, nonsocial insects, because the parent, the queen, remains the same. Rarely is there a succession of queens.

Before we examine insect societies, let's see what categories other species of insects fall into (table 9.1). Most species are solitary; that is, they live independently of other members of the same species. Solitary insects each gather their own food and find or accept a single mate (sometimes they mate more than once). Interaction with other members of the same species is usually by chance, and encounters are usually only at a common feeding area when reproduction takes place. This category includes the majority of insect species. Yet, there seem to be trends in the direction of social organization, as one might expect, because of the need to form associations for mating, and because of similar tastes in food.

Subsocial insects

Some insects are subsocial because of attention to the young they produce. This is a loose grouping because of varying degrees of subsocial organization, which is best explained by a few examples. The earwigs (Dermaptera) (fig. 9.1) are good examples of

Figure 9.1. *Earwig female guarding her eggs.*

this. These paurometabolous insects are nocturnal. A wide variety of both living and dead plant material is eaten by the nymphs and the adults. The female, at least of certain species, will deposit her eggs in a sheltered chamber that she has constructed in the soil and guard them from predators. She remains with the eggs until they hatch and then cares for the nymphs to some extent, at least until they are able to fend for themselves. Once on their own, she leaves them without any further concern or even recognition.

Figure 9.2.
Tent caterpillars in their protective tent.

Another insect exhibiting a loose subsocial organization is the eastern tent caterpillar, *Malacosoma americanum* (Lepidoptera). This species overwinters in an egg mass formed by the female in the fall on tree twigs. These hatch in the spring and feed on the new leaves of cherry, apple, and other trees, primarily species of the rose family. The larvae are gregarious; that is, they live in large numbers in silken tents jointly constructed by the caterpillars (fig. 9.2). They leave this nest during the day to feed on the leaves of the tree and return at night. After eating their fill and growing to maximum size, they leave the tent, pupate, emerge, and lead an independent adult life. There are many other tentmaking larvae, but these are among the most striking.

Some web-spinning and leaf-rolling sawflies (Hymenoptera) are also gregarious in the larval stage, but beyond this single action of constructing a joint tent, there is little else to indicate a subsocial organization.

A particularly surprising subsocial life style exists in species of the wood roach, *Cryptocercus* (Dictyoptera). Wood roaches live gregariously, as do other cockroaches, in small colonies. The difference is that these live in cavities in rotten logs. They feed on wood, which is digested by special symbiotic protozoa living in their digestive tracts. This is similar to termite digestion, and, hence, we believe termites evolved from cockroaches similar to this wood roach. With each molt, the nymphal cockroach loses the symbiotic protozoa, without which it would soon die of starvation. They replenish their supply by feeding on excrement of members of the cockroach colony. Only by the formation of a colony is it possible for these roaches to use this type of feeding and the symbionts they harbor.

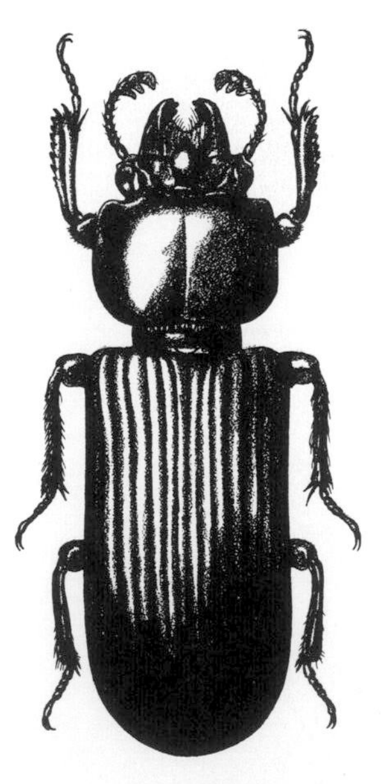

Figure 9.3.
Peg beetle (Coleoptera: Passalidae).

The bess or peg beetles (Coleoptera, Passalidae) (fig. 9.3) form small colonies, also in decaying logs. The adults rub roughened areas on the underside of the hind wings against similar areas on the dorsal side of the abdomen to produce sound. This is used, as we discussed briefly in chapter 8, as a means of communication, possibly an alarm signal, between adults and larvae. This is loud enough to be heard outside the log, a good way to locate the beetles. Tap a log, and then carefully listen for these sounds.

Other beetles showing subsocial behavior are certain species of dung beetles (Coleoptera, Scarabaeidae). The male and female work in pairs collecting and rolling dung into a compact ball. The ball is buried in the ground, and the female deposits her eggs on or in it. The adults of some species may remain with the eggs until they hatch, displaying the beginnings of maternal care, the rudiment of social life.

Other degrees of social life approach true, eusocial behavior. Communal behavior requires that adults of the same generation share the nest or hive, but do not cooperate in the care of the offspring. Bees of the family Halictidae often nest in soil in vertical banks along a stream or in old gravel pits. The adults feed on nectar and pollen and provide larvae with this food. They nest in a common gallery, usually with one entrance. Some species "appoint" one bee to guard the entrance. Bees returning to the nest with impressive loads of pollen are given the right of way by the guard bee to enter the chamber; others are warded off.

Quasi-social behavior is found in those species where the adults use the same nest and, in addition, take part in the care of offspring. Some carpenter bees fit this category. These giant bees (*Xylocopa* spp.) are over 25 mm in length. They resemble bumble bees. Extensive galleries are excavated in wood, including porches and building walls, as well as dead trees. The colonies that form approach those of eusocial insects, making it difficult to draw clear distinctions between these various nesting organizations.

Still another category of behavior is that termed semisocial behavior. A worker caste develops comprised of individuals of the same generation as the reproductive caste. These care for the young. Some species of sweat bees (Hymenoptera, Halictidae) display subsocial behavior.

Eusocial insects

The remainder of this chapter is devoted to some examples of true social life as represented by the termites, wasps, ants, and bees. Space permits only a brief introduction to this subject, and you should realize that many books have been written describing the ways of these interesting insects.

Termites. More than 1900 species of termites live throughout the world, most species in tropical and subtropical regions. Forty-four species have been recorded in the United States, but few of these range north of the extreme southern part of the country, and only one lives in the snow belt. Without doubt, termites evolved from cockroaches, but they do not resemble them to any great extent. This is the only group of eusocial insects with incomplete metamorphosis. Their immature stage is a nymph, and, of course, no pupal stage exists.

Termite colonies have several features uniquely different from those of the hymenopterous species. Members of the colony usually are equally males and females, each playing an equal role in social structure (table 9.2). In addition, they feed on the cellulose of dead, often decaying, wood, rarely on twigs, leaves, grass, humus, or even dung. Their chewing mouthparts and the stock of symbiotic protozoans or bacteria make them capable of digesting cellulose, a feat impossible for other animals.

The termite colony itself is a complex of interrelationships between primary reproductives, secondary reproductives, workers, and soldiers (fig. 9.4). Primary reproductives have fully developed wings during their nuptial flights. These flights occur as a swarming action taking place usually once or twice each year. Sometimes thousands of reproductives take to the air, setting out to establish new colonies. Often this occurs just after the first rain of the season. The flights are short, and, when they touch the ground or a building, they drop their wings. These wings are described as dehiscent, because they have a fracture

Table 9.2 *Differences between termites (Isoptera) and Eusocial Hymenoptera.*

Termites	***Eusocial Hymenoptera***
Paurometabolous, with egg, nymph, and adult.	Holometabolus, with egg, larva, pupa, and adult.
Worker caste consists of males and females.	Worker caste of females only.
Caste determination is based primarily on pheromones and, in some species, it involves sexual differences; some factors still unknown.	Caste determination is based on nutrition; pheromones play a role in some species.
Primary reproductive, the king stays with the queen after the nuptial flight, helps in nest construction, mates with her a number of times after the nuptial flight.	Male fertilizes the queen during the nuptial flight and dies soon after. Male does not assist queen in nest construction.
Definite soldier caste, with some highly specialized individuals known as nasutes.	Soldier caste mainly one of size differences of the workers, the majors.

Figure 9.4. *Termite castes. top, winged reproductive; lower left, worker; lower right, soldier.*

zone near the wing base, which permits them to break off. We have experienced swarms in the tropics when these insects were so abundant that the floor of the building was covered with the shed wings, and the air seemed to be full of more swarming insects.

Supplementary reproductives lack fully developed wings, but retain wing pads similar to their nymphs. They are capable of producing sperm and eggs when they must replace dead or declining primary reproductives. The bodies of the supplementary reproductives are not as fully developed as the primary reproductive caste. Two wingless, sterile castes, the workers and soldiers, complete the colony membership.

Most individuals in a termite colony are workers. They perform a variety of chores for the colony, including foraging for food, construct nests, and care for the young. Chewing tunnels in the wood to form these nests causes severe, structural damage to buildings. One estimate claims that termites cause more damage than fire in the United States. Workers are pale, sightless, wingless and obviously equipped with chewing mouthparts, but these are especially adapted for wood boring. Soldiers are also wingless and sterile, but differ by having enlarged heads and mandibles used for defense. Some soldiers have mandibles designed for use in actual combat; other soldiers (nasutes) (fig. 9.5) have a nozzlelike projection from the head used to squirt

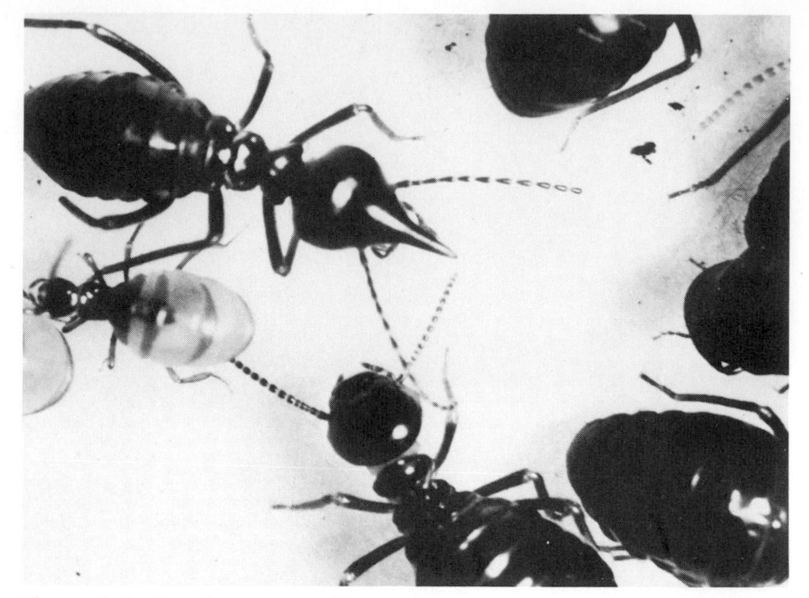

Figure 9.5. *Termite nasute (soldier).*

defensive fluids at enemies, and these help protect the colony from invasion from predators, especially ants.

Years ago there were accounts of battles taking place in the deserts of Asia Minor. There termite colonies are abundant. An officer, later to become the head of one of these countries, was injured in battle by a lance. To prevent infection, hot butter was poured into the wound, which included slashes through the stomach wall. These cuts needed sutures, and medical help was unavailable. Instead, a termite colony was opened, and several dozen soldier termites were captured. The termites held their mandibles open to try to capture the colony invaders. Taking advantage of this, the men sewed the wall of the stomach by getting the termites to close their mandibles on the cut edges of the stomach. When they bit, their bodies were torn from the heads. The mandibles remained closed, and, of course, eventually this organic material was absorbed by the officer's body, but the stitches held until the stomach healed.

The termite's life cycle is best illustrated by the pest species, *Reticulitermes flavipes*, the eastern subterranean termite. Winged primary reproductives are produced in the warm months. The emergence of winged primary reproductives signals the beginning of a nuptial flight involving thousands of individuals. When a female completes her flight, her fragile wings drop off when she touches ground. She runs about in search of a male, who also has lost his wings. When a pair meet, they run along the ground together in a pattern known as tandem running. The queen leads this run, which continues until they find a suitable piece of wood in which to start a nest. This, at first, consists of a tunnel with an egg chamber at the end. The new king and queen plug the entrance with chewed wood. Eggs are produced. When the nymphs emerge, they are fed by the king and queen, an act important not only for the survival of the nymphs, but also for the fundamental development of the colony. Feeding consists of regurgitating food in liquid form from the crop, a sack-like portion of the digestive tract. This type of feeding is termed trophallaxis and seems to be characteristic of all eusocial insects. The nymphs are soon capable of feeding on wood on their own. The colony grows by the continuous production of eggs, and the nymphs hatch into workers and soldiers. After several years of growth, when the colony is firmly established, new primary reproductives are produced. These leave the colony and establish new colonies, thus repeating the cycle. The original pair may live in the colony for over ten years. If the original queen dies, a secondary reproductive queen will develop into a primary queen, and she takes over the duty of egg laying.

The mechanism of caste determination is not fully understood, but entomologists believe that one type of pheromone (as pointed out in chap. 7) is released by the royal pair, which suppresses sexual development in the nymphs. This substance is transmitted by feeding. Another pheromone is produced by soldiers, which inhibits the development of additional soldiers beyond a certain proportion per colony. When the population of soldiers is reduced due to death (as may occur after an invasion of ants), more soldiers develop. Thus, a critical balance is kept between the number of workers and soldiers, apparently due to the concentration of this soldier-suppressing pheromone in the nest. The regulation of castes in social insects is a morphogenetic phenomenon. Changes in the colony count depend on the activity of the insect's endocrine system's secreting growth hormones. The activity of the corpora allata, a gland that secretes juvenile hormone, keeps the soldiers and workers from developing into reproductives. Some experimental evidence now exists that suggests that soldier formation from workers can be induced by administration of termite juvenile hormone. Since worker castes cause the damage to wood of buildings, fences, and so on, upsetting this balance in the colony by inducing pheromones that would cause the production of soldiers, instead of workers, might be a way of controlling these pests. This seems to be an interesting avenue for research, especially when seeking means of biological control, now of high priority in today's chemical-drenched world.

The eastern subterranean termite, a major pest in eastern and southern United States, starts its colonies in wood in the soil.

If the wood is not in contact with a building, a tunnel of chewed wood and saliva is constructed through the soil to a nearby building. To prevent these infestations, two things are required. First, builders should not bury wood cuttings near a building. This will cut down on the chance of infestation. Second, a barrier of metal and concrete should be constructed between the soil and the wood in a building. This baffle will keep the exploring termites from reaching the building. Fortunately for owners of wooden structures in temperate regions, this species of termite requires a soil connection. However, once the termites reach the wood, it must be replaced, usually a costly procedure. At the same time, the soil near the building should be treated with an insecticide to kill the colony.

Many species in the tropics and subtropics live above ground without contact with the soil. Other species construct a variety of nests for protection. In parts of semiarid Africa and Australia, termites make above-ground termitaria (fig. 9.6), some of which are as much as 20 feet tall and 12 feet in diameter. These nests are made by workers cementing soil and fecal material together. They stand for years, even after the colony has died off. The design of these structures is such that it keeps out the effects of the heat of the sun and the loss of water as much as possible. Termites are capable of regulating the temperature in the colony to within a few degrees by the design of the termiterium.

Figure 9.6. *Termitaria (termite colony) found in the Australian "out back."*

Eusocial Hymenoptera

Social wasps, ants, and social bees are the best examples of eusocial behavior. Their social habit has evolved independently in at least ten groups, each with a variety of nesting habits, but with a social structure centered about a queen, just as was the social structure of the termites, along with a variety of workers. Males are barely tolerated, functioning only as studs and of momentary value to the colony. Since eusocial Hymenoptera are holometabolous, a larvae and pupal stage is present. The larvae are legless and require feeding by the mature workers. The colony always lives in nests (except certain ants) constructed and enlarged, first by the queen, then by the workers.

Figure 9.7.
Paper wasp (Hymenoptera).

Social wasps. The eusocial wasps include the common paper wasps, hornets, and yellowjackets, all members of the family Vespidae. Paper wasps (species of *Polistes*, fig. 9.7), common throughout North America and, in fact, worldwide, nest in trees, bushes of forests and gardens, and even in old buildings. In the spring, several females, survivors of the winter, work to construct separate paperlike hanging nests of wood pulp glued together by their saliva. Soon other females join the first group, but are subordinate to the foundresses. Eggs are laid and hatched, and the larvae are fed by the queens. The first summer generation are all workers (nonreproducing females). When these are mature, they carry on all the work of the colony. Unfertilized eggs produce fertile males. At the same time, sexually mature young queens appear. After these appear in late summer or early fall, the colony declines, and all individuals die off by the first frost, except the new, fertilized queens. These overwinter, the survivors that will start new colonies in the spring. They spend the winter under loose bark or in stone walls and similar hidden places.

These paper wasps do not usually sting humans. They are, in fact, positively tolerant of humans, but the reverse is not true. We humans spend much time burning out these wasps' nests.

Figure 9.8.
Yellowjacket (Hymenoptera).

Hornets. Hornets and yellowjackets are more aggressive. Several species of these wasps (fig. 9.8) occur in North America. The best known is the baldfaced hornet, *Dolichovespula maculata*, widely distributed in North America. This hornet is 16 to 20 mm in length, with distinct black and white patterns on the head, thorax, and abdomen and with smoky brown wings. Again, only young mated females survive the winter, the bulk of the colony dying off at the approach of winter. In the spring, the young females chew wood to form the pulp used to build pendant nests. Nests are constructed in the open on trees, buildings, and other exposed places. Eventually, each nest has many layers of cells covered on the outside by several layers of paper with a doorway at the bottom. The first generation is exclusively female workers. Males eventually develop from unfertilized eggs, and the colony follows the same pattern found in the paper wasps.

Yellowjackets. Although there is no real distinction between hornets and yellowjackets, this name usually is applied to members of the genera *Vespula* and *Vespa*. They cannot be distinguished from, and are confused with, hornets. They all have fierce stings, which can be used repeatedly, always without advance warning. Some species, particularly *Vespula maculifrons*, are the scourge of patio parties, picnics, and campgrounds throughout eastern and midwestern North America. *V. maculifrons* and others make subterranean nests.

Mated females survive the winter. In the spring, females construct small nests and bring food to the larvae. The first brood of females consists of workers, who help expand the nest and tend the young. Males develop from unfertilized eggs and mate, as in all eusocial Hymenoptera. Yellowjackets usually nest in the ground or in old stumps or logs. These social wasps feed their larvae daily with a mixture of masticated dead insects, while the adults feed on nectar and picnic goodies. Often adults are seen around garbage cans foraging for food and even feeding on soda from discarded soda cans and bottles. Trophallaxis between larvae and adults occurs, which has a similar function in the control of the colony caste, as does the same process in termites.

Ants. Over 10,000 species of ants (Formicidae), all completely eusocial, are found throughout the world. Ants are easy to identify as a group, but are difficult to identify to species. They are probably the most familiar insects to many individuals and are often the delight of children, unless, of course, they get stung. They can be easily distinguished from termites, with which they are often confused (table 9.3). Ant and termite swarms are similar, in fact, so much so that termites have been called "white ants." Then, too, ant swarms alarm home owners, who fear that they may have a termite infestation. Ants have a slender waist, known as a pedical (fig. 9.9), which consists of one or two bead-like abdominal segments located between the thorax and the remainder of the abdomen. Also ants can be distinguished from termites by their antennal structure. Ants have elbowed antennae, whereas termites have straight antennae. The wing vein patterns of the two groups are distinctly different and, on this alone, swarms of one or the other can be distinguished after a little practice.

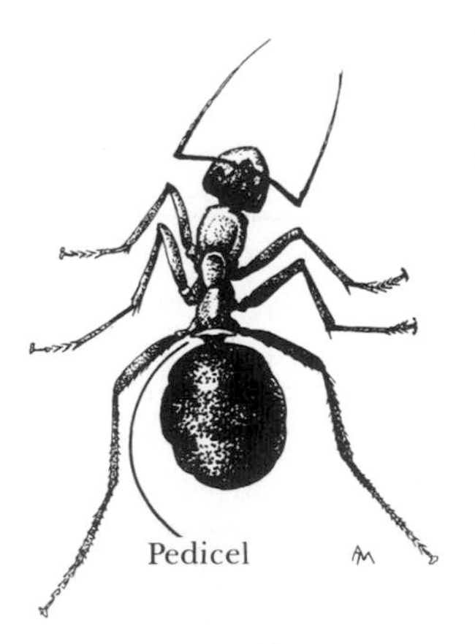

Figure 9.9.
An ant showing abdominal pedicel (Hymenoptera).

Table 9.3. *Comparison of the morphological features of termites and ants.*

Termite (Isoptera)	*Ants (Hymenoptera)*
Paurometabolous metamorphosis (egg, nymph, adult)	Holometabolous metamorphosis (egg, larva, pupa, adult)
Antennae straight	Antennae elbowed
Two pairs of wings of equal length; venation simple, without closed cells	Two pairs of wings of unequal length; venation with closed cells
No "waist" between the thorax and abdomen	Distinct "waist" (pedicel) between thorax and abdomen

Ants live in a wide variety of habitats. Their colonies develop in underground tunnels in soil or in elaborate galleries in dead or rotting wood, soil, branches of trees and shrubs, and even hollowed-out cavities in thorns. Ants have a caste system parallel to that of other eusocial Hymenoptera. These are queens, males, workers, and soldiers. At various times during the warmer months, winged reproductives, both males and females, emerge from the nest and disperse to establish new colonies, the swarms discussed previously. The foundress queen and mother, on reaching the site of a new colony, performs all duties until new workers mature and take over. Later her duties are restricted to egg laying. As we see among the social Hymenoptera, males contribute little to the colony, except to fertilize the queen's eggs, after which, they usually die. The workers, sterile females, specialize as laborers or defenders of the colony (workers and soldiers, respectively). In some ant species, the soldiers are of a larger size and shape, especially with larger heads and mandibles. Members of this subcaste are often called majors. Because there may be a range of sizes, other ants are termed medias and minors, with smaller heads and mandibles. These workers assist the colony by gathering food, nursing the young, and maintaining the nest.

Each species of ant has a specialized life style, among which are predators; many others, including the much-feared fire ant, are scavengers. Some harvest seeds; others raise fungus for food in underground gardens; and still others tend aphids for their honeydew excretions. Among the many species that culture honeydew excretions. Among the many species that culture fungus for food is the Texas leaf-cutting ant, *Atta texana* (see chap. 12, for more details). This species occurs in the woods of Texas and Louisiana, but many other species occur in the southern United States, south throughout the tropics.

The black carpenter ant of the eastern half of the United

States, *Camponotus pennsylvanicus*, is found in dead wood of trees, logs, poles, and beams of houses. The nests are constructed in dead wood, but the insect feeds on other insects, honeydew, sugar, and other sweets. They are unable to digest cellulose, but the damage done by their tunnels inside the wood of buildings is sometimes confused with termite damage. All stages are found in these galleries. Although it takes many years to establish a large colony, eventually these ants can cause enough damage to require control measures. Carpenter ants are capable of inflicting a painful enough bite with their large mandibles, but they do not sting. These ants are often found in houses near wooded areas. They almost never actually infest the house, but it is well to locate their colony and eradicate them, lest they decide that your house beams are closer to their food source.

Ants noted for their vicious stings, however, are the several species of the fire ants. One, *Solenopsis geminata*, inhabits fields and woodlands along the Gulf Coast from Florida to Texas. They are scavengers feeding on dead insects, seeds, and plant debris. This ant constructs a mound, usually near shrubs or trees, but also in open pastures and even in cultivated fields. They sometimes damage young plants, but they are not considered to be a serious crop pest. Domestic animals and humans frequently are stung by these ants when either accidentally steps into a nest. This causes a painful burning sensation (hence, their common name) and the development of a pustule in the area of the sting. Some may even suffer an allergic reaction to the venom. Fire ants are so troublesome in some parts of the south that chemicals dropped by airplanes have been used in an attempt to control the ants, but with little success.

Of all the described species of ants, probably the most spectacular is the army ant of the tropical rain forests of North and South America. These ants do not construct permanent nests. Instead, the entire colony migrates in search of food, establishing temporary bivouacs in a sheltered area, logs, hollow trees, and so on. Worker ants form a solid mass, probably a million or more, ranging up to a meter across, with each individual linked to another, hooked together by their tarsal claws. The queen, larvae, and pupae are carried in the center of this mass as they move along. Nights are spent in the bivouac, but at the first light of dawn, workers, guarded by soldiers, extend out from the bivouac in search of prey. Workers deposit trail pheromones to guide other workers to the prey and to help them return to home base. When prey is discovered, it is stung to death, dismembered, and transported back to the bivouac area for all to eat. The helpless larvae and the queen are fed. Their prey may include insects, small reptiles, birds, and even small mammals. At the end of each day, workers return to the bivouac area to

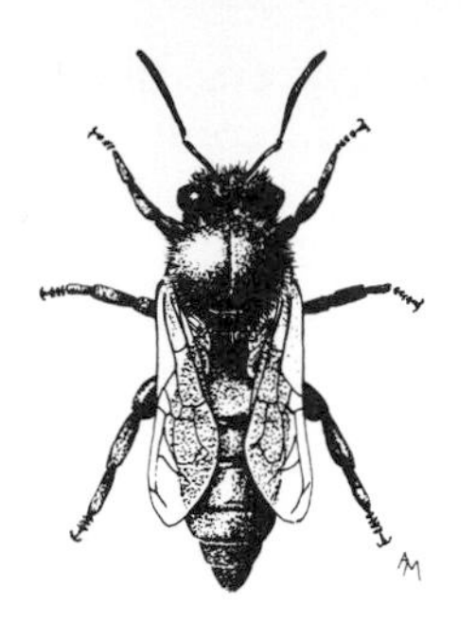

Figure 9.10.
Honey bee (Hymenoptera).

spend the night with the others. Trailmarking pheromones are an important means of communicating information to other members of the community telling them the location, as well as the quantity, of the food. Most trails are weak, and the scent lasts only for a short time. Stronger trails are established when a larger source of food is found. At any rate, when the available food is used up in one place, the ants are off again on another foraging expedition. They live their entire lives in this manner, the older members being replaced by younger ones, and on they go, probably for years before the colony dies of disease. But there are as many variations of this as there are species. Ants are a highly developed group of social insects, having a greater social organization than that of the social wasps or termites, second only to the social bees.

Social bees. The many species of bees (Hymenoptera, Apidae, and other families) are closely similar, and not all are social. Those that are include the well-known honey bee, *Apis mellifera* (fig. 9.10) and species of bumble bees *(Bombus* spp.) (see fig. 9.11 for *Bombus pennsylvanicus*, as an example). A total of fifty-seven species of Apidae are found in the United States; all but two are bumble bees or carpenter bees.

Bumble bees are robust, with bodies covered with hairlike setae. They range in size from 17–27 mm in length and are marked with conspicuous black and yellow, sometimes red, bands. Bumble bees nest in the ground. In temperate zones, the colonies die off each year to be started anew in the spring by the fertilized overwintering queens. These leave their winter hiding places and select a nesting site in a cavity in the ground for the new colony. The first eggs to hatch are workers. They enlarge the nest, collect food, and care for additional eggs laid by the queen and the new larvae. This continues all summer, with the new workers bringing in nectar and pollen to feed the young. In late summer, males and new queens are produced. Mating takes place, and the males die, the same familiar pattern we have seen throughout our survey of eusocial life. The cycle repeats itself each season. Bumble bees are pollenators and, therefore, are beneficial.

Figure 9.11.
Bumble bee (Hymenoptera).

The last, and most familiar, of all bees is the honey bee. In North America, the single species, *Apis mellifera*, is raised throughout the continent. This species was brought with the Europeans when they settled the New World. Here the honey bee lives only in man-made hives, except for a few that have escaped "captivity" and live in hives in hollow trees, the so-called bee trees, logs, or tree stumps.

The colony is composed of diploid (with two sets of chromosomes) females, haploid (with a single set of chromosomes) males (the drones), and sterile female workers. The life cycle starts with the egg placed in a brood cell in the comb of the hive. The egg hatches, the resulting larva staying in the cell where it is fed by workers. It pupates, and eventually an adult emerges.

Workers spend their lives at a succession of jobs. They spend their first three days cleaning the cells of the hive and feeding on pollen-honey mixtures. For the next five days, they feed larvae, first on salivary secretions and then on a substance termed bee bread. After this, they devote six days to building new cells from wax secreted from their wax glands. Three days are used to convert nectar into honey by fanning and also fanning to regulate hive temperature. A day is spent patrolling in front of the hive and attacking intruders. Beginning on the twenty-first day, they collect nectar and pollen until they die a few days later. Thus, the workers we see are near the end of their lives, and these are the ones that sting. When they sting, they leave their stinger in the wound and soon thereafter die. The workers assist the queen initially in setting up the colony and are, therefore, critical to its success. These sterile females comprise the most numerous members of the colony.

New colonies are formed when a swarm, led usually by a young queen that is accompanied by a group of workers, leaves the old hive and flies to a new nesting site. If the beekeeper anticipates this event, he is able to capture the young queen, place her in a new hive along with her following of workers, and establish a new hive. Bees are the only insects capable of maintaining themselves continuously with a store of food and in such large numbers throughout many successive generations. Other social insects, as we have seen, die off about the time of the first frost, leaving behind only a few fertile queens to start a new colony in the spring.

The queen is the center of the honey bee hive. She provides all the eggs, the majority of which develop into workers. Some of her offspring also develop into drones (males) and a few into queens. Queens are produced only from unfertilized eggs, this under three circumstances: when the present queen dies; when the colony becomes too large and must swarm; or when the queen is too old to produce the necessary eggs. Production is accomplished by feeding newly hatched larvae a special diet of protein, a rich secretion from the salivary glands of the workers, the well-known royal jelly. If a queen dies, only one queen takes her place. The workers quickly begin to construct larger-than-normal cells for replacement queens. More than one replacement is developed, but the first one to emerge from the pupal stage becomes the new queen. The others are killed by the workers. If two queens emerge at the same time they battle until one queen is killed.

Workers have various body modifications suited for the tasks they must perform. Legs are used for cleaning their antennae, using specially adapted structures for this purpose. The hind legs have pollen baskets and pollen scrapers. Pollen is removed from the pollen combs by a row of still hairs at the end of the tibia and is then pushed upward into the pollen baskets by a projection located just below the tibial comb at the junction of the basal tarsal segment.

Honey bee communication, first reported by Nobel prize winner, Karl von Frisch, is at its highest development in these bees. They are capable of transmitting detailed information by a dance pattern. Both the distance to the source of food and its direction can be reported by an elaborate method of bee dancing, turning in circles, and figure eights. A round dance (fig. 9.12) is performed if the source of food, pollen, and nectar is close to the hive. Workers report this information to other workers to show them where to locate the nectar flow. If it is over fifty meters from the hive, a scout bee, the worker who first found the food, will perform a waggle dance to communicate this information. The direction is signaled by the angle of the dance in relation to the direction of the sun. Bees are also able to detect shapes and colors used in recognizing honey-flow plants. The research of von Frisch opened an entirely new area of insect behavior or insect ethology, some of which we described in chapter 7.

Drones (males) have been mentioned several times in the discussion of social Hymenoptera, though, as stated, they have a minority role in an otherwise matriarchal hive. The drones' sole function is to mate with the queen. They are usually found clustering to one side of the hive, neither working nor helping in the construction of combs nor caring for the young. They do fly around on warm summer days, but when they mate, they die. In fact, the male genitalia evert out of the abdomen on meeting the queen, and the ensuing shock kills the male. Males that never mate are stung to death by workers and cast outside the hive as winter approaches.

Honey bees and other social bees perform a critical service to agriculture, because they are the chief pollenators of many fruits, vegetables, and forage crops. Without their assistance, these plants could not reproduce, and these would be absent from our tables.

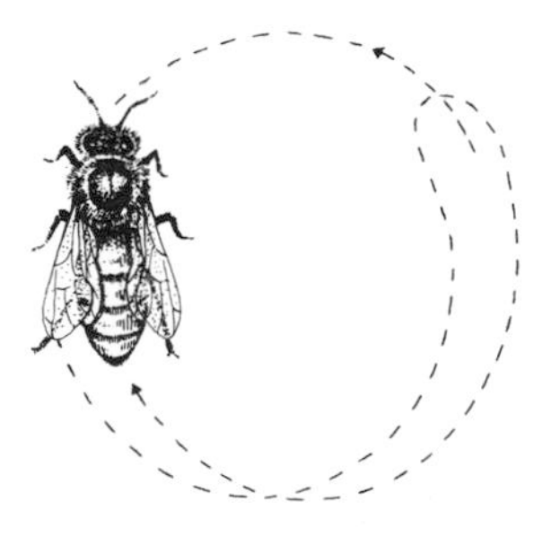

Figure 9.12.
Round dance of honey bee.

African honey bees. The honey bee species, *Apis mellifera*, is a native of Europe and Africa, including Madagascar. Three other species are native to other parts of the world. No species is native to North or South America, Australia, or New Zealand. Races of the European *A. mellifera* have been introduced to all countries on earth, other than tropical Asia. This species is divided into more than thirty distinct races. Some of these "strains" are usually referred to in the literature as subspecies. Those originally imported into North America were European, German, Italian, and Caucasian races. Each race has differing behavior features. The European strains are more gentle and easier to handle thansome of the other strains. For example, most beekeepers manage hives of the Italian bee, because it seldom stings. Some prefer the Caucasian bee, because it seems to gather more honey. The German bees tend to escape from the hive. Most of the wild bees in our woods are this strain. Two subspecies are now known in Africa, the African punic bee, *A. m. intermissa*, and the culprit African Honey bee, *A. m. adansonii*. It is this last bee that is presently the cause of so much concern. It is not generally understood by the "media" and exaggerated reports of danger from this bee have excited many.

The "Killer African Bee," so-called because it is extremely aggressive, deserves further discussion, because of the interesting triad of insect, bird, and man, a relationship that probably started millions of years ago, long before the "media" reports regarded this as "news." The African bird, a honeyguide (whose scientific name, appropriately, is *Indicator indicator*), has an appetite for wax. To satisfy it, this unpleasant, robin-sized relative of a woodpecker, whose young kill the young and take over the nests of other birds, at least two million years ago or probably more, started a war between the African honey bee and man. This war is still unsettled, and, seemingly, the bees are well ahead, as South American beekeepers know and North Americans may some day find out.

The Greater, or Black-throated, Honeyguides have learned that humans can be led around by their own sweet tooths. Incredible as it might sound, honeyguides will first locate a honey bee hive. They will then fly to a village and, circling and calling, attract attention to themselves. Then, pausing periodically to make sure they are being followed, they will lead a villager to the nest. There human honey-hunger takes over. The honeyguides wait patiently for the human to tear the bees' nest apart for its honey. Then the birds, whose gut contains colonies of bacteria for digestion of wax, gorge themselves happily on the remnants of wax honeycomb scattered about—snatching a few insects on the fly for dessert (the honeyguide's basic food is insects). The human gets most of the honey—and the stings! Many Africans are superstitious about this and believe that they must leave some honey for the honeyguides or, failing to do so, the birds will not lead them again or will lead them to a dangerous snake or a lion.

Chances are that the honeyguides evolved their ability to pit bees and humans against each other originally in relation to some other mammal, such as the honey badger, perhaps even before evolution produced man. Honey badgers will also tear into a bees' nest for patient honeyguides, and these animals may actually be the original work horse for these lazy birds. Once humans came along, their greed and the birds' trickery combined forces against the bee. Actually, we are not sure what the attitude of the bees is toward the birds, but it seems that this relationship is the cause of the excessive aggressiveness of the African honey bee.

After millions of years of having their nests torn apart by pre-human, and then human, creatures, African honey bees have evolved a healthy antipathy toward man and other large honey-eating mammals. They are, as one biologist puts it, "incredibly aggressive."

Ordinarily, that would be little more than an interesting fact outside of Africa. But in 1956, the man-hating African honey bee was introduced into Brazil for research purposes and escaped. It has been working its way gradually north, at approximately 200–300 miles a year ever since, overwhelming competing native honey bee races as it goes. It is creating problems for man and animals throughout most of South America, is now well established in parts of Central America, and probably will invade the United States in the late 1980s or early 1990s. According to Father Moure of Parana, a specialist on wild bees, it is next to impossible to work near the hives of these bees because of their stings. If a person approaches within fifty feet of the hive, the bees will swarm out and sting. One must take to water or thick jungle to avoid them.

With the invasion of the bee into thickly populated areas, it could seriously affect the practice of seasonally transporting large numbers of bees to pollenate fruit and nut crops, as well as legumes, oilseed crops, vegetable seed crops, and curcubits. Such bees pollenate crops valued from $1 to $6 billion annually. An intrusion by the African bees could also affect honey and wax production and create serious public health hazards.

On the positive side, the African honey bee is hardy and industrious. It flies longer hours and goes farther to get honey than the strains it replaces and works in a wider range of weather conditions. Many beekeepers report that it produces more honey than other strains. Little has been done, however, to study the genetics of honey bees. This could be a fruitful task in the attempt to solve the problem of hybridizing the good qualities of the African bee with the docile Italian honey bee.

We should emphasize that the African honey bee's individual sting is no worse than the Italian or other honey bees' stings, but the African bee is more likely to sting because of its former relationship with the African natives and the honeyguide. It is harder to dissuade—it takes something like four or five times as much of the smoke beekeepers use to befuddle these bees, so that they can get into the hives for honey, as it takes for other domesticated bees. The African honey bees are easier to stimulate to a stinging frenzy, because they are particularly sensitive to the chemical pheromone a bee releases as it tears loose from the stinger it has planted in the hide of some adversary. This tearing loose of the implanted stinger from the bee's body is fatal to the bee, but, at the same time, the scent chemical released triggers a chain reaction of stinging in other bees. A disturbed colony of Italian honey bees might be stimulated to sting at a rate of ten a minute by the pheromone, while the African honey bees' sting rate is anywhere from 200–300 a minute. As a consequence, despite incomplete reporting, records in some years show as many as 300–400 human deaths a year from bee stings, as compared to the one hundred or so reported in the United States.

African bees present real dangers, but they are not to be feared the way that they are presented in movies and TV programs. They are a problem to be attended to by entomologists, and they certainly are to be avoided by all. Most of our fear of these insects will be overcome by an understanding of the problem. We can hope that entomologists will be successful in crossbreeding and selecting their best qualities. Honey bees have many pests, predators, and diseases. Efforts to eradicate the invading African bees by pathogens and to immunize the other bees are, as yet, unsuccessful.

Termite, ant, and bee guests

In table 2.2, chapter 2, we listed various kinds of symbiotic insect associations. This includes hundreds of species found in termite and ant nests, and beehives. Some of these species have evolved to look like ants, while others are rounded, compact, and hard. The latter are capable of folding up into a round ball and withstanding the attacks of the soldier termites and ants and the sting of bees. Even army ants have an assortment of these guests running along with the foragers. Their legs are elongated, which enables them to keep stride with their symbionts. The many intricacies of these relationships could by themselves be the subject of an entire book.

10

Migration and Other Travels

"Lazy flying
Over the flower-decked prairies, West
Basking in sunshine till daylight is dying,
And resting all night on Asclepias' *breast;*
Joyously dancing,
Merrily prancing,
Chasing his lady-love high in the air,
Fluttering gaily,
Frolicking daily,
Free from anxiety, sorrow, and care!"

C. V. Riley (1843–1895)

As we have seen in chapter 6, the adult insect is the stage between two eggs. Most of the activities of adults are concerned with the production of eggs and their distribution. Population regulation is dependent on maintaining the size of the population throughout the range of the species—it must not be too large in the immediate vicinity of the egg layers, nor too sparse to inhibit the spread of the species. Just as flowers, bushes, and trees must extend their seed shadows, that is, the area in which the seeds are dispersed, so too must eggs be distributed.

Insects have complex distribution patterns. Often this is related to plant distribution, their food source, or the distribution of their hosts and prey. Plant distribution is partly related to their adaptation to environment, and partly due to their means of seed dispersal. It is physiologically necessary for seeds to be distributed some distance from the parent, otherwise offspring and parents would be competing for their places in the sun. Same problems face insects. Some species of butterflies "count" out the number of eggs they lay on a single plant, which prevents food competition among siblings. Thus a study of insect distribution is, in part at least, a study of the ecology of insects. Insects occur where they do because they are adapted to the particular habitat they select for egg laying, or because they are in search of a new habitat. Although most insects do not migrate in the sense of bird migration, some do, and these special cases of habitat selection are always of great interest to naturalists.

All individuals belong to a species, but, as we have seen, species are divided into populations—all members of a species are not in direct contact with each other. It is these populations that together comprise the species as it is now, and as it was in the past since its origin, and as it will be until it becomes extinct. The samples we study (see chap. 15) are population samples representing the species. As we learn when collecting these samples, some may be found in abundance, while others are scarce (table 10.1). No general agreement exists as to how properly to describe this variation in the number of individuals. One of us (Arnett) coined the terms "underabundant" and "overabundant" populations to help determine the nature of this varying number of individuals in a population. An underabundant population is one that does not inhabit all of the ecological habitat to which it is adapted. Thus, when collecting in a particular habitat, one finds specimens only in portions of the habitat. Although sometimes this is the result of specialization of the species for certain microhabitats only, generally it is the result of not enough individuals to spread into all areas. Much the same as the early years of the history of the New World when vast regions were uninhabited by man, so too are many areas uninhabited by a particular insect species, because it hasn't, as yet, penetrated, for whatever reason, into all areas it would otherwise be able to inhabit. Later, as with the human population, when numbers increase, migration to new areas takes place. The complex factors controlling this contribute to the fascination of the study of distribution. However, you have already observed,

Table 10.1. *Classification of insect abundance.*

(Letters in brackets are symbols that might be used in notes on the species.)

1. Abundant: easily collected in long series (100 or more) [A]
2. Frequent: often collected, but the number is not enough for long series (15 or more) [F]
3. Scarce: found only by using special methods to attract or to find the species, or by persistent collecting of an area known to harbor the species (under 15 specimens) [S]
4. Broods: in flight as adults for a short time and then disappear as adults for long periods (seasons or years) [B]
5. Migratory and of temporary occurrence in most localities [M]
6. Local: locally abundant in very restricted habitats [L]
7. Economically important species: pests of crops, forests, ornamentals, man, and domestic animals, often very abundant [Ec]

some species seem to occur everywhere. The house fly seems to be a splendid example of this. Here is an overabundant insect, which we usually call pests, but being a pest is not a part of our definition of overabundance.

To classify these differences in abundance, the following categories have been devised to describe the condition for any population. We have tried to avoid the ambiguous terms "common" and "rare," but actually a common species is usually an overabundant species, while a rare species is an underabundant species. It should be noted that the condition of any given population may change considerably from year to year. Hence, the use of any of these terms is predicated on the condition at the time of collecting. This also means that when the collector finds an abundant species, samples should be taken, because, it may be learned later, what was thought to be common was actually a "rare" species. Sometimes, luring insects to UV lights attracts "rare" species from nearby localities. Those not collected stay to breed in the new location and become "common." Thus a "rare" species may become a "common" species through the action of collectors (see chap. 17). Conversely, and paradoxically, a pest species may be "rare," simply because at any particular time and place, the species is represented by a few individuals due to a combination of factors—parasites, predators, lack of food plant, adverse ecological conditions, and so on.

Factors affecting insect populations

The environment may be divided into the physical environment and the biotic environment. These factors are termed, collectively, as the effective environment. The subject of environment and population dynamics fills volumes, but to succinctly state the subject, we may say that any force that impinges on the individuals of a population contributes to the total effect it has on the population. The reaction of the individual is predetermined by its genetics. The physical factors are temperature, light, oxygen, water, and available chemicals. The biotic factors are organisms of the same species and organisms of different species (these two are very different, if you stop to think about it).

Physical factors

Temperature. Change in season brings about extremes in temperature. The activity of insects, being coldblooded animals, varies with the temperature. The warmer it is, the more active they are; the colder, the slower. Each individual has an optimum temperature and fixed extremes, beyond which death occurs, either because it is too hot or too cold. It is a good thing that this variation among the individuals exists, otherwise, an extreme in the weather would kill the entire population. As it is, survivors almost always are left alive to reproduce.

Various adaptions help insects to survive climatic change. In extremely hot areas, they may be active only at night, or, they may be white, to reflect heat. Some dig into the soil to escape the sun's direct rays, and others seek shade during the day and fly only at dawn or dusk. Extreme cold is usually easier to survive than heat, because the insect becomes inactive, uses little energy, and requires virtually no oxygen until warm weather returns.

Light. Some insects are characteristic of open areas with bright sunshine and others are typically found in shade. Probably no direct correlation can be made between the amount of light and its effect on the insects. More likely, it is a matter of the preference for sun or shade by their food plants or their preys' host plant. For example, species of *Collops*, a beetle (Melyridae), feed on the eggs of species of the beetle genus *Leptinotarsus* (Chrysomelidae). These eggs are laid on the leaves of plants that grow in bright sunshine. *Collops* spp. are not partial to sunlight, but to the eggs that are easier to get in the sunshine. Scorpionflies prefer the shade of woods, probably only because they prey upon woodland species.

Oxygen. All organisms depend on oxygen to use in the metabolic processes of growth and energy production. Almost always, lack of oxygen is not a limiting factor. The atmosphere contains an abundance of this gas, released by plants during photosynthesis, enough to provide the needs of all plants and animals. A few insects are adapted to live in anaerobic (without atmospheric oxygen) habitats. Flour beetles penetrate bags of flour and reproduce deep in the interior away from atmospheric oxygen, or at least where it is so scarce that it can't be used for respiration. These beetles are able to survive, because they can release oxygen from the carbohydrate (flour) on which they are feeding. This oxygen, termed metabolic oxygen, is absorbed into the cells of the insects, where it is ready to be used in other metabolic processes.

Water. All terrestrial organisms must be able to retain water in their bodies in order to stay alive. It is used in all biochemical reactions of living organisms on this earth, these being water-soluble reactions. Drying out of the insect body through exposure to sunlight or excessive heat is a frequent threat, overcome only because of the nature of the insect integument. The more or less hard covering is impervious to water, except at the intersegmental membranes and the tracheae. This is mainly because of the waxy or oily nature of this covering material. Drying can occur through open spiracles, and these must be kept open for metabolic gas exchange when the insect is active. We have the same problem when we exercise. Water escapes through our sweat pores and out our mouth and nose. This must be replaced, or we will die. Insects are frequently seen drinking water condensed on leaves, at puddles, and so on. Confined insects will soon dry out and die, unless they have access to water.

Chemicals. Apparently insects receive all of the minerals needed in their diet through the food they eat. Salt, calcium, phosphorous, the various trace elements, and vitamins (somewhat different from ours) are obtained from the plant and animal materials they ingest (as do we). Some of the vitamins we require are synthesized by insects, and they require some we don't. Beetles known to feed on highly refined carbohydrates, such as flour, cannot survive without additional food. When these beetles are cultured, yeast is added to the flour. When they infest stored flour, they apparently contaminate the material with yeast, which grows and provides the needed nutrients.

Biotic factors

The marvelous range of foods used by insects and the range of habitats they occupy accounts for their great diversity. All of the physical factors so briefly discussed here are basic to their adaptive ranges. But equally effective are the biotic factors discussed below.

Species of the same kind. To some degree, every species must be able to recognize its own kind, which is probably a biochemical recognition. To the extent that this statement is true, this is another way of saying (at least among many animal species) that the behavior of an individual is affected by its fellow members of a population, and, ultimately, the entire species is affected. Obviously, mating is the most important of these relationships. Food getting can be competitive between individuals of overabundant populations. Insects show no compassion. They eagerly feed when food is present and share none of it when it is scarce. Even those social insects busy feeding their young will eat first if food is scarce. The third intraspecific relationship, protection, is a curious, innate factor that seems to be lacking in most species. It's every insect for itself. Even though each seems to be able to recognize each other, this does not protect it from cannibalism, if the species is predatory. We have even seen plant juice-sucking bugs suck the juice from another bug, and bed bugs suck blood from other bed bugs. Among the social insects, certain castes inherit the ability to protect the nest, which includes fighting between one nest and another, human style. The female praying mantid does not waste any food by sparing the life of her mate once she is fertile. The male is just another piece of useful protoplasm and is quickly consumed when he is within reach of her lightning quick claws. In crowded conditions, many species are cannibalistic and will eat their own young, and even each other.

Organisms of a different species. Interspecific relationships involve the remainder of all biological phenomena. Symbiotic partnerships are the subject of such studies as parasitism, phytophagous and trophozoic feeding, coevolution, and so on. Much of what is written in this book is concerned with the relationship of organisms of one kind with those of another kind. All of these factors are involved in the development of a species and its ability to disperse. Each species has a definite, but not permanent, geographical distribution. It seems clear that no species is "content" with its present state in life, and it persists in trying to expand its distribution range.

Dispersal

Returning now to the main subject of this chapter, the scattering of eggs, we see that many factors are involved in the placing of eggs where they have the best chance of developing into adults to produce more eggs, the only excuse for the adult stage. It seems to be the mission of each species to continue its existence. We know, of course, that species, the same as individuals, have a life span. This starts with the birth of the species and ends with its extinction, be it natural or through the effects of man's activities. All species eventually become extinct through natural causes, if not otherwise. The time span of a species may be short (perhaps even a matter of seconds, hours, or days; who knows how often a species is "born"). Or, as we like to say, the "well"-adapted species last for millions of years. While an insect species exists, its role in the world seems to be to distribute eggs. How this is done differs in as many ways are there species.

Most insects (in terms of numbers of species) fly. Thus, we will discuss egg dispersal as if all species flew. But we are not overlooking the fact that wingless insects also disperse, either by walking, crawling, or by being transported by air currents or by other organisms. For example, many tiny insects ride air currents whether or not they can fly. Some insects are dispersed in the seeds they infest in the same way that the seeds are distributed. Ectoparasites are transported from nest to nest by their hosts.

Winged insects are able to fly from place to place. It seems that a certain proportion of the fertile females of every species migrates to new habitats; that is, they leave the confines of their own population range and move into new areas. These forays may be successful, but generally all perish and the effort seems useless. Species are, for the most part, rigidly adapted to the area in which they breed and only rarely can they escape to new areas. Then, too, most species are underabundant, as we said previously, and simply do not have the personnel for these experiments. All dispersal flights are at random. Where they go cannot be predicted. Yet, rarely it does happen. Witness the range extension of various species of economic importance, or the distributional changes of many species of birds during the past fifty years (compare the first edition of Peterson's *Field Guide to the Birds* with the most recent edition to see this range change).

Migration

As generally defined, true migration occurs more or less regularly at a particular time of year. The flight (we usually think of insect migration of flying species) is usually persistent; that is, it lasts for several days or even weeks. It is continuous in one direction and for a long distance. This definition eliminates flights from the breeding area to the feeding area. For example, mosquitoes develop in aquatic habitats, where they mate. But the female usually must fly some distance to obtain her required blood meal, and then return to the egg-laying site. This is not considered to be migration.

The difficulty in defining or determining true migration comes from the many different types of insect movement taking place. Some entomologists apply the term migration only to masses of insects traveling together in the same general direction, and, after reaching their destination, the same individuals return to the same general original habitat. This restricts true migration only to a few species, a view adopted here. Before these few cases are discussed, however, we must look at what some other insects do that might be confused with migration.

Some insects have a one-way migration during part of their life cycle. This occurs in many species of aphids. Many of these sap-sucking insects regularly change hosts and migrate or disperse in great numbers from the habitat of one host to the habitat of the new host.

Passive migration occurs when insects are swept up and carried high into the atmosphere, sometimes thousands of feet into the air, and are passively transported by air currents to new areas. Collections made by nets attached to airplanes show that hundreds of species are found many feet in the air, some as high at 30,000 feet. Few doubt that many species find new habitats in this manner, including some that become pests in their new location. We do not know how many species are carried across continents and oceans in this manner. However, there is no direct evidence of this happening. No one has actually observed this taking place and, to the best of our knowledge, no experiments have been conducted in an attempt to show that this happens.

Sometimes the habitats of immature stages are so different from those of the adults that the newly emerged adults must "migrate" to their proper adult habitat, and then the fertile female must go back to the original habitat of the immatures for egg laying. Sometimes the immatures are left to find their own habitat, but, obviously, this is not common. A notable example of the latter is found in the peculiar actions of cicadas. The nymphs of these insects spend their life deep in soil feeding on rootlets. This may be as long as 17 years in the case of the seventeen-year locust. When mature, the nymph digs its way to the surface of the soil, crawls from there to a nearby tree, ascends the trunk for some distance. Then the nymphal skin ruptures and the adult crawls out, leaving the old skin behind. As soon as the wings are inflated and dried, the adult flies to the top of a tree where it loudly stridulates to attract mates. After mating, its eggs are inserted under the bark of delicate twigs of trees where they develop. The first instar nymph is now high in the air, far away from the roots upon which it must feed. It must crawl or drop to the ground and dig into the soil to reach its feeding habitat. One wonders what change took place during the evolution of these species that seemingly leaves these delicate, newly hatched nymphs "high and dry," so far away from their food. Migration to their proper habitat must result in the loss of a high percentage of the new population.

Migration flights

True migration takes place in three stages. First, the insect flies from its habitat, usually before the area becomes uninhabitable through a change of season. The adaptation factors that initiate this usually are related to the length of day or changes in temperature or humidity, factors that minimize the chance of the insect's returning until the habitat again becomes suitable for breeding. Observers of the departing flights note that there is an orientation flight at the time they leave. This consists of a circular flight around the area, which seems to imprint landmarks in the nervous system of the insect.

Next, prolonged flight takes place in one direction, either to reach the area where diapause is to take place, or, when returning, to reach the breeding site. Actually then, the second stage is the leaving of the breeding ground. The flying is usually aided by wind currents, but sometimes the insects must actually fly against the wind.

The third stage is the return of the same adults, after diapause, to the original breeding grounds. Presumably no orientation flight occurs at this stage, because these individuals, so far as it is known, never return to the diapause area. No one knows how the new generation of adults knows how to find the diapause area, yet, as we will see in our discussion of the Monarch Butterfly, the first stage flights are to specific locations year after year.

The energy required for these flights is not as great as one might expect, because insects take advantage of, and, to a considerable extent, depend on, wind movement. The insects ride up on heat convection currents and passively glide much as motorless airplanes, to sustain their flight for hours.

Once the insects arrive at their destination, new behavior patterns are triggered to help them find host plants or resting sites. Migration depends on complex factors, many of which are controlled by the environment—temperature, food supply, and crowding seem to dominate in most species. The migrating population always is led by females. Males may migrate with the females, separately, at some distance from the females, or not at all, depending on whether the females have mated or not. Since the development of the ovaries parallels the development of wings, the migrants are always sexually mature. In the immature stages, food has been stored and energy is obtained by using fat stored in the fat body, a structure similar to our liver and a center of metabolic activity. Insects arrive at their diapause site with little of this reserve used, because of their ability to glide.

Professor C. G. Johnson of the Rothamsted Experimental Station, England, in his classic study of insect migration, proposes three classes of migration, but points out that there are no clear-cut differences. Insects behave more than one way and cross the lines defining these groups. Johnson's classes are as follows:

Class I: Adults, with their lifespan limited to within one season, that emigrate from a breeding site, disperse, oviposit, and die.

Class II: Insects that emigrate from a breeding site to a feeding site: the same individuals then return later in the season, after oogenesis, to oviposit in the former breeding site, or in other sites. Some species repeat such back and forth flights several times in a season, and, except when some flights take some individuals beyond the area within which the back-and-forth flights normally occur, these are probably not to be regarded as truly migratory.

Class III: Insects that emigrate from a breeding site to a place where they hibernate or aestivate, usually aggregate; the same individuals disperse next season, some ovipositing in the territory from which they came.

Johnson's classes are based entirely on insect flight migration. It might be well to add another class which includes immature stage migration.

Class IV: Migration from one feeding site to another or stream drift migration.

Many species of moth larvae routinely migrate in groups from one host plant to another. They have definite behavioral patterns, as described in chapter 7.

Some time ago, researchers became aware of the phenomenon now termed "behavioral drift." Streams that are the breeding site of aquatic insects have a continuous drift of invertebrates, now known to be a natural feature of these streams. This drifting is not a simple washing downstream, but is the response of organisms to the twenty-four-hour light/dark cycle of each day. The amount of drift by bottom-dwelling insects increases greatly at night, usually just after sunset. Failing light triggers increased activity by these insects. It seems that most of these drifting naiads, members of the orders Ephemeroptera and Plecoptera, and larvae of certain Diptera and Trichoptera, have two nightly peaks. For some, the greatest amount of drift takes place just after sunset, followed by a minor peak just before dawn. For others, the peaks are reversed, the minor one at sunset and the major at dawn. The naiads and larvae are distributed throughout the column of water. Although predator naiads of Plecoptera maintain feeding territories, as do Odonata naiads, and, therefore, do not drift, the detritus feeders pass by and may drift for miles throughout the season.

Studies of this phenomenon are not complete. Therefore, not much can be said about why this drifting takes place. Some researchers say it is because of crowding, while others claim it is simply a testing of new environment. All seem to agree that it is a regular part of the behavior patterns of many species, but they fail adequately to explain the role darkness plays in triggering the action, which can be experimentally stopped by shining artificial lights on the stream at night. Interest continues in this subject, because the drifting naiads and larvae are food for fish.

Nonmigratory flights

Obviously, wings are used for many flights other than true migration. As already pointed out, most insects do not migrate, but to help recognize true migration, it is necessary to know what flights are not considered migratory.

Many insects make repeated flights within a season from oviposition to feeding sites. That is, the adults feed in one area, but the immatures in another, as discussed above. Therefore, egg laying takes place in one area, adult feeding in another. This is characteristic of many beetles. For example, pollen-feeding beetles, such as species of Oedemeridae, fly to flowers as adults, feed on the pollen, and then lay their eggs in the ground or under bark some distance away. Since these flights are not outside the area occupied by the population of the species, the flights are neither for dispersal nor migratory.

Post-teneral flights, flights taken after full development of the wings, may be made from the immatures' habitat to the feeding grounds of the adult. Only if this is for a long distance and the result of an inherited habit can this be considered migration. Usually these flights are for only a few meters, although some species may mate before the flight and disperse at this stage to widen the area occupied by the population.

Examples of true migration

Migration has been described mainly in the Lepidoptera, where species with the most sustained migratory flights occur. Certain Orthoptera migrate; some beetles, particularly the Ladybird beetles (Coccinellidae), and certain leafhoppers (Homoptera). The following are examples.

Monarch butterfly. The migratory habit of the Monarch has attracted much attention, and it has been widely studied by lepidopterists. These butterflies belong to the family Danaidae, a moderate-sized tropical family whose larvae feed exclusively on milkweed. Only one species has invaded the temperate region, *Danaus plexippus* (fig. 10.1). Only this species is known to migrate.

For many years, individual butterflies have been marked (fig. 10.2), released, and recovered in various parts of southern United States and Mexico. In this manner, much has been learned about their flight patterns. When they migrate, they have the habit of congregating in trees at night in large numbers to "sleep." Some of these trees have been used year by year, making them famous as "butterfly trees." If marked specimens are found in the trees, then it is possible to learn where they came from. Only recently, however, entomologists have been able to discover the location of the major final winter resting grounds in Mexico. There they cluster in large numbers, millions, perhaps billions, of individuals packed closely together. They take advantage of this saturation as protection from predation by birds.

 Figure 10.1. *Monarch butterfly (Lepidoptera).*

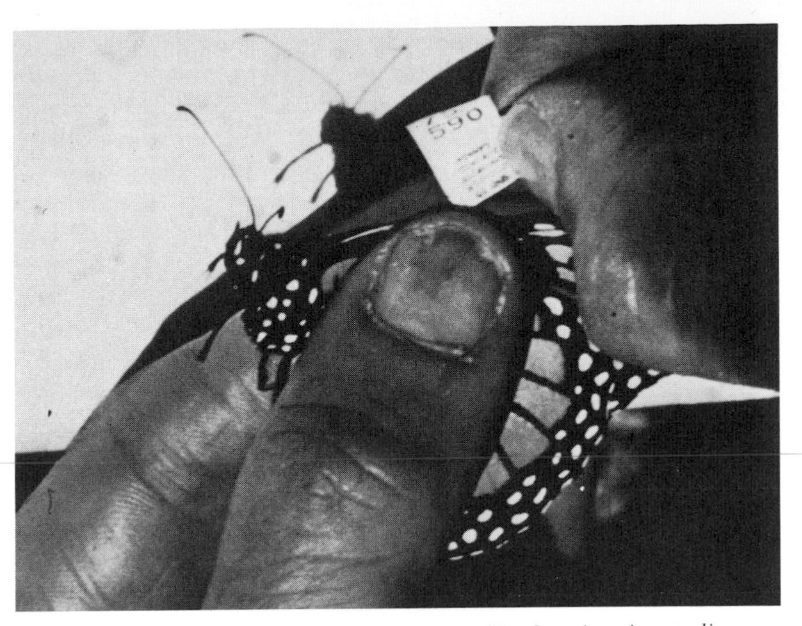

Figure 10.2. *Marking living Monarch Butterflies for migration studies (Lepidoptera). (Courtesy of Lee D. Miller)*

The larvae of this species feed during the summer on milkweed plants growing in northern United States and southern Canada. Milkweed is a poisonous plant, but the poison does not affect the larvae. Instead, this material is stored in the body of the larvae, and, eventually, the adult. Eventually a chrysalis is formed and, in late summer, when the days become noticeably shorter, the adults emerge. They congregate in small groups and then head south. After many days of travel, a major portion arrives in the lowlands of Mexico, and then flies to one specific location in a highland forest. The cold of the high altitude prevents gonad development throughout the winter. Winter storms bring snow to the area, but relatively few individuals are lost by freezing. Some birds have learned to feed on these poisonous butterflies. Although most birds will not attack them, some will, and apparently these birds have learned how to scoop out the inner, nonpoisonous tissues. It is estimated that they have learned this procedure during the past two hundred years.

In the spring, these adults again migrate, flying north laying eggs as they go. Two or three generations, each lasting from four to six weeks, are produced, the old adults dying off and the young taking over. In the fall, it is the young replacements that migrate south, flying entirely by instinct.

Other migrating Lepidoptera. The Snout Butterfly, *Libytheana bachmannii*, is a familiar migrant in the southern states. It breeds from New England and southern Ontario, south to Florida and west to the Rocky Mountains and Arizona. Their larvae feed on *Celtis occidentalis*. Large flocks of these butterflies may be seen in mass migration. This appears to be true migration, but not as precisely carried out as that of the Monarch.

Various species of Noctuidae migrate northwardly in the spring and breed throughout the northern part of the United States and southern Canada. One, the Cotton Leafworm *(Alabama argillacea)*, does considerable damage to cotton in its path. None of these species overwinters there, however. They die out and are replaced the following year by new migrants. Some species migrate north from Mexico and may be captured as far north as New York state, but they, too, do not survive the northern winters.

The larvae of the Armyworm *(Pseudaletia unipuncta)* feed on various grasses. At irregular intervals, they become so numerous that they eat everything in one area and then migrate as a group to new feeding grounds. Their vast numbers attract the attention of the general public and cause great concern.

Many species of butterflies captured in Texas, Arizona, and as far north as Kansas are not known to breed in the United States, yet they are relatively common and are permanently listed in books on the butterflies of the United States and Canada.

Migratory locusts. Since biblical times,[1] grasshopper plagues have been described and are frequently reported in the news. Mass migration of various species of these insects occur throughout the world. Volumes have been written on this subject. In many places, particularly in Africa, species of the spur-throated grasshoppers (Acrididae) (fig. 10.3) cause severe damage to vegetation, because of the tremendous numbers and widespread migrations that occur at irregular intervals. In 1873–1875 in Europe and 1874–1877 in the United States, the outbreaks were extremely severe, and many other similar swarms have appeared since. One such scourge had an estimated 124 billion insects in it. In five days, a swarm consumed

[1]"But if thou resist, and wilt not let them go, behold I will bring in tomorrow the locust into thy coasts: to cover the face of the earth that nothing thereof may appear, but that which the hail left may be eaten: for they shall feed upon all the trees that spring in the fields. And they shall fill thy houses, and the houses of thy servants, and all the Egyptians; such a number as thy fathers have not seen, nor thy grandfathers, from the time they were first upon the earth, until the present day. And he turned himself away and went forth from Pharao." (Exodus 10:4–6).

Figure 10.3. *Adult migratory locust (Orthoptera).*

7,000 tons of oranges. This is more than the entire yearly consumption of oranges in France.

Of the many species of grasshoppers (over 570 species in the United States and Canada alone and several thousand throughout the world), only nine are known to enter these migratory phases. In the United States, the principal species is *Melanoplus sanguinipes*. Most species of the large genes *Melanoplus* are solitary feeders, avoiding each other as much as possible. This is also normal for *M. sanguinipes*. Those that do this are said to be in the solitary phase, a phase that may last for many generations. When conditions become crowded, an entirely different phase, the migratory phase, develops. In this phase, the insects differ in shape, coloration, behavior, and metabolism. They tend to aggregate and the "urge" to travel sets in. As a rule, these ravenous hordes come only from certain districts, which are apparently favorable for the development of this phase. The females mate and lay large masses of eggs. Billions of eggs are produced, and fifty to sixty days later, depending upon the temperature and humidity, the eggs hatch, and the ground is covered with thousands upon thousands of small insects (the nymphs) creeping toward any growing plant (fig. 10.4), all in one direction, crossing ditches, over hills, and even across ponds and rivers. As they march, they molt and grow from the original tiny nymphs to the winged adults. Still, they go on, flying in swarms (fig. 10.5) so extensive and dense that they shut out sunlight. It is reported that Tucson, Arizona had to turn on street lights during one such flight, even though it was the middle of the day in that sunny, desert city.

Figure 10.4. *Marching nymphs of the migratory locust (Orthoptera).*

Figure 10.5. *Migratory locust adults flying in swarms (Orthoptera).*

Ladybugs. One of our most beneficial insects, the beetle we call "Ladybug" (Coccinellidae) congregates during the winter month, most noticeably, those in mountainous areas, high in mountain meadows. For some reason, probably to keep cool and not break dormancy before there are other insects for them to eat, they collect in large masses (fig. 10.6) on vegetation in these mountains. They swarm in the fall, flying ten to fifteen miles or even more to reach these winter homes.

Figure 10.6. *Aggregates of ladybird beetles clustering on weeds on Mt. Lemon, Arizona, where they will hibernate for the winter (Coleoptera: Coccinellidae).*

Woollybears. Grasshopper nymphs and armyworms are not the only immatures to migrate. One species of moth (Lepidoptera: Arctiidae), the Woollybear larva *(Isia isabella)*, has become famous as a weather prophet in the northern part of the United States. Woollybears are covered with a coat of hairlike setae. The anterior and posterior ends of the body are black; the middle, brownish or orange-brown. It is believed that the width of the central band forecasts the length of winter. A narrower band, which means that there is more black, predicts a longer winter; less black, less winter. Strangely enough, there is some truth to this prediction.

The larvae of this moth feed on various low, herbaceous plants, including plantain and a variety of weeds. When feeding is completed, the caterpillars migrate some distance to find a suitable, protected spot. There they spin a cocoon in which to pass the winter. It is well known that a wet summer usually means a longer winter. Also, the color of many insects is determined to some extent by humidity as they are growing. It follows, then, that insects tend to be darker, if they grow during a rainy season, lighter during dry seasons. Thus, it is likely that dark Woollybears mean long winters, even though there is no direct correlation between these populations and weather forecasting. Color is, of course, controlled for the most part by genes. Even so, watch for these interesting larvae. They may be seen crossing highways in the fall. Collect some, provide them with a place to spin their cocoons in a box or jar. Keep them outdoors all winter. In the spring a beautiful moth will emerge.

11

Disease Spreaders

"insects have
their own point
of view about
civilization
a man
thinks he amounts
to a great deal
but to a
flea or a
mosquito
a human being is
merely something
good to eat"

Don Marquis (1878–1937), "archy and mehitabel"

Of primary concern in the minds of many people are the diseases transmitted through the bites of insects, or fancied other means. There are few principal ways that disease is transmitted by insects. Usually it is from man through insect back to man; or domestic animal through insect back to domestic animal; or domestic animal through insect to man. Most of the insect-borne disease organisms pass part of their life cycle in the insect. Some are transmitted by direct contact only.

Medical entomology is a branch of entomology concerned with the annoyance, injury, or spread of disease to humans and domestic animals by insects and mites.

Insects affect man's health, either directly by stinging or biting, causing great discomfort, or indirectly as vectors of disease. For an insect or other arthropod to be a vector, it must be able to transmit successfully the pathogen (disease-causing organism) to humans or animals.

Throughout history, insects and mites have been the plague of man. Biblical accounts tell of "a grievous swarm of flies that entered the house of Pharaoh and corrupted the land" (Exodus 8:24). References to lice date back to Homer in 900 B.C., and Aristotle referred to mites as "those eight-legged arachnids" that cause such problems as scabies and mange.

Very little scientific knowledge about insects and their relationship to disease was available until the invention of the microscope, the rejection of the theory of spontaneous generation, the introduction of the germ theory of disease, and the knowledge that all organisms are cellular. In 1848, Josiah Nott published his belief that mosquitoes caused both malaria and yellow fever. A milestone in medical entomology occurred in 1878 when Patrick Manson observed the roundworm, *Wuchereria bancrofti*, in the body of a female mosquito. It was eventually proven that a mosquito was the vector (carrier) of this roundworm, the cause of the serious diseases in humans, filariasis and elephantiasis. In the late 1800s, several important discoveries in medical entomology were made, which helped establish medical entomology as a distinct discipline. Some significant discoveries include Laveran's discovery in 1880 that the causal organism of human malaria was a parasitic protozoan, *Plasmodium malariae*. In the same year, Carlos Findlay proposed the theory that female mosquitoes transmit an arbovirus that causes yellow fever. In 1895, Bruce found that the tsetse fly carries a pathogen, the causative agent of nagana in animals in central Africa; and, in 1889, Theobald Smith discovered the organism that caused Texas cattle fever, a disease transmitted by ticks. In 1900, Walter Reed and his co-workers proved that the female mosquito, *Aedes aegypti*, carries yellow fever.

The problems insects cause to human health have changed the course of history, although many historians have overlooked their decisive role. Plague rates as perhaps the most serious disease in the history of human populations and has claimed millions of lives. The great pandemic in Europe in the 14th century resulted in twenty-five million deaths, or nearly 25 percent of the entire population of Europe.

Napoleon's army in Russia was decimated as much by louse-borne diseases as by hunger and exposure to the cold winter. Typhus and trench fever, both carried by lice, were disasters to both the European and Russian armies.

African sleeping sickness, transmitted by the tsetse fly, has prevented the development of many areas of central Africa. Malaria has played a major role in retarding the development of many areas of the world; it is still a major disease in some parts of the world and claims thousands of human lives every year. Malaria and yellow fever, both diseases transmitted by mosquitoes, prevented the construction of the Panama Canal until the American doctor, Gorgas, realized the need to control mosquitoes. During World War II, the Korean War, and the war in Vietnam, mosquito-transmitted diseases were a constant threat. Troops exposed to the bites of mosquitoes in these areas often contracted malaria. They could not fight in mosquito netting and mosquito-resistant clothing. Antimalaria drugs were not effective. Not until airplane spraying with DDT was perfected (at the Army School of Malariology in the Panama Canal Zone) was it possible to kill the mosquitoes prior to landing party invasion.

Closer to home, mosquito-transmitted St. Louis encephalitis and eastern-equine encephalitis occur in many areas of the midwest and eastern United States and have caused disease and death to many. In 1977, the state of Florida reported over one hundred cases of St. Louis encephalitis with eight fatalities; usually there are only a few cases, but, on occasion, serious outbreaks of both diseases occur.

Insects themselves cause numerous problems to humans. They annoy us by their mere presence. Some people suffer from entomophobia, fear of insects. The bites are injurious, as well as annoying. When they bite or sting, they inject either saliva or venom, and the wound may result in dermatosis. Sometimes eggs or larvae are deposited on the skin. The larvae bore into the body through the skin, causing myiasis (meaning the "disease" is caused by the larvae of flies). The problem is more common in animals, though there have been reports of myiasis in humans. Many, if not most, people are allergic to the bites and stings of insects, rarely so much so that medical attention is required, but sometimes even hospitalization and occasionally death results if the allergic reaction is severe.

Entomophobia may be caused by the mere presences of an insect. Often this fear develops even when no insects are present. Howard Hughes suffered from this affliction for many years. Insects can induce both annoyance and worry that may lead to nervous disorders. This problem may be more common than realized and, of course, is the result of the patient's complete ignorance of what to fear and what not to fear. Fear of spiders and other archnids is known as Arachniphobia, a term used only because spiders and mites are not insects.

Annoyance from insects is difficult to evaluate and determine. Flies at home or work, ants at a family picnic, a bee in the car: all are examples of annoying insects. The loss in employment time, worry, and fear is nearly impossible to determine.

Injury to delicate membranes occurs when insects or arachnids enter the eyes, nose, or ears, where they cause mechanical injury. This is especially prevalent, of course, during the summer months when many flying insects are about.

Envenomization, the pain and injury caused by stings, bites, and defense secretions of insects, is a serious problem. Allergic reactions from envenomization can be caused by direct contact, bites, and stings. Bees, ants, and wasps are especially harmful. More individuals die each year from stings by hymenopterous insects than from snake bites and scorpion stings combined. Venom from insects enters a victim by a stinger. The most serious syndromes associated with envenomization are allergic reactions, which may include hives, eczema, and asthma, or the more severe reaction, anaphylaxis.

Dermatosis, skin irritations caused by the injury of insects through their bites, secretions, or actual skin invasion, and pediculosis, caused by lice, are common forms of skin lesions. Scabies, mange, and acariasis are common diseases caused by mite infestations on humans. Fleas, mosquito, and bed bug bites are all capable of producing dermatosis in humans.

Besides direct injury by insects, disease transmission is the most serious problem facing medical entomologists. Insects and mites are vectors of pathogens. They transmit these organisms from one vertebrate host to another. The various ways insects serve as vectors depend on the arthropod. To complete the disease cycle, the pathogen must pass part of its life cycle in the vector. The disease is further compounded by reservoir hosts that can maintain the pathogens until suitable vectors are present. Yellow fever and Chagas disease are examples of diseases that have numerous reservoir hosts.

Besides their role as vectors, some insects (such as fleas) serve as intermediate hosts for some pathogens. Fleas are the intermediate host for dog tapeworms; the dog, in turn, can transmit these tapeworms to humans.

The number of species of insects that are human health problems is small compared to the nearly 750,000 described species of insects in the world. The number affecting human health is less than five hundred. But despite the small number, the problems and suffering to humans is overwhelming.

Mechanisms of disease transmission

Before looking at each group of insects that is a human health problem, it is necessary to look at the methods of disease transmission by these pests. Pathogens capable of being transmitted by insects include protozoans, bacteria, viruses, and such helminths as tapeworms, flukes, and numerous roundworms. Generally, two ways are possible for an insect to transmit a pathogen: either by mechanical transmission or by biological transmission.

Mechanical transmission by insects is the carrying of the pathogen from its source to the host. This may be by mouthparts, legs, or other parts of the body. Biological transmission is when the pathogens enter the body of the vector. There it undergoes morphological change, multiplies, or both. In these cases, the pathogen is dependent on the vector for its survival, development, and transportation to a new host. The most serious human diseases transmitted to humans by insects are those that are biologically transmitted, including particularly, malaria, yellow fever, encephalitis, and epidemic typhus.

Cockroaches

No other insects in the world are so hated and dreaded as cockroaches. These insects evolved over 350 million years ago, and, during all of this time, they have changed little in appearance. They were contemporaries of the dinosaurs, but, unlike them, they persisted to the present day, certainly an indication of their adaptability.

More than 3,500 species of cockroaches live in the world, most of them in tropical regions. There, they come in all sizes, some as small as a house fly, others as large as a bat. Some fly around at night in search of food. Most are green, orange, pink, or dull brown and black; only a few could be called attractively marked with color. Some are horned, and some can hiss. Less than twenty-five are known to be pests in human habitations. These pesky species are often called "domestic roaches," and, in some places, they are termed "water bugs," perhaps as a way of detracting from the less-than-acceptable presence of these strange insects among "polite society."

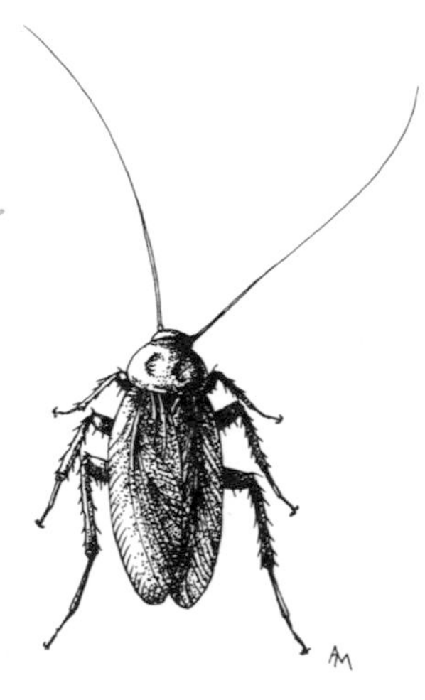

Figure 11.1. *An adult cockroach (Dictyoptera).*

All cockroaches are flat, with long filamentous antennae, chewing mouthparts, and two pair of wings, although some species may have vestigial wings. The "hood" that appears over their head is actually a part of the thorax, known as the pronotum, and is a distinctive feature (fig. 11.1). Cockroaches belong to the insect order Dictyoptera, along with the beneficial preying mantids. Their common name, cockroach, is supposedly derived from the Spanish name for this insect, "cucaracha."

The life cycles of roaches are similar, with variations from species to species. The eggs of cockroaches are contained in a bean-shaped capsule known as the ootheca (see fig. 6.7). The number of eggs per capsule varies from sixteen to forty, depending on the species. The ootheca may be carried by the cockroach for some time or may be deposited in cracks and crevices. Oothecae can be glued to all types of surfaces, from clothing to paper bags, thus permitting the excellent dispersal of the cockroaches; this is one main way the cockroach makes its way into your home, office, or other place of business, particularly warehouses, restaurants, and grocery stores. The structure of the ootheca is specific for each species and offers identification features useful when determining the species infesting the premises.

Eggs hatch within twenty-five to thirty days, depending on temperature and humidity in the habitat. The immature cockroach is wingless and light colored when it molts. Adulthood is reached in two months in some species, or may take two years in others, again depending on the temperature, humidity, and food supply. Adults may live for a year, or more, during which time a single pair may produce as many as 400,000 eggs.

Domestic cockroaches are omnivorous, eating sugar, chocolate, grains, cheese, meat, pastry, in short, those things humans eat. They also feed on a wide variety of other materials, such as bookbindings, blood, their own dead, and, under human stress conditions (as in slums), sputum, fingernails, and toenails of sleeping or sick humans. They have rightfully earned their legacy as being one of nature's most disgusting creatures. Yet, notice that all the things they infest are human debris. We might ask: Are roaches pests or beneficial detritus-cycling insects? Can we blame them for being attracted to unsanitary human habitations?

As nocturnal insects, they are most active at night and hence scamper for shelter when lights are turned on. Therefore, we do not see enough of them to know the true extent of their populations. They are also gregarious, gathering in large numbers to feed and mate, but they are not social insects. Some species also migrate from building to building, which adds to the control difficulties.

The four main domestic cockroach pests are found in the United States, with several other pest species that occur locally in certain sections of the country, mainly in the southern states. Table 11.1 gives some details about the common cockroaches, including their habits and life history.

Their feeding habits raise some suspicion that they might spread disease pathogens. Although cockroaches have some disgusting habits, they have never been proven to be a vector of pathogens, except perhaps in the most stressful conditions of

Table 11.1. *Summary of cockroaches of economic importance.*

Common Name	***Scientific Name***	***Habitat***	***Life Cycle***	***Distribution***
German cockroach (12–18mm)	*Blattella germanica*	Kitchen, bathroom	3–4 months	Cosmopolitan*
American cockroach (30–40mm)	*Periplaneta americana*	Basements, sewers, alleyways, steam tunnels	1–2 years	Cosmopolitan
Oriental cockroach (22–28mm)	*Blatta orientalis*	Sewers, basements, outdoors	1 year	Cosmopolitan
Brown-banded cockroach (12–18mm)	*Supella longipalpa*	TVs, furniture, living room, closets	6 months	Cosmopolitan
Surinam cockroach (18–25mm)	*Pycnoscles surinamensis*	Leaves, debris, homes	1 year	Southeastern U.S.A.
Madeira cockroach (5 cm)	*Leucophaea maderae*	Outdoors, can invade home	2 years	Tropical America

*Cosmopolitan generally refers to living throughout the world; in the case of cockroaches, they are most often found associated with humans in dwellings.

war, poverty, or other disasters. It is known that they can harbor and carry a wide variety of pathogens on their bodies and mouthparts, but it has never been proven that they can transmit these pathogens from human to human under natural conditions. In fact, there is evidence that the pads on the feet of roaches actually produce a bactericide. The elimination of cockroaches from homes, places of business, restaurants, and any place where food is stored or consumed is required by sanitation codes, mainly because their presence indicates otherwise unsanitary conditions.

To control an established infestation of cockroaches is difficult and requires first the correct identification of the species. Although general control methods are helpful, knowledge of the life cycle and habits of the species will enable more specific control measures. Home sanitation is the first step toward control. Keeping the house free of food particles and open food containers and protecting stored products are absolutely necessary. Preventing access to water, although difficult, is required. Guard against bringing roaches into the house in bags and cartons. Movement from building to building cannot be prevented; therefore, chemical control is necessary. This may be obtained from reliable pest control companies. Information on the proper use of insecticides may be obtained from your county extension agent.

Even so, cockroaches are difficult to control with most of the wide variety of pesticides now on the market, because many species have developed resistance to these chemicals. Part of this resistance has developed because of the poor techniques employed by the homeowner in trying to control these pests, usually because the owner feels that if a little is good, a lot is much better. The folly of this is apparent when it is known that a few individuals always are hardy enough to survive. Thus, a percentage of their offspring also are resistant to these chemicals. After a few generations, the entire population within the dwelling may have developed immunity.

One method of control that seems to give good results is the application of boric acid to cracks and crevices. Cockroaches resistant to other insecticides have not yet developed resistance to this material. After walking on it, the roach ingests some of the material when it attempts to clean its body parts. This chemical is lethal even in small doses.

Entomologists are hard at working on a wide variety of control measures and do not intend to give up the battle in the constant war against this pest. Urban entomologists are looking some recent research to outsmart this master of invasion. In 1979, a Columbia University chemist, W. Clark Still, synthesized the sex pheromone produced by the female American cockroach. This material, called periplanone B, entices males from hundreds of yards away to enter traps baited with this material. Entomologist William Bell of the University of Kansas tested this pheromone, saying that it holds promise of being effective in these bait traps. The sex pheromone for the more troublesome German cockroach is not, unfortunately, effective for any distance. Entomologist Eugene Wood, of the University of Maryland, is working on another pheromone called Aggregate. Both sexes of the German cockroach emit this pheromone, effective in bringing hordes of the insects together, but, as yet, no practical application of the substance has been devised.

At the University of Florida, Ann Trambarula, a graduate student, found a banana spider that likes to eat cockroaches. To be sure, this would be a unique method of biological control; however, does one settle for a houseful of cockroaches or a houseful of banana spiders? One thing seems certain, humans will never entirely eliminate cockroaches. The best we can do is to keep the population in check and keep our homes free from this pest as best we can. Any insect that has been around for over 350 million years with little change will undoubtedly be around for another 350 million years.

Bugs

Bug, a name often applied to all insects, is restricted by most entomologists as a common name for members of the insect order Hemiptera, although to insist on this restriction verges on pedantry. The order Hemiptera contains many plant pests, but also some species are bloodsucking insects, and, whenever blood is taken, you can expect that, one way or another, disease is transmitted. Two offending families in this order are Cimicidae, the bed bugs, and Reduviidae, the assassin bugs. The latter family contains the conenose or kissing bugs.

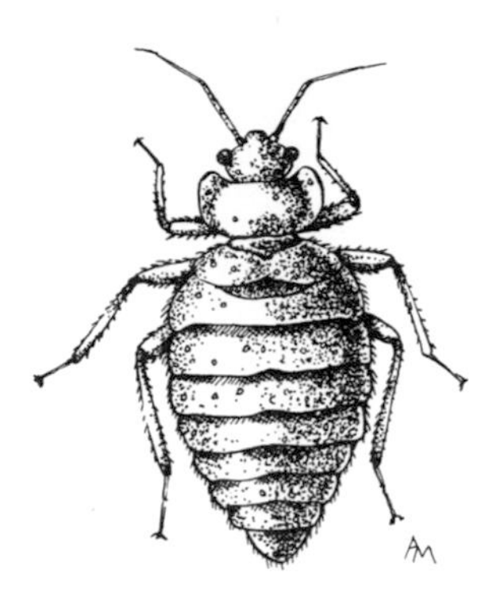

Figure 11.2.
A bed bug (Hemiptera).

The bed bug family includes many species, commonly called swallow bugs, poultry bugs, bat bugs, and the bed bugs themselves. Bed bugs are broad, flat, wingless insects with piercing-sucking mouthparts and incomplete metamorphosis. The human pest bed bug (there are other species on other animals) is *Cimex lectularius* (fig. 11.2), a cosmopolitan species. They feed on human blood at night and hide in cracks and crevices during the day. Although they are gregarious, they are not social. In situations where there is a heavy infestation, they may be found in large numbers hiding in the mattress, cracks in the bed, or even under loose wallpaper. They occur in some rooming houses, campus dormitories, hotels, and motels and are especially common among members of the armed forces during times of stress. They are most prevalent in run-down and unsanitary public dormitories.

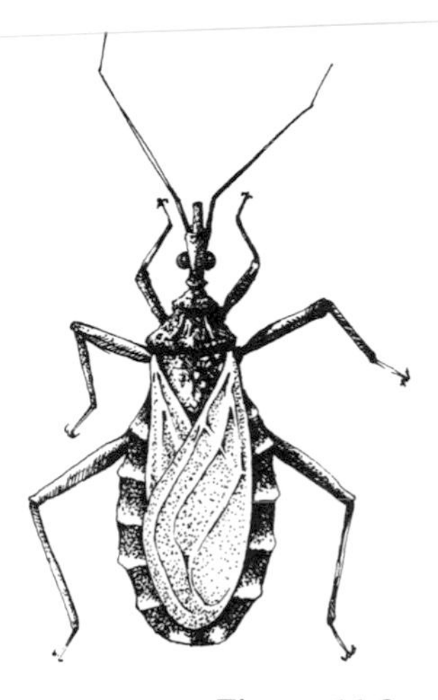

Figure 11.3.
Kissing bug, transmitter of Chagas disease (Hemiptera).

Bed bugs usually bite the face, and the area around the hole made by the insect's proboscis swells and itches. Allergic reactions may occur due to the bug's saliva, injected at the time of biting to prevent blood coagulation. Bed bugs are not vectors of any diseases, although experimentally, they can carry pathogens. They have never been incriminated as natural vectors. However, heavy infestations of these bugs can result in a considerable loss of blood. Persons being bitten night after night may show signs of anemia and weakness.

The best control of bed bugs is personal protection. Avoid unsanitary conditions and be sure all bedding is clean. Infested premises must be fumigated by professional exterminators using poisonous gas, such as cyanide.

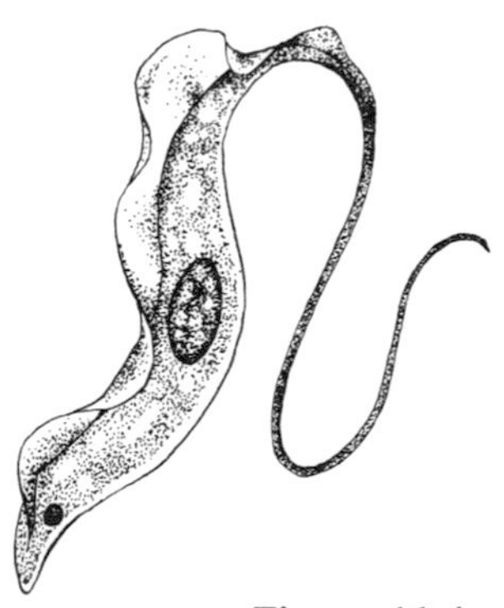

Figure 11.4.
Trypanosoma (Protozoa), the causitive agent of Chagas disease in Central America.

Kissing or conenose bugs (fig. 11.3) affect human health in Mexico, Central and South America. These insects, in America, transmit American trypanosomiasis or Chagas disease, a disease caused by the protozoan, *Trypanosoma cruzi* (fig. 11.4), which is a parasite similar to the causative agent of African sleeping sickness. It lives in the blood in the same way as the malaria parasite. It also invades the tissues of the heart. At times this is a serious disease in tropical areas and must be guarded against being introduced into the United States.

Conenose bugs feed at night on their sleeping victims. Even though they are about 15 mm long, they are able to consume a considerable quantity of blood. The bite is made with the insect's piercing-sucking mouthparts. Although the bites of some members of this family are extremely painful, those of the conenose (species of *Triatoma*) go unnoticed at night while the hapless victim is sleeping. Chagas disease has stricken over ten million people in rural areas of the tropics where it enters unscreened homes. The disease can be fatal. Therefore, when traveling in these countries, protect yourself at night from the bites of this bug. It can enter tents and even trailers, if the screen and doors are not tight.

Lice

If the cockroaches caused you to look twice around your home, the lice will cause you to itch and become somewhat hysterical over ways you might pick up these little monsters. Lice belong to the insect order Anoplura, a group of insects that are commonly called sucking lice. Only three of the some 500 species directly attack humans, but these three are certainly responsible for a variety of problems. Lice are bloodsucking ectoparasites; all species are wingless with a more or less flattened body. The legs are adapted to hold onto hair or feathers. The three species infecting humans have interesting life histories, but, unfortunately, there is considerable myth surrounding their lives.

The head louse, *Pediculus humanus capitatus* (fig. 11.5), is gray in color and from 2–3 mm in length. This species occurs on the head, about the ears, and, in severe infestations, may be found on all hairy parts of the body. The greatest myth about the head lice is that they are a sign of uncleanliness. Actually, how often one shampoos has nothing to do with head lice. Children are most susceptible to head lice, or at least they are the most frequent patients. Estimates indicate that approximately 10 percent of young girls and 7 percent of young boys are infested with head lice. This lower rate among boys may be due to the length of their hair and the use of hair oils. Children commonly acquire these little pests through combs and brushes infected with nits (eggs) after use by an infected child. They also may acquire lice from hats, helmets, towels, ribbons, and, yes, even the earphones of "walkman" radios!

Figure 11.5.
Head louse (Anoplura).

Head lice spend their entire life on the human head. They glue their small, white eggs, the nits, to the hair shaft (see fig. 6.6). Hatching occurs in five days, with nymphs joining adults to feed on the skin or scalp of the head and at the base of hairs. Within three weeks, the entire life cycle is complete. A large population of lice on one's head can develop, if the problem is not properly addressed. Adult lice survive for about forty-eight hours away from the head. If they fall from the body, chances are good that they will not find a new host. But the area where they may fall is critical. Should the lice end up in bedding, or better still, in a hat, they transfer to a new head and will soon establish a new colony.

Infestation by head lice calls for proper action. First, of course, it is necessary to determine for sure that the invaders are lice, and not just a stray booklouse (Psocoptera). Many times the latter have been confused with true lice and have caused great distress among school officials when no louse infestation actually occurred. The easiest way to be sure is to find the nits in the hair. Once nits are found, either by a competent nurse or a doctor, then they may be easily controlled. It is wise to avoid using any of the home remedies recommended by well-meaning friends and neighbors. Oil, gasoline, and kerosene are all dangerous substances and should never be used. The best treatment for all three species of lice infesting humans are the various medicated shampoos, especially pyrethrin-based shampoos, which may be purchased at drug stores. The most effective can be obtained only with a doctor's prescription. Always follow directions explicitly; avoid getting the shampoos in eyes and mouths. Never use more than the recommended dose. Some shampoos may not kill the nits; hence, repeated application may be necessary, keeping in mind the length of time it takes for the nits to hatch and the nymphs to mature. Nits can sometimes be removed by a fine comb, sometimes called a "nit-picker" (hence, the use of the term for those who look for tiny mistakes). Articles of clothing and bedding should be laundered in water at a temperature above 125° F.

The body louse, *Pediculus humanus humanus*, looks almost exactly like the head louse; in fact, it is nearly impossible to tell the two subspecies apart, except by noting the locations and the life cycle. As their name implies, they are found on parts of the body other than the head. They are most common on clothing where the nits are attached. While head lice are common on clothing, the nits are never attached to it. Head lice are common in all segments of the population, but body lice are usually associated with unsanitary, stressful conditions.

The life history of body lice, also called "cooties," is the same as that of the head lice, except for their location and deposition of the nits. Individuals with heavy lice infestations are said to be lousy. They often suffer from pediculosis, a rash caused by the irritation from the bites. When the population of lice on the body is large, the individual then suffers from a general tiredness and irritability. This louse is the only one capable of transmitting epidemic typhus, trench fever, and epidemic relapsing fever to humans. These are not found in North America, but are distributed in parts of Central and South America, tropical Asia, and Africa. One should follow the same recommendations given for head lice to get rid of body

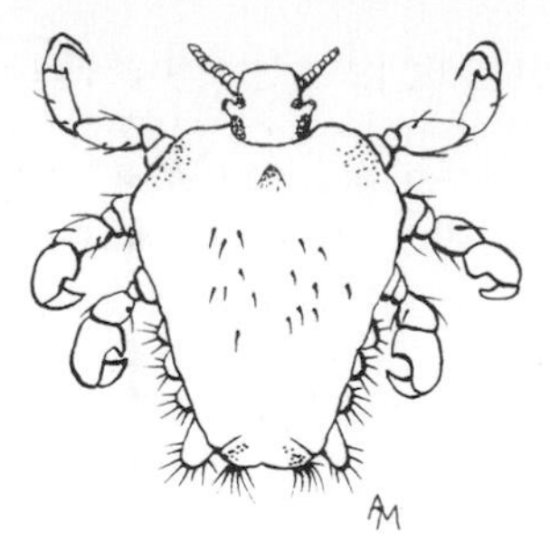

Figure 11.6.
Crab louse (Anoplura).

lice, with special consideration of infested bedding and clothing.

Crab lice, *Pthirus pubis*, are also known as pubic lice. Their common name, crab lice, refers to this insect's miniature crab-like appearance. They are grayish white, about 1.5–2 mm in length (fig. 11.6). The lice are almost always acquired by physical contact with infected persons. It has been claimed that infections can be obtained from toilets and showers, but this has never been verified. It would be difficult for this insect to be acquired this way. The only proven means of acquiring crab lice is through sexual contact, contact with infected bedding, or from clothing of an infected person. These lice are rarely found other than in the pubic region, although occasionally beards are infected. The control of these lice is the same as that used for head lice, which is effective if directions are carefully followed.

Fleas

Fleas (order Siphonaptera) are bloodsuckers as adults (fig. 11.7), ranging in size from 1.5-4 mm in length, with piercing-sucking mouthparts, laterally compressed, hardened bodies, no wing, and hind legs fitted for jumping. The larvae are considerably different, being elongate and soft-bodied and having chewing mouthparts. They do not suck blood, but, instead, feed on organic debris found in animal nests.

Adult fleas are temporary, but obligate, ectoparasites. Their eggs are scattered in the nest of the host, never attached to hairs. Hatching occurs in two to twenty days, depending on temperature and humidity. The immature stages develop in the nest wherever the host sleeps or rests. Flea larvae are active, feeding as scavengers in the nest. After about two weeks, the larvae spin cocoons, and the pupal period begins. This stage lasts for about three weeks. The newly emerged adults jump onto the host, where they remain for a short period while obtaining a blood meal (fig. 11.8).

The two common fleas in the United States are the dog flea, *Ctenocephalides canis*, and the cat flea, *Ctenocephalis felis*. Both species attack dogs and cats, as well as humans. Humans often get bitten by these two species of fleas, because of neglect of their pet. Dogs and cats permitted to run free in the warmer months without flea collars or treatment with flea powder often pick up fleas. The population of fleas can become established in homes with pets, and, before long, immature fleas can be found in the animal's box or bedding. If the pet leaves home or dies, the fleas that once fed on your pet suddenly move to the humans in the household for a blood meal. Flea bites itch, and a body rash develops, often accompanied by allergic reactions.

Figure 11.7.
Flea (Siphonaptera).

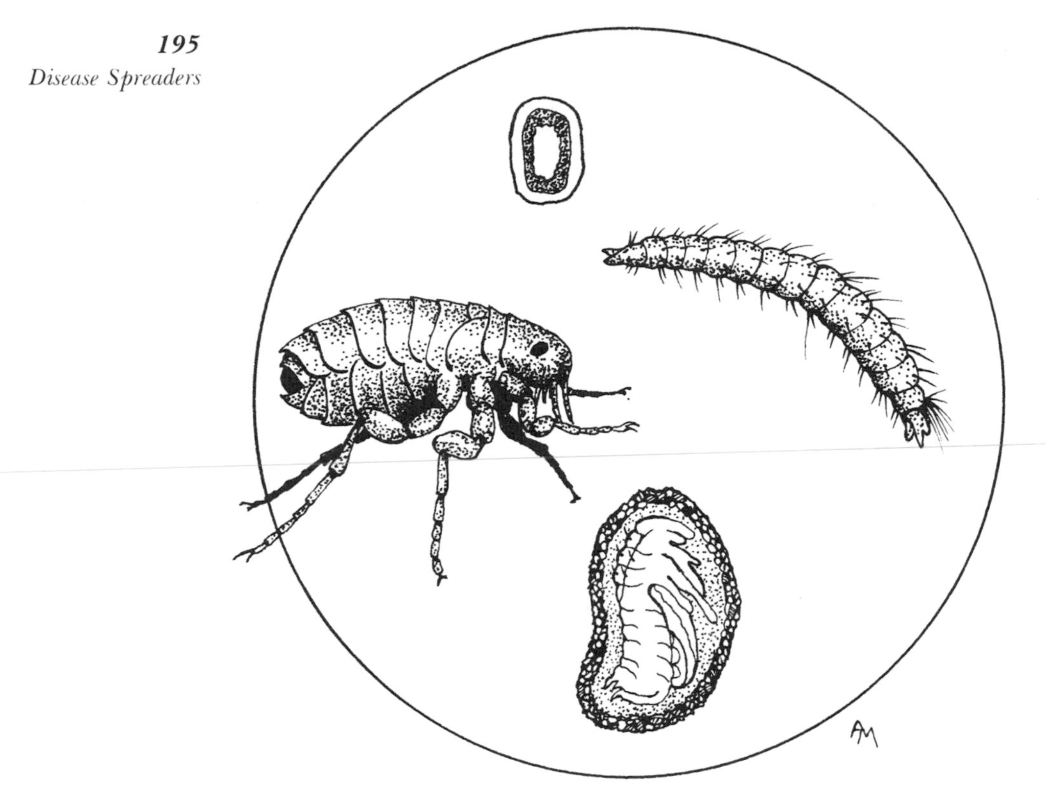

Figure 11.8. *Flea life cycle (Siphonaptera).*

One potentially serious problem caused by dog and cat fleas is their role as the intermediate host of the dog tapeworm, *Diphylidium caninum*. Dogs and cats acquire this internal parasite by eating fleas as they attempt to prevent them from biting. Children often come in contact with the parasite in much the same way, by accidentally ingesting a flea. The parasite will develop in children as readily as in dogs. This, of course, requires careful testing by a competent physician. Unfortunately, many cases go undetected, resulting in serious internal injury.

Fleas have a much more notorious reputation as the vector of plague. The Oriental rat flea, *Xenopsylla cheopis*, is responsible for the spread of plague. This flea is found wherever the rat, *Rattus rattus*, is found. Plague is a bacterial disease and appears in several forms. One, bubonic plague, is acquired by humans through the bite of fleas. Another, pneumonic plague, is the form that can be spread easily through droplets of saliva, instead of an animal vector. Once plague is established in a community, all forms can and usually do develop, making both control and prevention difficult. Plague was responsible for past pandemics, and history shows that it has killed millions of individuals in

Europe, Africa, and Asia. Even today, over five thousand cases occur each year in the world, with as many as a dozen cases in the western part of the United States, especially in New Mexico, Arizona, and western Texas. There it is a rural disease associated with the fleas of wild rabbits and other wild animals that are often hunted and eaten. Camping in areas where this disease is endemic may be dangerous. In other parts of the world, the major concern is to keep the rat population down, especially in urban areas of northeastern United States.

Fleas are usually not difficult to control. Again, it is a matter of proper sanitation of homes and pets. They must be protected from fleas by the available chemicals. The use of flea collars seems to be, at least now, an effective control measure.

Flies

Without doubt, the most important order of insects accused of disease transmission is the order Diptera, the true flies. This includes mosquitoes, the notorious killers throughout the history of man. Malaria, which is transmitted only through the bites of certain mosquitoes (rarely by blood transfusions), remains the number on disease killer in the world today. There are nearly 100,000 described species of flies throughout the world. Although a large and diverse group with more beneficial than harmful species, they are notorious disease vectors.

Flies, without exception, have only one pair of wings or none at all. The second, or hind, pair is reduced to balancing organs called halteres. The bodies of many flies are beset with a series of spines. These are uniform in their location. The variation of these spines is useful as classification features, and these characteristics are used in identification keys. Their mouthparts vary connsiderably from group to group. Some are mandibulate, with modified chewing mouthparts, while others have greatly elongate parts modified for piercing-sucking. It is this type that comprises the groups that feed on blood, and, thereby, are disease vectors.

All flies undergo complete metamorphosis with eggs, larvae, and pupae. Fly larvae are legless, but have chewing mouthparts. The pupae are enclosed in a puparium formed from the last larval skin. Some pupae, such as those of mosquitoes, are active, but, as might be expected, they do not feed. The food of larvae and adults ranges from simple detritus feeders to predators, and, as already mentioned, adult blood feeders.

The flies of major medical importance include black flies (Simulidae), mosquitoes (Culicidae), horse flies and deer flies (Tabanidae), house flies (Muscidae, which also includes the tsetse flies, vectors of sleeping sickness), blow flies (Calliphoridae), and warble flies and bot flies (Oestridae). Some flies cause myiasis (see the introduction to this chapter and the glossary). Others, including midges and gnats, are annoying, as well as possible disease transmitters.

Black flies

A major annoyance to both humans and animals in North America, black flies (Simulidae) are important disease vectors in various tropical regions, although they are most often found in north temperate and subarctic zones. Adults are small, 1–5 mm in length, with piercing-sucking mouthparts and a somewhat "humpback" appearance (fig. 11.9) with broad, iridescent wings and short antennae. Many are yellowish brown instead of black; hence, their common name is somewhat misleading.

Black fly larvae are found in water, and, in North America, they prefer swift-running, well-oxygenated streams. Pupation occurs in the water and adults emerge in enormous swarms near the stream.

Figure 11.9. *Black fly (Diptera* Simulium *sp.).*

Only female black flies feed on blood, a requirement for ovarian development. Males feed on nectar. As many as six generations may occur per year, depending on the climate. They are notorious for pestering campers and fishermen and are also traumatic to both wild and domestic animals, due to the annoyance and loss of blood. In tropical Africa, southern Mexico, and Central and South America, black flies are vectors of a filarial nematode, *Onchocerca volvulus*. This pathogen causes a severe disease, Onchocerciasis, or river blindness, a condition that affects some twenty million people worldwide. The most serious problem associated with onchocerciasis is blindness. The worm destroys the sight of up to 30 percent of the affected population. Onchocerciasis does not occur in North America, but annoyance from black flies is a serious problem in northern areas to the point where humans are limited to fly-free seasons for outdoor activities.

Mosquitoes

Mosquitoes are the most important family of disease-carrying flies. Not only are they a major annoyance to humans, but they are responsible for the spread of several fatal diseases. They attack and spread diseases to man and all warm-, and some coldblooded, animals. Table 11.2 shows the major diseases transmitted to humans; some of these will be discussed further in this section.

Table 11.2. *Major mosquito-borne diseases of the world.*

Disease Name	***Causal Organism***	***Vectors***	***Geographic Location***	***Human System Attacked***
Malaria	Protozoa: *Plasmodium* (4 species involved)	*Anopheles* (wide range of species)	All tropical, subtropical regions of the world	Circulatory system, red blood cells
Filariasis (elephantiasis)	Nematoda: *Wuchereria bancrofti*	*Culex* sp. *bancrofti*	Humid tropics	Lymphatic system
Eastern equine encephalitis (EEE)	Arbovirus	*Aedes* sp.	Eastern seaboard, Canada to Argentina	Nervous system
Western equine encephalitis (WEE)	Arbovirus	*Culex* sp. *Culiseta* sp.	North Central USA, Western USA	Nervous system
St. Louis encephalitis (SLE)	Arbovirus	*Culex* sp.	Continental USA	Nervous system
Dengue	Arbovirus	*Aedes* sp.	Caribbean, SE Asia	Fever, rash
Yellow fever	Arbovirus	*Aedes aegypti*	Tropical Africa, South America, Central America	Fever, pain, jaundice, nausea

The study of the metamorphosis of mosquitoes is of great importance, since it is through a study of the various stages of their life cycle that entomologists have learned to control them. Eggs are usually laid in water; adult females may deposit them singly or, sometimes, in rafts formed of a mass of eggs stuck together, floating on the surface of the water, depending on the species. Some species even deposit eggs in areas that will become flooded during the spring and early summer rains. Once wet, these eggs hatch, but otherwise, they will remain viable, sometimes waiting for several years before conditions become suitable for hatching. All mosquito larvae, known as wigglers (fig. 11.10), feed and develop in water. Wigglers have chewing mouthparts modified just for gathering and filtering out microorganisms, particularly algae. A few are predacious, often feeding on other mosquito larvae. It is possible that the propagation of these species could become a means of biological control in certain situations. Wigglers are found in all types of water: fresh, slightly saline, and that in containers, treeholes, and flower and leaf bracts. They are well-known inhabitants of pitcher plants. Wherever standing water occurs, except in the ocean itself, you are apt to find mosquito larvae. We have observed the larvae of pest mosquitoes in small holes in rocks along the coast of the Seychelles, bathed by waves of sea water incessantly beating these rocks. The larvae must cling to the rock itself to escape being washed out to sea.

Pupae, called tumblers, are also found in the same water as the larvae. Both wigglers and tumblers breathe through projecting siphons. A pair of siphons known as trumpets (fig. 11.11) is found on the thorax of pupae, while a single tube placed at the

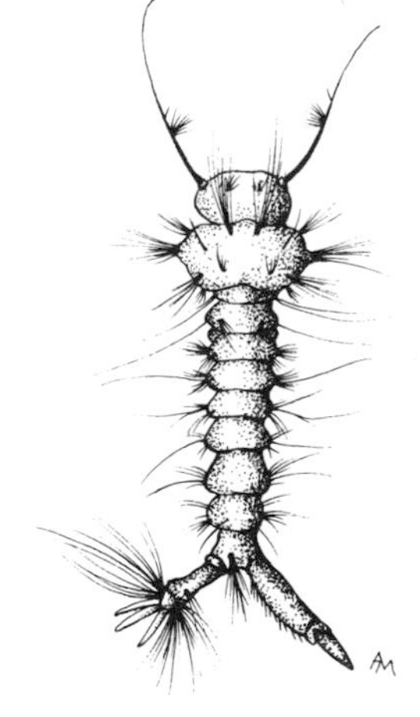

Figure 11.10.
Mosquito larva, a wiggler (Diptera: Culicidae).

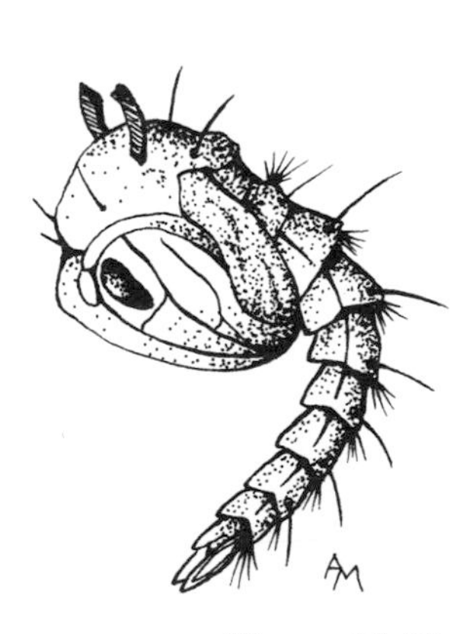

Figure 11.11.
Mosquito pupa, a tumbler (Diptera: Culicidae).

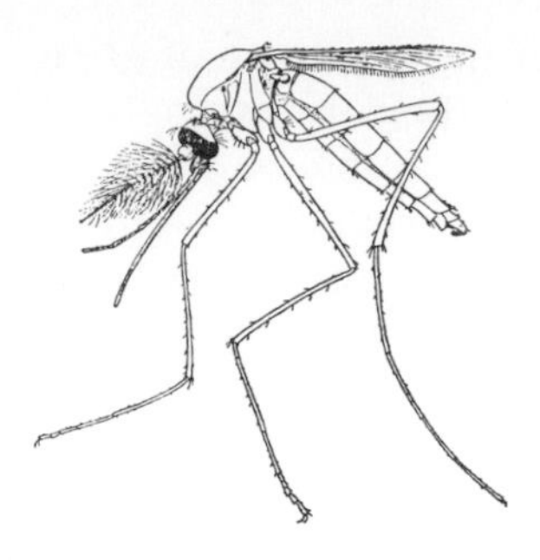

Figure 11.12. *Adult male mosquito; note the long palpi and bushy antennae (Diptera: Culicidae).*

end of the abdomen of the wiggler serves these larvae. When the water is disturbed, both larvae and pupae quickly drop (wiggle, or tumble) to the bottom, where they remain for a few minutes. Eventually they must return to the surface for air, even though the larvae, at least, have gills through which some oxygen is absorbed from the water. These structures are not efficient.

Larvae take about seven days to complete their feeding (this, of course, varies with the species and the available food) and remain about three days as pupae. Again, this varies with the species. We have observed the complete life cycle, from egg hatching to emergence of adults, of a species of *Psorophora* found in a temporary rain puddle in Florida, to take place in three days from the time the area was flooded by rain.

Adult male mosquitoes (fig. 11.12) feed on nectar and plant juices; females (fig. 11.13) feed on blood, besides nectar. This is necessary in most species for ovarian development.

Adults superficially look similar, with long legs, an extended abdomen, a short, humpback thorax, and medium length antennae. Mouthparts are sexually dimorphic, the males and females easily distinguished by the shape and length of the palpi and the structure of the antennae. Of the 2,500 species that occur throughout the world, primarily in the tropics, only about 150 species occur in the United States and Canada.

Annoyance from mosquito bites is well known. Great swarms are not uncommon during the summer months, especially around recreational facilities near water. This makes it difficult for campers and others engaging in outdoor activities to enjoy themselves fully. Precautions must be taken to prevent being bitten. Seashore towns and inland cities have been plagued with the salt-marsh mosquitoes, preventing not only recreational activities, but real estate development, as well. Annoyance to domestic animals and nesting birds is well known. In fact, wild birds suffer greatly from bird malaria, a disease of birds similar to that of human malaria.

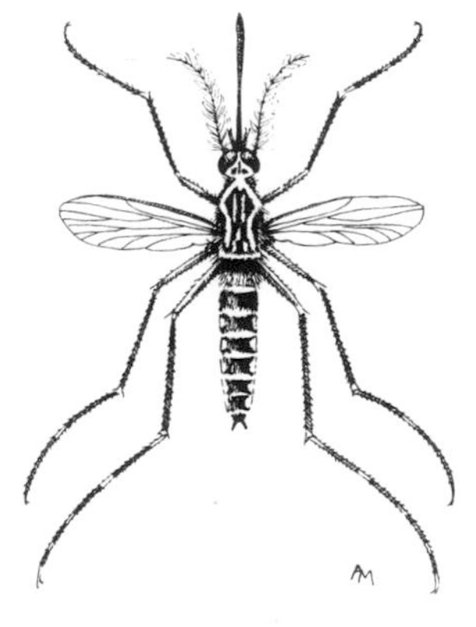

Figure 11.13. *Adult female mosquito; note the short palpi and slender antennae (Diptera: Culicidae).*

As if this annoyance were not enough, mosquitoes also transmit malaria, yellow fever, and dengue, as well as some forms of encephalitis. These diseases are not confined to tropical regions, but may occur throughout southern United States, and, in the past, as far north as southern Canada. In fact, during the building of the Erie Canal, malaria and yellow fever were as deadly to the American builders, as they were to the builders of the Panama Canal. Transmission of filariases, especially elephantiasis, is now a serious problem to the natives of American Samoa and throughout many tropical regions.

Malaria, the world's most serious disease, continues to strike down thousands of humans each year. The death toll from this dreaded disease has run into the millions during major outbreaks. Some regions have almost 100 percent infection, either as endemic or epidemic cases. Over two million cases occur annually.

Malaria is caused by protozoans belonging to the genus *Plasmodium*. There are four species that infect man, always transmitted by mosquitoes of the genus *Anopheles*. Other species of plasmodium infect birds and are transmitted by other kinds of mosquitoes.

The parasite develops internally in red blood cells, which they destroy. The complicated life cycle of the parasite is only partially completed in human blood; the main, sexual part of the cycle takes place within the mosquito. The mosquito itself does not seem to be affected by the parasite.

Even now there are a few cases of this disease annually in North America, north of Mexico, and many cases south of the border. Usually this occurs in individuals who have travelled in malarious areas and failed to take proper precautions. The danger of getting the disease is always present in areas where anophelene mosquitoes live. Two potential vector species occur in North America. It is doubtful that any endemic cases occur in the United States.

Mosquito-borne human filariases is a debilitating disease of humid tropical areas caused by two filarial nematode roundworms. *Wuchereria bancrofti* and *Brugia malayi* are pathogens that cause lymphatic filariasis and obstructive filariasis (elephantiasis) respectively. Several vectors, including mosquito species of the genera *Aedes*, *Culex*, and *Anopheles*, are widely distributed in these areas. The disease is serious in underdeveloped countries, where 200 million individuals have contracted filariasis. Vector control and chemotherapy are the main weapons in the attack on this disease.

Viral diseases are easily carried by mosquitoes. More than 90 viruses have been isolated from wild mosquitoes and ticks. Arboviruses, as these are known, are widespread. Probably the most famous of all is yellow fever, a serious, often fatal, disease of wide distribution. Now it is confined to tropical regions of Africa, Central and South America, but it once affected millions of people throughout the world, wherever yellow mosquitoes occurred. The principal vector is *Aedes aegypti*, a mosquito that breeds in artificial containers in urban areas. Yellow fever was found in the United States as far north as New York State as recently as 1878, when the annual death toll was as high as 20,000. Today the disease is much less widespread, with less than five hundred cases a year worldwide. One word of caution must be extended to travellers in areas where yellow fever is endemic: the disease can be easily acquired, and travellers are advised to get immunization, since it is now available for the prevention of the disease.

In the United States, three arboviral diseases have, at times, caused great concern to health officials, namely, eastern equine encephalitis, western equine encephalitis, and St. Louis encephalitis; all are present and have caused human mortality when populations of the vector and disease carriers are present. All three of these arboviruses affect the nervous system; symptoms may be mild or can cause serious damage to the central nervous system. Health officials are constantly monitoring populations of the vector and the various reservoir hosts, such as nestling birds, to keep the spread of viral encephalitis in check.

Still another arboviral disease, dengue, can cause serious complications in children and infants, but is usually mild in adults. Dengue is found primarily in the islands of the Caribbean, Central America, and in Southeastern Asia. *Aedes aegypti* is again the main vector of the disease, but it can also be transmitted by other culicine mosquitoes. Epidemics have been reported in various endemic areas in recent years. Control of the vector is the main way to prevent the disease.

One cannot overestimate the importance of mosquitoes in any community. Mosquito-transmitted diseases occur throughout the world, and high populations of mosquitoes cannot be tolerated. But control is not easy, especially now that so many effective insecticides can no longer be used. Considerable effort has been devoted to the biological control of mosquitoes, but none is, as yet, effective; hence, mosquito bites continue to be as prevalent in the summer as the common cold is in the winter! Many professional entomologists are employed to manage mosquito abatement projects and to conduct research toward finding effective mosquito controls.

While mosquitoes may top the list both for annoyance and the spread of human disease, the tabanids (horse and deer flies) win the award as the most painful and vicious bloodsuckers. Female horse flies have bladelike mouthparts that can inflict a deep, painful wound. After the initial puncture, the fly will lap up the blood with a sponging labella. The mouthparts cutting into the skin are somewhat scissorlike. They open the wound until blood flows freely. Often when large numbers of these flies attack horses and cattle, the pain causes them to stampede.

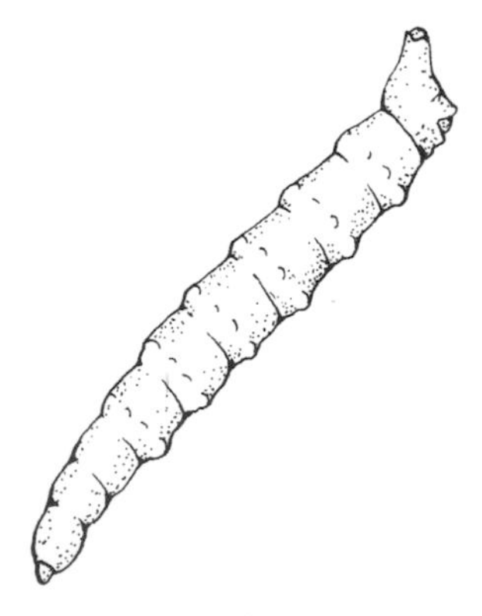

Figure 11.14.
Larva of a horse fly (Diptera: Tabanidae).

The larval stage of tabanids (fig. 11.14) is usually aquatic or semi-aquatic. The larvae feed on organic matter. When they are mature, pupation occurs in moist or wet soil on the shore of their feeding area. Besides the annoyance caused by these flies, they are responsible for the mechanical transport of pathogens to humans and animals. Horse flies are known to transmit anthrax,

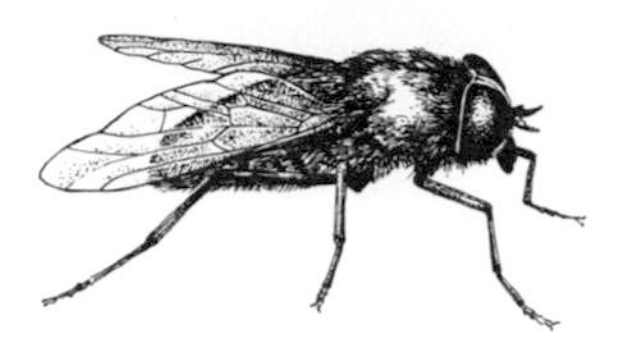

Figure 11.15. *Adult horse fly (Diptera: Tabanidae).*

a bacterial disease, also called lockjaw, to domestic animals and humans. Tularemia, another bacterial disease of humans and various animals, including chickens and wild game, and loa loa, another nematode worm found in Africa, infect humans and animals through the agency of these flies.

Control of tabanids (fig. 11.15) is difficult, since the adult stage is not usually near human dwelling, but in less accessible marshy and swampy areas. Therefore, personal protection from their bites is necessary by the continuous use of repellents. Tabanids have made some campgrounds, lakes, and beaches unuseable during certain seasons of the year. It is alleged that the Declaration of Independence was signed on July 4, 1776, because the horse fly population in Philadelphia was so great at that time of year that the delegates refused to consider the wording of the document further, signed it, and left for home to avoid further discomfort from the fly bites. Domestic animals also can be protected with repellents. Spraying of larval habitats is impossible, because of their location.

The house fly, *Musca domestica*, is probably the most familiar of all flies to humans. This gray fly, 6–9 mm in length, with four dark stripes on the thorax (fig. 11.16), is the most common fly found in dwellings in North America, especially in the home. The adults have mouthparts of the sponging type, an extremely complicated structure (see fig. 5.4). There are no stylets, and they do not suck blood. However, the sponging mechanism is ideal for the mechanical transmission of pathogens.

House flies multiply at a fantastic rate. There may be as many as twelve generations per year. The female deposits her eggs in a wide variety of places, including excrement, garbage, and organic matter. Eggs require only eight to twelve hours before they hatch. The larval stage, known as a maggot, requires about five days, and the pupal stage, about five days. The life cycle is affected by temperature and may take a bit longer in cooler areas.

These flies are attracted to a wide variety of materials, flying from one to another. First, they may visit human feces and then go on to a sugar bowl. Each time the fly lands on an object, it examines it with its mouthpart, which is beset with various sensory setae. If the material is something suitable, such as food, the fly regurgitates some of its stomach contents on the substance and then sponges it up, along with the new material, to add to the contents of the stomach. It does the same at its next stop, thus contaminating our food material with excrement. If the excrement contains pathogens, such as those that cause typhoid fever, dysentery, and cholera, it is easy to see how these diseases are transmitted.

Figure 11.16. *Adult of a house fly (Diptera: Muscidae).*

Control of house flies involves chemicals, traps, baits;

sometimes, certain biological control methods have been helpful. What is probably more important than control is prevention; that is, keeping flies away from food and, particularly, from substances containing disease organisms. This should include destruction of breeding sites, screening of buildings, and the proper management of the feces of domestic animals as well as of humans. House flies should not be tolerated, for even one presents a potential danger, because of their demonstrated ability to transmit disease.

Many other species of flies both annoy humans and carry pathogens. These include blow flies, blue bottle flies, and flesh flies that lay their eggs on various food products. The larvae that hatch out, of course, destroy the food. Stable flies are capable of biting and sucking blood. This fly, along with the dreaded horn fly and screwworm, are serious pests. One tropical fly, the tsetse fly, species of the genus *Glossinia*, transmits African sleeping sickness, a disease caused by protozoans, *Trypanosoma* spp. This disease is more or less isolated in central Africa, but it renders vast areas uninhabitable and takes a death toll of over 7,000 humans each year. Volumes have been written about this fly, the problems it causes, and attempts to control it. It still remains a serious problem today.

Several species of flies cause myiasis, the infestation of organs and tissues of humans and domestic animals with fly larvae. Although more common in wild and domestic animals, at least one species infects man. The most commonly diagnosed myiasis in the United States is that caused by the primary screwworm, *Cochliomyia hominivorix*. The adult fly is attracted to wounds or sores on animals, where the female deposits as many as 300 eggs. The eggs hatch within twenty-four hours, and the maggots (larvae) begin to feed in the wound. The injury, wound, or sore is made larger and, depending on the location, other complications may set in. The wound is, of course, subject to bacterial invasion, and it is this secondary infection that causes the death of the animal. The maggots feed for about five days, before they drop off the host and penetrate the soil to pupate. After another seven days, adults emerge.

The screwworm problem is serious in the southwestern United States and Mexico. The problem once extended into southeastern United States, but complete control in this area was achieved by what is known as the sterile fly technique devised by F. Knipling of the United States Department of Agriculture. Millions of flies are mass reared, and the males are treated with low-level radiation, enough to destroy the sperm, but not enough to prevent them from mating with females. These males are released throughout the infested area. They mate with the females. These females accept males only once. Since they are not fertilized, the eggs they lay do not develop. Once the population is swamped with these sterile males, it rapidly declines. Over a period of years, the number of cases of screwworm myiasis was greatly reduced and virtually eliminated in the United States. However, complete eradication is not possible, because of repeated infection from across the border where the problem still exists. Both governments continue their efforts to completely eradicate this pest.

Besides the diseases caused by flies, many species are agricultural pests. These are discussed in chapter 12.

The problems caused by flies are overwhelming in some parts of the world. Their destructive activities are of great concern and control measures are often ineffective. At the top of the list as a successful means of control is personal protection through screens, repellents, and protective clothing.

Bites, stings, and insect allergies

Aside from the threat of disease transmission, bites and stings from insects or other arthropods can be painful, and often patients show an allergic reaction to varying degrees. The organic chemicals produced by these animals are of three types, classified on the basis of their function: 1) venoms that are introduced into the skin and produce an immediate and apparent response, for example, the sting of a honey bee; 2) irritating defense secretions that are secreted on contact, such as the urticating defense hairs of certain caterpillars; and 3) defense substances expelled by insects to ward off would-be predators. Humans react in various ways to these substances. As is well known, every individual reacts to foreign proteins and related chemicals that are introduced into the body through the skin, then into the circulatory system. These reactions are termed allergies, some of which are passive, and others, violent, even resulting in death. As explained previously, whenever an insect inserts its proboscis into our body, it injects saliva, which acts as a blood anticoagulant. This causes the itch of mosquito bites. Stinging insects use a more powerful substance, which induces immediate and severe pain, for example, that of bees, wasps, and some ants. They inject this material from a stinger at the end of the abdomen (fig. 11.17). Contact with the larvae of such insects as the saddleback caterpillar and io moth caterpillar can be painful because their hairs inject irritating urticating substances into the skin. Finally, there is active projection of substances "sprayed" as a protection from aggressive attack. Seldom does one come in contact with the latter, unless, of course, the victim is an insect collector. Various beetles and a host of arachnids employ this means of

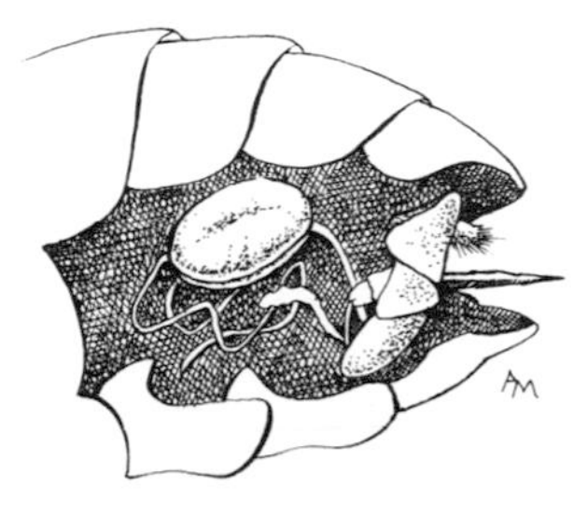

Figure 11.17. *The stinging apparatus of a hornet.*

protection. Some persons experience the itch or pain from these substances, and that is the end of it. Others exhibit more symptoms, including considerable swelling, dizziness, rash, and even anaphylaxis, requiring immediate medical assistance. Sometimes these reactions result in anaphylactic shock and death (table 11.3).

Only recently has attention been given to the effects of these bites, stings, and their allergic responses. Allergies can occur from many different kinds of insects, ranging from cockroaches, ants, and bees, to spiders, mites, and scorpions. The subject is complex, and our current knowledge about it is limited. As is true of the effects of various chemicals in the environment, in food, and in water, arthropod allergies have opened an entirely new field of medicine and public health.

Table 11.3. *Reaction to bee stings; nonsusceptible and susceptible individuals.*

Nonsusceptible:	Sting produces pain, swelling, some redness. Pain should pass within a few hours. An ice cube or meat tenderizer can be used to help ease the pain and discomfort.
Susceptible: I.	Sting brings on malaise, anxiety, pain, and redness at the site of the sting. Consult your physician; you may wish to be tested. Don't panic, but act.
II.	If the reaction to the sting causes chest constrictions, wheezing, nausea, abdominal pain and/or dizziness, you should consult a physician at once. If these reactions intensify, CALL YOUR LOCAL FIRST AID SQUAD OR GET TO A HOSPITAL EMERGENCY ROOM IMMEDIATELY.
III.	If the sting produces even more serious reactions, such as collapse, confusion, or speech impairment, the only recourse is to get to a hospital as quickly as possible. Collapse, drop in blood pressure, cyanosis, and unconsciousness are all signs of serious shock reaction and are life threatening. Do not delay.

If you notice any physical difficulties after a sting, a visit to your physician to discuss the symptoms should be a priority. An allergist may later be able to develop a program of desensitization for you.

Noninsect problems: spiders, mites, ticks, and scorpions

When a medical entomologist refers to other arthropods that are of concern (those who study these terrestrial arthropods are generally included in the field as entomologists), he is generally thinking of the vast array of arachnids, the spiders, mites, ticks, scorpions, and related groups comprising about 75,000 species. Of these, more than four thousand occur in the United States and Canada, with several of them coming in contact with the general public, especially during outings, camping, and so on.

Arachnids, also arthropods, differ from insects (table 11.4) by several distinct anatomical features. Almost all species have poison glands or defense secretions.

Spiders are the largest group of arachnids, with about 35,000 species in the world, followed by ticks with 31,000 species, and scorpions, a relatively small group, with 2,000 species. Spiders are easily identified by their four pairs of legs and the lack of a distinct head, its being joined to the thorax to form what is termed the cephalothorax (table 11.4). All spiders are predacious and use venom to kill or paralyze their prey. They digest the contents of their prey by injecting digestive juices into the victim and then suck out the liquidfied contents. Spiders also have six pairs of special glands, located on the abdomen,

Table 11.4. *Morphological differences between spiders (class Arachnida) and the insects (class Hexapoda).*

	Spiders (Class Arachnida)	*Insects (Class Insecta)*
Body regions	Two (cephalothorax and abdomen)	Three (head, thorax, abdomen)
Antennae	Lacking	One pair
Legs	Four pairs	Three pairs
Pedipalps	One pair	Lacking
Poison apparatus	Opening on fangs of chelicerae	If present, at posterior end of abdomen
Wings	Lacking	Two pairs in most species
Eyes	Simple eyes, 6–8	Compound eye, plus two or three simple eyes
Silk apparatus	Opening at hind end of abdomen	Only present in some larvae; when present, opening is on the lower lip of the mouthparts
Development	Direct, no larval stages; immature spiders resemble parents	Many species undergo metamorphosis; larval and pupal stage common; some insects have nymphal stage

used to spin silk. These spinerets, as they are termed, are capable of producing webs by some species, but other use this silk to line their burrows, form funnel-like traps, or even to form a balloon used to travel by air from place to place (fig. 11.18).

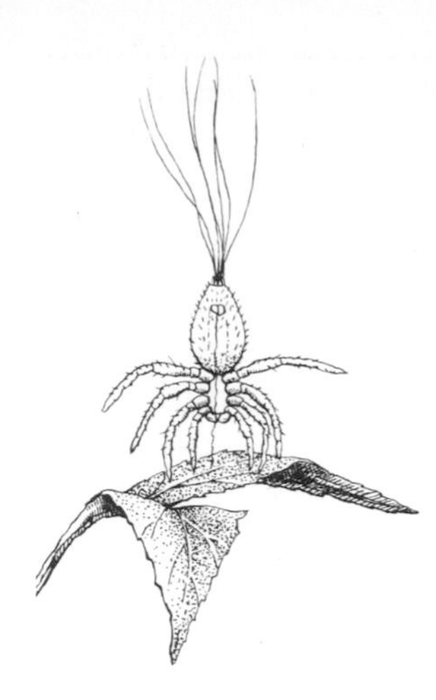

Figure 11.18.
A spider ballooning.

Spiders are beneficial, because they help to regulate insect populations, one of their principal foods. Although all spiders have poison-injecting fangs, few of them use these to bite humans. Two well-known species are known to cause serious bites, the brown recluse of the south and the infamous black widow spider of general distribution. A third spider with a bad reputation is the tarantula, but mostly because of its extremely large size. As any of the people who keep these spiders as pets will tell you, their bite, if ever inflicted, usually is not, or only slightly, painful.

The brown recluse spider, *Loxosceles reclusa* (fig. 11.19), is considered a highly venomous species. It is found indoors in basements or behind furnishings and outdoors in loose debris in sheltered areas. The spider is comparatively easy to recognize by the distinct violin pattern on the cephalothorax and is thus sometimes called the violin spider. The female is about 11 mm in length; the male smaller, about 6 mm. Their cephalothorax is orange yellow with a dark violin pattern. The rest of the spider is a dull grayish to dark brown in appearance. It occurs as far north as Ohio, south to Georgia, and west to Nebraska and Texas. As with most spiders, the puncture marks from the fangs develop a crust, redness, and swelling characteristic of all such wounds. The crust remains for some time, and, after it falls off, a deep ulcer is evident. The dry ulcer may persist for several months with permanent disfiguration. Although their bite is potentially fatal, few fatalities have been reported.

Figure 11.19.
Brown Recluse Spider.

The black widow spider, *Latrodectus mactans* (fig. 11.20), has a highly toxic venom, and there have been reports of some fatalities from the bites of this spider; however, usually this has occurred in small children or persons with weak hearts, or other physical weakness. Males are small, 3–4 mm in length, and are seldom seen. The infamous female is 8–10 mm in length, with red-orange hourglass markings on the ventral surface of her abdomen. The species occurs from southern New England to Florida, west to Kansas, eastern Oklahoma and Texas.

The black widow is seldom, if ever, seen indoors. It may be found in outbuildings, including old-fashioned outhouses where, in former days, bites were frequent. Now it is encountered under trash, outdoor furniture, and in protected areas on the sides of buildings. Despite her reputation, the female is rather shy and rarely attacks, unless she is disturbed while guarding her egg sac. As with many species of spiders, the female will attack and eat the male, unless she is well fed. Males frequently escape and live to mate again.

Figure 11.20.
Black Widow Spider.

Figure 11.21.
A tarantula.

The largest spiders known are the tarantulas (fig. 11.21). These spiders are close relatives of the smaller, common trapdoor spiders. In fact, many of these species also construct trapdoors. Two dozen species of tarantulas occur in the southern part of the United States, mainly from Florida to Texas, New Mexico, Arizona, and southern California. These species have bodies about two inches, or even more, in length; and, with their long, hairlike setae-covered legs, they may measure up to seven inches. They are nocturnal animals, hunting insects and other small arthropods found on the ground at night. We have observed some of the large tropical species chasing and capturing mice indoors. During the summer months, tarantulas average about one meal a week, but they may go for weeks without a meal when in captivity. They have been known to live as long as 20 years. They may be kept in captivity without danger. They molt up to four times a year while still sexually immature, and once each year after reaching maturity. Many species do not mate until after their eighth year.

Considerable superstition surrounds the tarantula; the name itself is of superstitious origin. It came from the village of Tarentum (Taranto) in southern Italy. According to legend, an epidemic of "tarantism" broke out in the village. The resulting spider bites caused victims to have an irresistible desire to dance wildly. From this legend, we have the European folk dance, the tarantella. Immigrants brought tales of the spider to the New World. Our species seem to be only slightly poisonous, but it is reported that some species in Central and South America are considered to be as dangerous as the black widow and brown recluse spiders. These spiders are sometimes found on stalks of banana imported from the tropics. They must be handled with care.

North American tarantulas have few enemies, but the tarantula hawk (a large wasp) is an exception. This insect, a species of *Pepsis*, attacks and paralyzes the adult spider with its sting and places the immobilized spider in the wasp's nest in the ground with its eggs laid on the spider's body. This provides food for the developing wasp larvae.

It is safest to avoid all spiders, but should you be bitten by any spider, the lesion and symptoms must be carefully watched. Medical treatment is advisable. If possible, capture the offending spider for later identification.

Mites and ticks

Mites and ticks, also members of the class Arachnida, are smaller than most spiders. Mites are extremely small, usually less than 1 mm. They, too, have four pairs of legs (immature ticks have only three pairs), a cephalothorax, abdomen, and chelicerae for feeding.

Most mites are free-living, many feeding on plants, and others are predacious, living in soil and in water. Other species are ectoparasitic on animals and humans. These bloodsucking species cause dermatitis with considerable tissue damage, and often cause strong allergic reactions in humans, pets, and domestic animals. Some species transmit tropical diseases.

Mite infestations are termed acariasis. Although most of the bloodsucking species are ectoparasitic, some, such as the scabies mite, burrows into the skin and causes the disease sometimes termed barber's itch. Scabies is a serious disease among domestic animals. The human scabies mite, *Sarcoiptes scabiei* (fig. 11.22), is extremely small, 250–450 microns in length. Intense itching occurs for about a month after infection in the area where the mite burrows in. This is caused by the toxic secretions and excretions from the mite. A serious rash and severe allergic reactions result in some individuals. Treatment for this disease is possible only after correct diagnosis is obtained from a physician.

Chigger mites (fig. 11.23) are the larval stage of a mite of the family Trombiculidae. They are often encountered in meadows and woods where hikers and campers pass. They are found in most regions in the United States, but are characteristic of the southern states. During mite season, which lasts most of the summer, precautions must be taken to prevent these bites. Several kinds of repellents may be used, especially around shoe tops, on ankles, and around the waist. Mites crawl up the body and stop to feed where they encounter a constriction in the clothing. By the time the mite is felt, it has already fed, done its damage, and dropped off.

Follicle mites (fig. 11.24) are among the smallest mites known, ranging in size from a scant 0.1–0.4 mm in length. They live in the sebaceous glands and hair follicles of humans, especially in the region of the nose, eyelids, and other parts of the

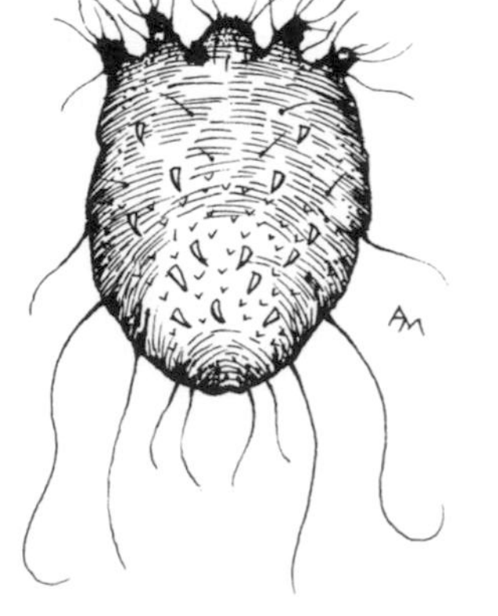

Figure 11.22.
Scabies mite.

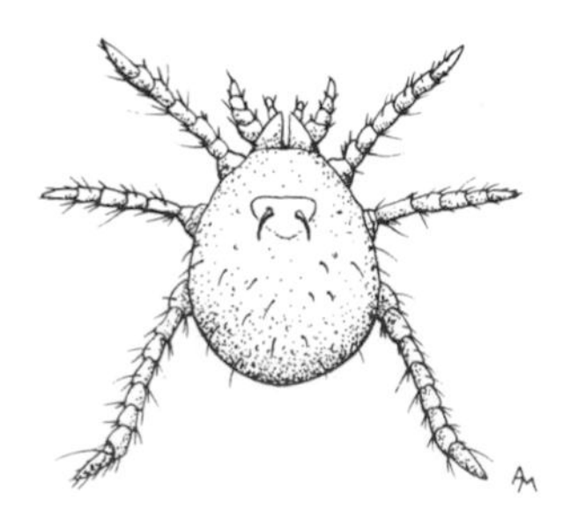

Figure 11.23.
A Chigger mite.

Figure 11.24.
Follicle mite.

face. Infestations are usually benign and are not of great concern, since it is alleged that every human has a considerable fauna of these creatures. Mites of this group belong to the family Demodicidae and are present in dogs and cattle. In these animals, the infestation may develop serious lesions and need to be treated.

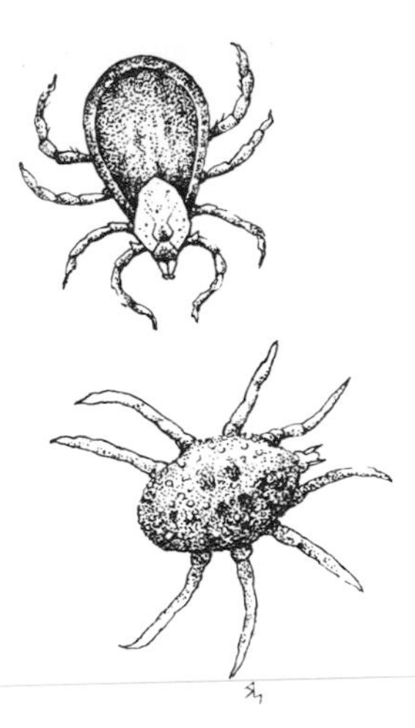

Figure 11.25.
Ticks, top, a hard tick; bottom, a soft tick.

Ticks, also members of the Class Arachnida, are the largest of this group. They are of two types, hard ticks (fig. 11.25, top) and soft ticks (fig. 11.25, bottom). Females engorged with blood may be 20 mm in length, but this is unusual. Most terrestrial mammals and birds are subject to attack by ticks. The hard ticks are common on dogs, deer, and other mammals. The soft ticks are most commonly seen in a chicken house, where they feed on the blood of chickens. Both male and female ticks, including immature stages, feed on blood. Tick bites cause a variety of symptoms, including inflammation, itching, swelling, and, sometimes, anaphylaxis, which can result in death. Tick paralysis, sometimes caused when a tick attaches itself to the base of the head, especially in children, results in partial or total paralysis of the victim which sometimes is permanent. Ticks also are vectors of the serious disease, Rocky Mountain Spotted Fever.

One should not panic at the appearance of a tick attached to the body. They are common in certain parts of the country and are easily removed by the following steps. First, one should never pull the tick out, as it is apt to leave its mouthparts imbedded in your skin, a potential source of infection. Instead, place some rubbing alcohol on the body of the tick. If this is not available, try clear nail polish remover. Both substances will cause the tick to disengage its mouthparts and drop off. Ticks should be removed with care, especially from areas in the ear where they may crawl. Since ticks are hard and difficult to kill by crushing, it is best to drop them into a vial of alcohol or to burn them with a match.

Dogs and cats frequently become infested with ticks. They later drop off in the house and crawl around on the floor. Pets should be treated with flea and tick powder or have a combination flea and tick collar. Remember that cats lick themselves and can be poisoned with some of the available insecticides.

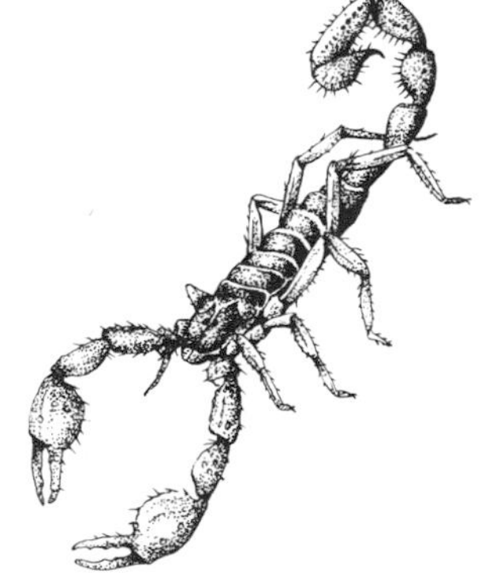

Figure 11.26.
A poisonous scorpion.

Finally, among the poisonous arthropods, scorpions are dangerous in many parts of southern United States (fig. 11.26). Scorpions have the usual four pairs of legs characteristic of the arachnids. In addition, they have pincerlike pedipalps and small chelicerae. Two venom glands are located in the tail or telson of this animal. With this stinger the scorpion stings its prey, and it will sting you without hesitation. The pain from these stings is severe, accompanied by considerable swelling. Since these animals are able to crawl into bedding, tents, cars, and trailers, great care must be taken to assure not being stung. They frequently enter buildings and hide under furniture, clothing left carelessly about, or in dark corners. If you are stung, immediate medical attention is advised. It is best to capture the animal to enable a specific identification, because treatment may differ according to the species. The best first-aid measure is to follow the same procedure followed for snake bite. Pack the stung portion with ice; do not use a tourniquet. Proceed to the nearest doctor or hospital when possible.

Well, this seems to be a chapter of sickness and gore, doom and gloom. There is no doubt that many insects have caused great suffering and death to humans for as long as there have been humans. But, compare the death toll of the highways with that of insect-borne diseases. Forewarned is forearmed, and that has been the main intent of this chapter. It is unfortunate that many of us are afraid of the wrong things. We generally do not fear a house fly or a mosquito, but scream at the thought of being attacked by a giant beetle! Maybe not after reading the above.

12

Coevolution Among Insects and Plants

"*'And what does it live on?'*
'Weak tea with cream in it.'
A new difficulty came into Alice's head.
'Supposing it couldn't find any?' she asked.
'Then it would die, of course.'
'But that must happen very often,' Alice remarked thoughtfully.
'It always happens,' said the Gnat."

Lewis Carroll (1832–1898), *Alice in Wonderland*

We might add to this chapter title, "or the friendly competitors," because, like Alice's Gnat, sometimes insects can't keep up with the plants and they go hungry. Plants "strive" through the marvelous process of selection, to outwit pesky insects by forming many kinds of poisons, thorns, hairy structures, and other things to keep ahead of the herbivores, but, at the same time, they also use them for their own survival.

Plant-feeding insects, that is, phytophagous species, comprise about 50 percent of the known insects. As we have seen from the chart showing the basics of ecosystems, the first energy loss in the transfer of solar energy is that used by the phytophagous species, both plant (e.g. fungus) and animal. Phytophagy, the act of feeding on plants, must have evolved just before predation, but probably the two went hand in hand. At any rate, the process of coincidental change of plants and the animals that feed on them, termed coevolution, is now well documented, even though the mechanism for this phenomenon is not yet clearly understood. Coevolution occurs as a continuing process involving plants and phytophagous insects, vertebrates, invertebrates, and pollinators of all kinds. Especially among the insects, many interesting relationships have evolved. In general, we may conclude that insects have had much to do with the changes and evolution of plants.

In this chapter, we will examine many different types of insect-plant relationships. Pollination is obviously vital to the plant and of benefit to the insect, but some of the other associations, such as the fungus gardens of ants, carnivorous plants, galls, and so on are of diminishing mutual return, and, eventually, one or the other suffers. In other words, not all interactions are in balance. Because of this, in chapter 13, we end up with a look at the war that goes on between insects and plants. Insects are, at times, highly destructive to those plants humans wish to cultivate. But then, we turn the tables on some by using insects for the biological control of weeds.

Pollination and pollinating insects

We generally take flowers for granted, as they seem to be available for the taking or buying. They play an important role in our lives, bringing joy and satisfaction on special days such as Mother's Day, Valentine's Day, and at weddings, and they serve as an expression of sorrow at the loss of a friend or loved one. But what is a flower? The flowers we observe are actually only modified leaves, but these leaves have a special function, a role in the reproduction of the plant. They combined to form a protective envelope around the reproductive organs of the plant and produce the sex cells, actually tiny plants within the plant, the pollen and the ovules. The flower of most angiosperms is composed of the perianth or floral envelope, composed of the calyx (sepals) and the corolla (petals), the androecius (stamens with their anthers) and the gynoecium (pistil and ovary). At the base of the

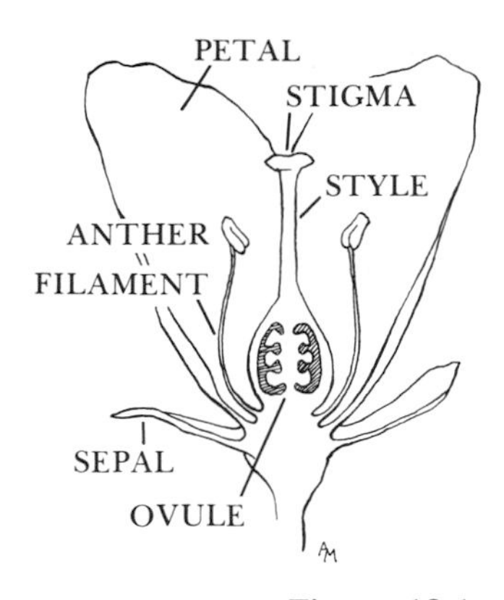

Figure 12.1.
Parts of a flower.

pistil is the ovary or the ovaries from which the style connects to the stigma (fig. 12.1).

Pollination is the prelude to the union of the male and female gametes. It is accomplished by the transportation of pollen grains from the anther of the stamen to the stigma in the flower. The fusion of the gametes will produce an egg, which will develop into an embryo, and, along with other parts, will form the seed. The final product, the fruit, is the ripened ovary or group of ovaries containing the seeds. It almost goes without saying that fruits, like flowers, play an important role in our everyday lives and in the economy of the world. Without pollination, the world would be a much duller place in which to live, and we would not have the opportunity to enjoy many of our favorite fruits and vegetables (table 12.1).

Insects are not the only pollinators, but they are by far the most important. Pollination agents include: the wind, anemogamy; the water, hydrogamy; the insects, entomogamy; the birds, ornithogamy; and the mammals, therogramy. Insects in several orders pollinate flowers. Flies (Diptera), beetles (Coleoptera), butterflies, skippers, and moths (Lepidoptera), and, most of all, bees (Hymenoptera) are the orders with the most pollinators. Bees certainly are, by far, responsible for the pollination of the greatest number of our common fruits and vegetables.

Besides the obvious pollinators, such as the honey bee, *Apis mellifera*, there are numerous other insects involved. Some remarkable adaptations have evolved, structures to induce or entice, if you will, insects to the flower to gather pollen. Two of these occur in the fig plant (*Ficus* spp.) and species of yucca (genus *Yucca*).

Table 12.1. *Some cultivated plants that require insect pollinators.*

alfalfa	carrots	onions
apples	cauliflower	peaches
artichokes	celery	pears
asparagus	cherries	plums
blackberries	chicory	pumpkins
blueberries	chives	radish
broccoli	collards	raspberries
brussels sprouts	cucumbers	rutabagas
buckwheat	currants	squash
cabbage	eggplant	sunflower
cantaloupe	leek	turnips
	muskmelon	watermelon

Fertilization of the various species of figs is extremely complicated. The fig wasp, a species of *Blastophaga*, is one example of an insect which does not fertilize flowers for their nectar or pollen or because of any attractant such as perfume or nutritive tissue. Instead, the fig wasp lays its eggs in the ovaries of the plant. The mature female wasp bores into the fig inflorescence and lays her eggs in the ovaries of the sterile flowers. The males produced from these eggs fertilize the female wasp. When females escape, they are dusted by the pollen of the male flowers of the fig and carry the pollen to new fig tree flowers. Males are wingless; hence, it is the winged female that collects the pollen after she is fertilized and, in the process, deposits her eggs in a new location for the next generation. In this country, *Blastophaga psenes* is an introduced species that has been brought here for the sole purpose of pollinating the commercial fig. Without these wasps, it would be impossible to raise figs. But the story does not stop here. The Smyrna fig is grown extensively in California, but it produces fruit only when pollinated by pollen from the wild fig or caprifig. It is from these figs that the pollen is brought. The fig wasp develops a gall in the wild fig flowers. She, in turn, collects pollen from the male flowers of the wild fig. She only successfully oviposits in the shorter flowers of the caprifig, which is why these wild figs are necessary for the survival of the commercial Smyrna fig.

Yuccas, members of the family Agavaceae, are also dependent on insects for pollination. They are so specialized that only the small tineid moth, *Pronuba yuccasella*, can fertilize the flowers. The yucca flower gives off a perfume primarily during the evening, and this serves as an attractant for the moths. The female moth has a long ovipositor, as well as prehensile, spiny, maxillary palpi, structures unique to this genus. At night, with the aid of the palpi, the female collects some pollen and shapes it into a ball about three times as big as her head. She then flies to another flower and deposits some eggs in the ovary with her ovipositor. Climbing to the top of the flower, she places the pollen ball on the stigma. The ovules are then fertilized, thanks to the remarkable instinct of the moth. The pollen is so abundant that there is enough for the larvae to feed on as well. Without this method of pollination, the ovary would wither, and the plant species could not survive. And, of course, without the plant, the moth could not survive.

Why do insects visit flowers, and what stimuli attract them to certain flowers? Insects are attracted to pollen as a highly concentrated source of protein or to nectar as a source of carbohydrate. Some insects, as seen in the fig wasp, are also dependent on the flower as a site for oviposition and a food source for their larvae. Besides the odor of these substances, the color and shape of the flower guides them to the right place. Bees are capable of distinguishing the four colors, yellow, green-blue, blue and ultraviolet. Also, they select flowers that are radially symmetrical, whereas bumble bees seem to prefer flowers that are vertically symmetrical. The movement of the flower by wind, carrying with it the odor it produces, enhances the attraction.

The smell of flowers ranges from sweet and pleasant, to humans at least, to the unpleasant odors of such giant tropical plants as *Amorphophallas* spp., which smell of rotten fish and molasses. The amorphophallas are pollinated by beetles, while *Rafflesia arnodli*, a large Indo-Malayan plant, is pollinated by flies attracted by its color and pungent odor. The carrion flower, *Scapelia* sp. of Africa, brownish-purple in color, has the odor, as you might expect from the name, of rotting meat. This large tropical flower, therefore, attracts flesh flies and blow flies and, during their visits, effects pollination. Most flowers emit odors that are pleasant to both insects and humans, although these odors are not always present. They may be emitted at certain times of the day or at certain periods in the life of the flower. After pollination, many of the odors cease. Nectar production by a flower often follows certain rhythms, and there exists a synchronism between the activity of the pollinators and the secretion of nectar.

As discussed previously, some insects are attracted to flowers, because they are a potential breeding place. The other "rewards" for the insects' visit to the flower, besides those mentioned above, are certain waxes, resins, and aromatic or nutritive oils.

Many insects have a habit of "stealing" nectar without pollinating the flower. In other words, they reap the rewards without doing any work. They accomplish this task by piercing holes in the base of the flower, where the nectaries, the structures that produce the nectar, are located.

One group of flowers with enormous morphological, as well as esthetic, diversity are the orchids. About 35,000 species and varieties of orchids depend on insects for their pollination. Orchids can be either odorless or produce odors that range from pleasant to repugnant. This depends on the "customer" they wish to attract. Certain orchids are the food of Chrysomelid beetles, and they are pollinated, in the words of the great naturalist, Huxley, because "the flower offers its own flesh to satisfy her sexual desire."

One bizarre method of pollination occurs in orchids lacking nectar, oil, or edible tissue. These orchids, instead, mimic the female of certain bees. The male is visually attracted to the flower, which he mistakenly identifies as a female of his species, and begins to copulate with the flower. This method of pseudocopulation (false copulation) results in fertilization of the flower, because, during this futile act by the bee, pollen is transferred to the stigma of the flower. This is not a single case, but common among several orchid species.

It is interesting to note that, among the four major orders of insect pollinators, there exists a correlation with climatic conditions. Bees are, of course, the most common pollinators of tropical and temperate areas, but Lepidoptera tend to replace bees in the temperate mountains. Flies dominate in the arctic areas, and ants, in semidesert and desert areas. Where specially evolved pollinators are missing, beetles remain as the only pollinators. Indeed, there is strong evidence that beetles were the original pollinators and that much of the higher evolution of the flowering plants is a result of beetle activity in the blossoms of these plants.

Bees immediately come to mind when we speak of pollination. Therefore, you may be surprised to learn that not only the social honey bee pollinates our favorite plants, but many wild, solitary bees. For example, the alkali bees (a species of the family Halictidae) are important pollinators of commercially grown alfalfa (fig. 12.2). These bees are gregarious, but not social. They nest in the ground and overwinter in the pupal stage. Sometimes it is necessary to encourage the spread of these bees in alfalfa fields by importing them from other areas. Sweat bees, small species of the same bee family, are also important pollinators of some crops.

Figure 12.2. *Alkali bee (*Nomia melanderi, *Hymenoptera: Halictidae).*

The subject of pollination and the role of insects in this process is the subject of many books and articles. Much more could be said about the adaptations that have developed in this complex relationship. This is, of course, of great interest to gardeners and agriculturalists alike, but we must turn now to other relationships.

Fungus gardens

Humans are not the only organisms that cultivate plants. Many insects are associated with fungus. What are the fungi? These are species of a large group of plants that lack the ability to carry on photosynthesis, because they lack the photosynthetic pigment chlorophyll. These distinct plants occur in a wide variety of forms. They usually reproduce by spores, an asexual process, but most kinds also have a complex sexual stage. The vegetative body is composed of filaments similar to those of many agae. The cells may be multinucleated. Cellulose or chitin (a complex chemical also found in the arthropod body wall) or both surround the cells. Fungi, including bacteria, are the principle agents of decay and are the important part of the detritus phase of an ecosystem, which will be discussed in chapter 17. Among the familiar types of fungi we know are mushrooms, and many are acquainted with braken fungus, molds, smuts, and rusts.

Insects and fungus are sometimes closely associated. Some insects feed on fungus, the mycetophagous species, of which many widely separated groups are known. In the flies (Diptera), the family Mycetophilidae, or fungus gnats, are specialized fungus feeders. One species is a pest in mushroom cellars. Beetles (Coleoptera) have several families known to be feeders on fungus, the Myctophagidae, the fungus beetles, Endomychidae, handsome fungus beetles, and the Mycetaeidae, tiny beetles, all feed on fungus.

We will be particularly concerned here, not just with fungus feeders, but with those insects that actually grow it for food. Again, widely separate groups of insects are involved, including the fungus-growing ants and the fungus-growing termites. Atta ants, famous for their leaf-cutting ability, are highly evolved social Hymenoptera (see chap. 9). These ants, found in the New World tropics, belong to the subfamily Myrmicinae. The species of the genera *Atta*, *Trachymyrmex*, and *Cyphomyrmex*, especially, use a type of manure or compost made chiefly of cut leaves on which they plant fungus. The fungus is eaten and provides the only food for these ants. The 400 species of fungus-growing ants belong only to the one subfamily, and most of them to the genus *Atta*, the best-known genus. The workers of *Atta texana* cut leaves (fig. 12.3) from bushes and trees, carrying these particles back to their nests along well-beaten trails (fig. 12.4). A leaf particle is

Figure 12.3. *Atta ant carrying leaf fragment back to their subterranean fungus gardens.*

Figure 12.4. *Path worn by atta ants foraging for leaves for their fungus garden.*

held over the body as it is carried, earning these ants another common name, the parasol ants. The leaves are added to the "garden" inside the ants' nest. But how did these nests come about?

When the swarming atta ant female is fertilized, she drops her wings and looks for a nest in the ground. She digs a cavity in the soil and stays in it. She has carried to this cavity fungus hyphae collected on her body from her original nest, her birthplace, and stored in a pocket located near the mouth. Once these hyphae are planted in the soil cavity, they germinate, forming hyphae, or threads of fungus, the start of the garden. For several days, the queen will fertilize this developing fungus with her own excreta and even some of her own crushed eggs as manure for the garden. Eventually some of her eggs will develop into workers. They will be tended by the queen until they reach adulthood, after which they will collect leaves from nearby bushes and trees on which to grow the fungus. The compost is improved with various kinds of debris, including flowers, excreta of caterpillars, and other dead organic material. They continuously add to the garden, which finally takes on the shape and appearance of a sponge (fig. 12.5).

Under the constant care of the workers, the fungus garden produces special terminal bulbs on the hyphae. These are called

Figure 12.5. *Fungus garden of atta ants.*

bromatia, ambrosia, or kohrabia. These nonreproductive elements are the sole source of food for the adults and the larvae of the atta ants. The bromatia do not appear spontaneously in laboratory cultures of the fungus. It appears that temperature and humidity control is necessary for the proper growth of the fungus. The ants are able to control this by carefully positioning the entrances. If the ants cultivate their gardens, the fungus does not sporulate, and no fruiting structures are produced. The bromatia are produced by the constant pruning of the mycelia and by the antibiotic effect of the ant saliva and excreta.

Atta ant nests can reach an enormous size. Nests up to six meters long have been unearthed. The workers moving in long columns on leaf gathering forays can be destructive to some trees. They may completely defoliate a small tree in a few hours, although, generally, they remove only one or two pieces from each leaf, leaving enough to keep from killing the plant. Large nests may contain hundreds of separate gardens.

Atta ants are polymorphic with minute workers, the minimes, charged with the cultivation of the fungus and the feeding of the larvae. Medium-sized workers collect leaves, and the large workers act as soldiers, defending the colony.

Termites

Fungus-growing termites are found only in the New World tropics and belong to the subfamily *Macrotermitinae*. Organic matter, chiefly wood collected by the workers, is chewed and digested to form the media on which to grow fungus, fertilized by the fecal material from the termites themselves. These gardens may be dispersed around the nest. The inside cavity is covered with a mixture of saliva and dust and may be ventilated by a system of vertical conduits that reach the outside. The grayish to brownish gardens have a spongy aspect. The intestines of these termites lack the flagellate protozoans that other species have as symbionts to digest cellulose. The cultivated fungi predigests cellulose and provides these hosts with a source of food and vitamins. These species of termites cannot live on wood, as do other species, but use this fungus as their sole source of food.

This insect-fungal association appears to be an almost perfect example of a mutually beneficial symbiotic relationship. It is certain that, due to antibiotics secreted by these insects, only the particular species unique to the termite nest is allowed to grow. These antibiotics are also responsible for the production of species structures on the fungus, such as bromatia, and the prevention of the production of reproductive bodies. Levels of carbon dioxide, acidity, and temperature control also aid in the maintenance of the fungus gardens.

Carnivorous plants

Although we know plants defend themselves in various ways, through the production of toxins in the sap or by certain modified structures, such as leaf hairs, spines, and thorns, we marvel at those plants that are modified in such a way as to actually capture small insects and other arthropods to supplement their otherwise strictly inorganic chemical diet. The carnivorous plants, according to botanists, number over 500 species scattered throughout the world. They are found in nitrogen-poor soil, such as that found in bogs or other wet places, or in certain poor tropical soils. The water-holding and snap trapping structures of these plants are modified leaves; the sticky hair species, modified stems.

Carnivorous plants may be divided into two groups, those that trap insects actively and those that do so passively. Among the active trappers is the Venus's flytrap, *Dionaea muscipula*, one species that actively traps insects by snapping closed its modified leaves when trigger hairs are touched. This species is found in the wild in local areas of North and South Carolina in damp, but not sopping wet, largely sandy, peaty soil. Although the plants may be cultivated, they generally live only a short time. Attempts to grow the species in other areas have failed. The prey are chiefly small insects and spiders, animals too small to chew their way out of the trap of intersecting marginal spines. The excitation of the tactile hairs on the leaf surface causes a rapid loss of water in the cells of the hinge of the leaf. The turgidity of these cells otherwise keeps the trap open. Once this tension is relieved, the thin cell wall bends the two halves of the leaf together. The trap may stay closed for up to ten days after an insect meal. This trap, obviously, works well outdoors, but conditions are never right for the trap to work properly when the plant is potted indoors. Recently it has been suggested that the stimulus for springing the trap is uric acid, abundant in insect excreta.

Another active trapper is the bladderwort, species of *Utricularia*, water plants with attractive blue and yellow flowers. These plants are found in ponds and ditches worldwide. On their submerged, threadlike stems are numerous bladders (fig. 12.6) in which tiny aquatic invertebrates are caught. At the end

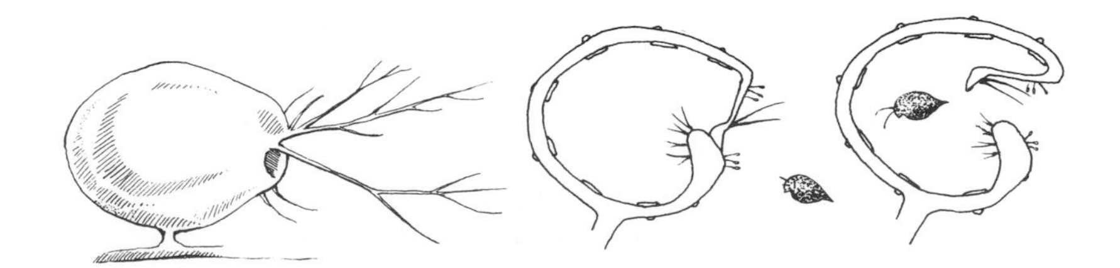

Figure 12.6. *Bladderwort, an aquatic plant, showing insect trap.*

of each bladder is a trap door, which suddenly snaps open when sensory hairs just outside the trap are touched. The prey is sucked in so fast that it is difficult to see the action. Once the door is closed, part of the water is pumped out and digestion takes place.

Passive trappers are those in which rapid plant movement is not an integral part of the trapping process. They include plants with pitfalls, the pitcher plants, and the "flypaper" or adhesive traps. Among the latter are species of the genera *Drosera* (sundews), and *Pinguicila*, (butterworts). The most common passive trappers are the pitcher plants (fig. 12.7), members of the genera *Sarracenia* and *Darlingtonia* in North America, and *Nepenthes* in the Indoaustralian region.

Pitcher plants, so-named because their leaves are modified into a water-filled vessel in which prey is captured and digested, are common in bogs in North America. These are members of

Figure 12.7. *Pitcher plant, a terrestrial insect trap.*

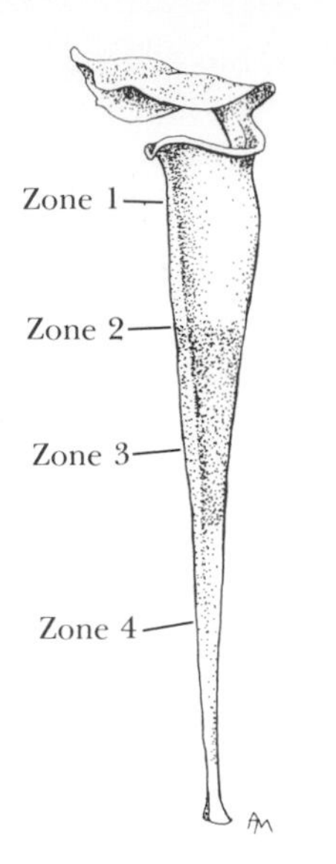

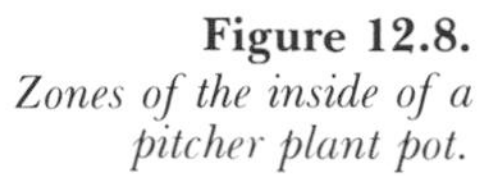

Figure 12.8.
Zones of the inside of a pitcher plant pot.

the genus *Sarracenia*, with seven native species. The funnel-shaped pitchers have a lidlike, but immobile, structure, the hood, at the top of the trap.

Both water and scents secreted by the plant attract flying and crawling insects to the rim of the pitcher. Ants and flies are common victims. The inner surface of the pitcher is divided into several zones (fig. 12.8). Zone 1 has scent glands and short hairs. Insects crawling in this area can get a foothold. Zone 2 has more scent glands, which attract the insects deeper into the trap, where they encounter a slippery surface and drop to the water in the trap and soon drown. Zone 3 has digestive glands, and zone 4 contains long downward-directed hairs that prevent any chance of the insect's crawling out before it drowns. Besides the digestive enzymes, botanists have recently discovered an intoxicant, coniine, which paralyzes the insect and increases the likelihood of its falling into the water. Insects thus trapped are slowly digested, and the nitrogenous compounds released are used in the metabolism of the plants for growth, thus making up for the lack of nitrogen compounds in the soil. At the same time, these plants carry on photosynthesis as their source of carbohydrates. Therefore, these plants are not truly carnivorous.

Galls

One of the many unique symbiotic relationships between plants and animals is the formation of galls, remarkable malformations, often as large as golf balls, which can develop on leaves, stems (fig. 12.9), buds, flowers, roots, and even fruit. Gall formation is stimulated by mites, nematodes, fungi, bacteria, and insects. It is difficult to see how this seemingly one-sided relationship can have any mutual benefit, but some biologists consider the production of a gall as advantageous, since it protects the plant against greater damage, the complete necrosis of the parasitized tissue. It seems to be far better to encapsulate the parasite and to provide it with food than to have the parasite moving around and destroying tissues. The formation of the gall, which is an hypertrophy (overgrowth) or dysplasy (abnormal growth) of tissue, limits the damage. It is a defensive and protective measure and seems to be one way to combat invasion and consumption of vital tissues.

Galls are extremely variable in shape and color. Some, to us, enhance the beauty of the plant, while others destroy the natural form of the plant and make it look sickly.

Insects are by far the largest group of gall-makers (table 12.2). Because of this, there is considerable entomological interest in these groups of insects and their plant hosts.

Figure 12.9.
Stem gall of horse chestnut.

Table 12.2. *Insects known to induce gall formation.*

Order	*Common Names*	*Families*	*Number of Galls Produced*
Diptera	Gall midges	Cecidomyidae Tephritidae	700
Hymenoptera	Gall wasps	Tenthredinidae Cynipidae	490
Homoptera	Aphids and psyllids	Aphidiidae Coccidae	60
Lepidoptera	Caterpillars of some	Tortricidae	15
Coleoptera	Beetles	Curculionidae (weevils)	12
Thysanoptera	Thrips	?	?

Galls are benign tumors. Usually an egg is laid in plant tissue, stems, and so on, which starts the irritation, but growth does not take place until the egg hatches and the larva starts to feed on the surrounding tissue. The growth of the gall stops when the larva becomes a pupa. The different shapes of the galls depend on the species of gall-maker. Experiments show that gall formation is due to a growth hormone, an auxinlike substance usually produced by the invading larva.

Obviously, with over fifteen hundred different kind of galls in North America and over eight hundred species known on oaks alone, it is impossible to begin to describe them here. One example will have to do, but we encourage you to collect galls associated with the gall-makers, as you will find this a fascinating study.

The goldenrod gall is a common species, easily located on the stems of goldenrod plants. These plants bloom in late summer in meadows, woodlots, along railroad beds, and in fence rows. The gall is easily seen as a swelling on the stem.

The goldenrod gall fly, *Eurosta solidaginis* (Diptera, Tephritidae), (fig. 12.10) is a fly, approximately 6 mm in length, with dark patches on its otherwise clear wings. The fly overwinters as a larva in the gall and pupates in the spring. The adult emerges from the gall and seeks a mate. After mating, the female locates new goldenrod stems just coming up in the late spring and early summer. She deposits her eggs in the stems, usually one egg per plant. The eggs hatch and the larvae burrow into the stem and create a chamber when they feed on the plant tissue. As explained above, this causes the gall to form.

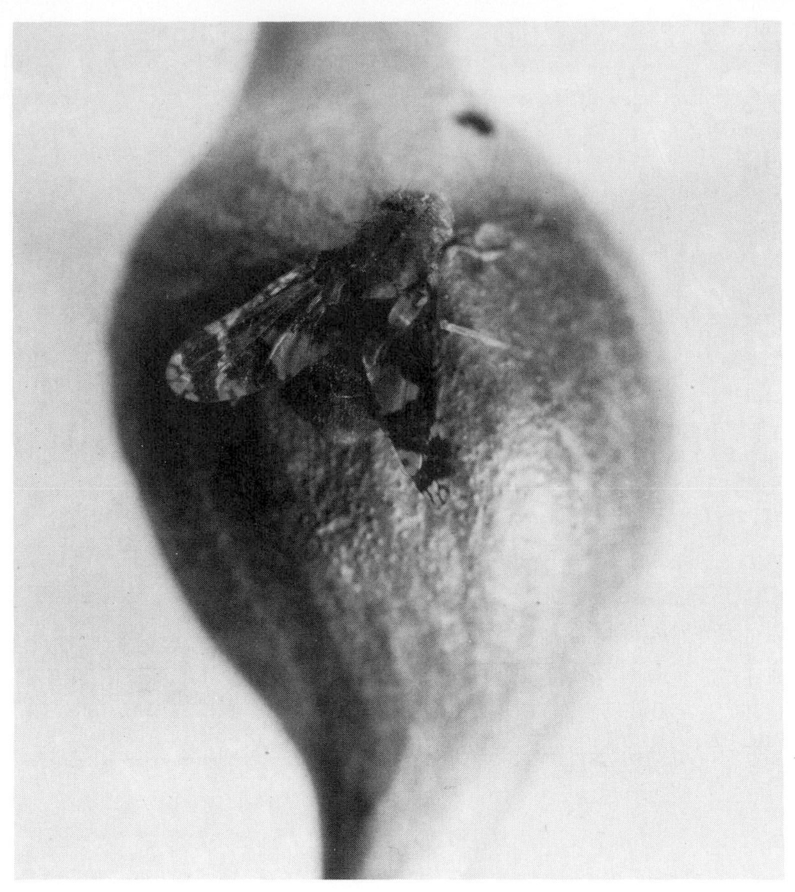

Figure 12.10. *Goldenrod gall with the gallfly.*

Similar life cycles occur for other gall-making species. Galls can be collected in any season and the contents may be reared out of the gall, provided they have matured sufficiently to complete their life cycle without further feeding. Many species of gall-makers can be identified from the gall alone. However, you may have a surprise in store for you, because sometimes parasitic species invade the gall and feed on the gall-maker. Thus, you may obtain a parasite instead.

13 Insect Wars

"So, naturalists observe, a flea
Hath smaller fleas that on him prey;
And these have small fleas to bite 'em;
And so proceed ad infinitum."

Jonathan Swift (1667–1745), *On poetry*

Out of balance

What happens when the symbiotic relationship we have discussed in chapter 12 becomes unbalanced? We have talked about populations, their distribution, their food, and their relationship with other populations. We know that all are not in perfect balance. The fact that we have such things as "pests" tells us that something is wrong. Who is wrong? Is this a war between species? We are constantly being told by the "media" that we are at war with the insect world. Nothing could be farther from the truth, but, at the same time, we cannot ignore that fact that some insects and man have overlapping interests. We try to protect our interests—so it is war. The best way to discuss this is to ask what happens when one side loses. The subject of economic entomology, the description of the insect war, is a vast one, and certainly beyond the scope of this book. There follows, however, a few examples of pests and pest control.

Unfortunately, many people believe that the chemical control of insect pests is all wrong. It is not the chemicals that are bad; it is the improper use of these chemicals that needs attention. Nor is it the entomologist that is, with some exceptions, using them incorrectly. The application of chemicals is exactly the same as the use of medicine. Only the right amount should be used. If every drug were available "over the counter," disaster would follow. Just because a certain dosage of medicine is good, it doesn't mean that more is better. Usually the converse is true. Many individuals, farmers, homeowners, and others were, in the past, able to use strong insecticides and herbicides however they wished. The result was the contamination of the environment. Now, through the action of the U.S. government, particularly the Environmental Protection Agency, these chemicals are regulated, not always for the best, perhaps, but at least now there is a way to regulate and deter further damage.

Insects as pests of plants

Less than one percent of the approximately 750,000 known species of insects are pests, and many of these seldom reach economic proportions. This is an indication of how far apart man and the insects really are. Yet, insects feed on almost every known terrestrial and fresh water plant in every conceivable environment. These pests destroy field crops, vegetable gardens, ornamentals, greenhouse plants, and every kind of stored organic product, even, including, boring through lead cables covering electric wires.

Most insects are pests only periodically and only when the natural balance of food and population suppressors is disturbed. Then, due to the availability of seemingly unlimited quantities of food, the number of individuals rapidly increase, become noticeable because of the damage they do, and gain our attention. The combination of weather conditions, reduction of natural enemies, either because they were left behind or because they were killed by other factors, migration, and even a behavioral change, all contribute to the rise of these pests.

A classic example of this is the periodic increase of the migratory locust, *Melanoplus spretus* in the Rocky Mountain region. This insect normally occurs in numbers low enough to pass unnoticed, or at least not great enough to make it worthwhile to employ control measures. But, under certain conditions of weather, combined with food availability, it increases in great numbers and migrates, causing serious damage to all crops, much the same as the locust plagues previously described in the Bible. In the 1930s, this insect inflicted losses to crops in the western United States, which have never been surpassed.

Sometimes insects are accidentally introduced into a new area through the aid of man. The infamous gypsy moth, *Lymantria dispar*, introduced from Europe into Massachusetts in 1868, is a good example, and it has recently been the subject of much concern by the news media. This insect feeds as caterpillars on forest and shade trees, especially oak, cherry, willow, elm, and apple (fig. 13.1). The insect is well established in Massachusetts and, recently, has spread throughout the northeastern United States and southern Canada and threatens to spread even further. It caused millions of dollars worth of damage to lumber trees, defoliating hundreds of acres of forest in some areas, especially in those areas at the fringe of its distribution. Few natural enemies are known, and they do not succeed in keeping the moth populations under control. Fortunately, the pest goes in cycles, and when it is at low ebb, new controls may be effective in reducing its spread.

Insect outbreaks are always a serious problem to farmers, but especially so when adverse weather conditions combine to ruin a crop. It appears that sometimes even genetic changes in the pest species, combined with changes in crop management, upset balances. When large scale plantings of a single crop, a monoculture, such as the planting of cotton or wheat on large farms, are invaded by a pest, the result can be financial disaster. A dramatic example of monoculture in American agriculture is the story of the cotton boll weevil, *Anthonomus grandis*, a beetle (Coleoptera, Curculionidae), a serious pest of the cotton boll.

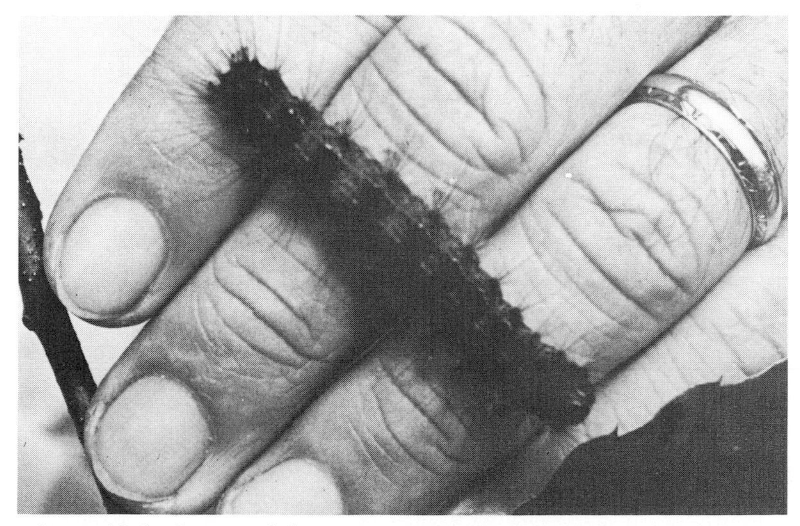

Figure 13.1. *Gypsy moth larva.*

This species entered the United States from Mexico in the late 1880s. At that time "king cotton" was a major financial investment in this country, and much of the economy of the South depended on the annual cotton crop. At first, farmers felt that they could control this pest. Many growers knew only how to grow cotton and did not diversify. They were suddenly completely at the mercy of this insect, because no effective controls were found in time to save their economy. With pure stands of cotton for miles on end, the weevil spread rapidly, and a major disaster was inevitable.

An interesting story about the boll weevil centers on the town of Enterprise, Coffee County, Alabama. This area once was a large cotton-growing area, but it was seriously affected economically by the boll weevil until 1915. The president of a local bank, also the owner of the Enterprise Cotton Seed Oil Company, decided to gamble and plant peanuts as an alternate crop to cotton. He persuaded a local cotton farmer to plant some of the peanuts he brought to Enterprise from North Carolina. The farmer planted over 120 acres of peanuts in 1916 and, while over 70 percent of the cotton crop was destroyed by the weevil, the peanuts grew well. When the farmer took the peanuts to the mill and netted over $8,000 (a large sum of money in those days), the days of "King Cotton" were over in Coffee County. So delighted were the inhabitants of the town that they erected a statue of Columbia (fig. 13.2) in the town square, and

Figure 13.2. *Statue of Columbia, Enterprise, Alabama, erected in honor of the boll weevil.*

Figure 13.3. *Closeup of the boll weevil atop statue of Columbia.*

on the top of this statue is a large boll weevil (fig. 13.3), the only monument to an insect pest. On the monument is an inscription that reads:

> In profound appreciation
> of the Boll Weevil
> and what it has done to the
> Herald of Prosperity. This
> Monument was erected by the
> Citizens of Coffee County, Alabama.

Obviously, the boll weevil taught them, and other American farmers, a lesson—that a monoculture is risky business. They no longer put all their "eggs in one basket." A recent drive through that area showed the authors that cotton has not gone from there. Government subsidies make it a paying crop, even if eaten by boll weevils. Both still thrive in the South!

The United States is still a major agricultural country, and we feed much of the world. Therefore, our economy depends on healthy, high-yield, insect-free crops. Our 224,000,000 people, and many millions of others, depend on our farmers for food. A collapse of any major field crop in the United States, or elsewhere in the world, causes worldwide problems. The major field crops in the United States are corn, wheat, cotton, soybeans, oats, barley, alfalfa, and clover. Each of these crops has many insect pests and plant diseases.

Corn has the highest value per acre among the world's major field crops. It is grown in every state except Alaska. Corn is marketed as cash grain, or kept for livestock feed. It ranks second only to wheat in worldwide acreage. Because of this abundance it is no surprise that insects attack all parts of the corn plant from roots to ears. In 1976 American farmers spent over 250 million dollars on insecticides to control corn pests.

Small grains, such as wheat, oats, and barley, are used for human consumption and feed for animals. Insects cause considerable loss to these crops—about $150,000,000,000 annually. All costs are passed on to the consumer as higher prices for bread, cereals, and meat. The story is repeated over and over again for each crop.

Control measures

More important than a discussion of either crop or pest is the method of controls, at least in general terms. Two major categories of insect control, chemical and biological, are apparent, but, recently, more emphasis is being placed on a combination of these two somewhat-diverged methods. Chemical control is the application of, usually organic, chemicals to crops to poison the insect pests in one or more stages. Biological controls make use of any natural condition of the environment or any organism that will inhibit the growth of insect populations. Integrated control is a combination of these two basic methods of control.

Biological control of insects is possible utilizing many different organisms. The use of parasites, predators, pathogens, and the environment is well documented and practiced in many places. Each of the many species of insects is a pest of crops, ornamentals, and stored products and, as we have already discussed, of man and domestic animals. Each "pest" is studied in detail in an attempt to find new control measures.

One obvious natural control is the weather. Weather conditions, including temperature changes, variations in moisture, wind, and sunlight, affect the populations of insects just as they do crop growth. To forecast insect outbreaks, a careful study of weather conditions through sophisticated analysis is essential. In fact, computerized models of insect populations enable the testing of weather changes and, hence, the forecasting of the activities of an insect pest as these weather factors change. This greatly affects decisions by the growers about when to plant, when to use insecticides, and when to harvest. Weather is a prime factor in the consideration of any control method.

All kinds of organisms, collectively known as "natural enemies," are potentially useful in the control of pests. The dream of all entomologists, as well as growers, is to find the super enemy, but this is a fantasy that will never come true for one good reason. No predator, no disease organism can kill its host, for, if this happens, it too dies. Natural enemies do, however, hold populations in check; they keep a "natural" balance. Every species has its enemies. If not, they reproduce and spread until the mere lack of food kills them back. This happens with insects, and it happens to man. When insect pests are introduced into a new area, they generally arrive without these natural controls. When they lack these, efforts are often mounted by entomologists to find predators and parasites to introduce into the infested area to help reduce the population. This can be done, but complete control is impossible using this method. Further examples of this type of control are discussed on following pages.

Applied controls

Applied control involves any method used to reduce the populations of a pest. This includes legislative, physical, mechanical, cultural, biological, host plant resistance, genetic, and chemical controls.

Legislative control calls for governmental action at both the state and federal levels. This includes the inspection of baggage and merchandise entering the country at airports and seaports. Certain materials are contraband. Others must be fumigated, if found to be infested. Certain products are quarantined and cannot be shipped into or out of certain areas. Whether any of these measures have ever prevented the introduction of pests is problematical, but certainly it has greatly reduced the chances of an unwanted introduction. These regulations are enforced by state and federal departments of agriculture.

Physical and mechanical controls are similar. The use of a light trap is one form of physical control. This works only in certain cases and, in general, is not effective. As mentioned previously, the so-called "bug zapper" certainly kills many insects, but none that are really pests, unless, of course, you consider night-flying moths pests. Mosquitoes, flesh flies, and few garden pests are attracted to these lights. Mechanical controls, such as fly screens on windows or baffles to prevent the invasion of termites, are effective. Even a fly swatter is an efficient mechanical control. Some years ago, an enterprising inventor of the "new mousetrap" variety invented a 100 percent guaranteed insect control method, if used properly. It was widely sold in a small cardboard box. The user was asked to follow the directions exactly, or the control would not work. It consisted of two wooden blocks. Any insect placed on one block and struck with the other would be killed!

Cultural controls refer to the cultivation methods used for growing a particular crop. Crop rotation is a common method used to prevent reinfestation of next year's crop by the hibernating pests in the old field. Corn cannot be planted year after year in the same field in some regions. This is due to the populations of corn pests that exist in the soil and build up each year. Soybeans are alternated with oats or corn for the same reason, as well as for a means of soil improvement. Soybeans add to the nitrogen content of the soil; corn reduces it. Other cultural control methods include modified methods of tillage, harvesting, trap crops, and crop-site modification.

The elimination of breeding sites is another way to control many pests. Plowing under stubble that might harbor eggs or pupae, or removing fence-row debris are two ways to lessen the chance of pests appearing the following years. Removal or containment of garbage and manure helps to prevent fly growth, and so on. In urban areas, garbage cans should be kept clean and covered. Although plastic bags are the bane of environmentalists, because they do not disintegrate, garbage that will otherwise become the home of many fly larvae can be kept free of insects by their use. Dumpsters should be emptied at least every three days, and their contents buried, burned, or at least sprayed with insecticides.

Fighting insect pests with repellent plants

Even in the home garden, the "everyday homeowner"—that Mr. John Doe—can employ a method of insect control without the use of chemicals. Although Mr. Doe may be employed or retired from any manner of job, if he is not a professional gardener or professional entomologist, he is invited to grow repellent or companion plants.

More and more gardeners are convinced that they can cease to use chemicals to control insect pests in flower and vegetable gardens. Part of this is a concern for one's own health and safety, and a large part is because of an awareness of the potential dangers chemicals may have to the environment. Hence, if a gardener is content with some loss of the crop and insect damage of the skin, flowers, and so on, these methods will work and should be encouraged, if, for no other reason, than to save money. Most gardeners will prefer to combine the common sense use of chemicals with these methods.

Numerous garden flowers, herbs, and even vegetables have qualities that repel certain insects (table 13.1). Some popu-

Table 13.1. *Insect-deterrent herbs and plants for your garden.*

Plant	***Used to Control***
Asters	Many insects
Basil	Flies and mosquitoes
Catnip	Flea beetles
Chrysanthemums	Many insects
Garlic	Japanese beetles, some others
Geranium	Many insects
Marigolds	Many insects
Mint	Cabbage butterflies and ants
Nasturtium	Aphids, squash bugs, striped cucumber beetle
Onion	Many insects
Petunia	Bean insects
Peppermint	Cabbage butterflies
Radish	Cucumber beetles
Rosemary	Cabbage butterflies, bean beetles
Sage	Cabbage butterflies, carrot flies
Tomato	Asparagus beetles
Thyme	Cabbage butterflies

lar flowers, such as marigolds and chrysanthemums, seem to deter many insect pests. They may be used as attractive borders around the flower and vegetable garden. Garlic has proven to be a potent killer of many insects, and researchers have found it to be effective in warding off mosquitoes, as well as pesky friends, and to help prevent heart trouble! Spice herbs, now popular in the home garden, also repel some insects. They can be planted in both types of gardens, since many of them are both ornamental and edible.

Host plant resistance

Host plant resistance is a natural attribute of every plant. It is well known that some plants are more susceptible to insect damage than others of the same species. Using this fact, plant geneticists are able to select hardy plants and develop insect-resistant strains. Three mechanisms make this possible: antibiosis, antixenosis, and tolerance, as discussed below. Antibiotic plant substances may kill, or at least greatly reduce, the chances of survival of an insect pest. Such protection may be by chemicals within the plant, such as those found in milkweeds or members of the potato family. Or, sometimes, the mere presence of spines or hairs on the leaves is sufficient to prevent feeding and oviposition. Host plant resistance is an important part of American agriculture, with many plants, such as wheat, corn, and soybeans, having resistant varieties. Some of these have been developed for tropical agriculture, with successful introduction into Latin American countries to help ease the food shortage.

Genetic control. Control by mass rearings is a new method of insect damage prevention. Lethal genes are selected, and the wild populations of pests are inoculated with these. For example, malaria or yellow fever mosquitoes with genes for the production of a high percentage of males are reared in large numbers and released to mate with the wild females. The resulting reduced number of females causes the population to diminish and sometimes die out. Genetic manipulation to produce sterile males of the screwworm has previously been discussed.

Antixenosis, which simply means "against guests," is the nebulous feature of some plants that appear unattractive for insect pests. Obviously, there is some as yet unknown reason for this. Although we say that everyplant has its own insect pests, certain plants, particularly ferns and horsetails, have only a few.

Tolerance to insect injury appears to be characteristic of some plants. They sustain considerable injury from attacks, yet they continue to grow and produce fruit and seed. This is, of course, the basis of host plant resistance and is a part of nature's grand selective process.

Biological control of weeds

Obviously, most of our efforts to control the loss of products from insect damage is directed to the elimination of the pest on the plant itself. We may forget that, besides fighting off these insects, plants are also battling against weeds. Every plant must find its place in the sun, and, to do so, it must compete with those "masters of shove," weeds.

The approach to weed control is exactly the opposite of that of insect control, yet, it too involves insects. Weeds too, have their insect pests. It makes little difference to an insect whether it eats something humans want or something they don't. Hence, it seems obvious that we can join with the insects here and enlist their aid in the control of weeds. This is exactly what is now being done.

The selection of insects for biological control of a weed is not a simple task. Insects that feed only on a single species of plant are most desired. Since many of our weeds, as are many of our insect pests, are imported, then it is necessary to find a monophagous insect in the native habitat of the weed. The following examples illustrate how this is done.

An important weed in waterways in the south is the beautiful floating plant, the water hyacinth, *Eichhornia crassipes*. Rivers, streams, and lakes may become completely clogged with this plant (fig. 13.4). It prevents navigation even of small boats, chokes out other plants, and prevents the oxygenation of the water, making it uninhabitable by fish and other aquatic life. The plant is found from Virginia south through Florida, where it is a major pest, west to Texas, and south into South America. Until recently, the only way to combat this plant was to drag it out of the water, dry it out, and burn it, or use it for compost.

For many years George B. Vogt and others of the United States Department of Agriculture travelled extensively in South America to find a suitable pest insect. Care had to be taken that the insect would feed only on the water hyacinth, not leave that plant for some other, and, eventually, become a pest of some desirable plant. To do this, extensive, careful, and thorough study had to be made. Carefully controlled experiments using reared specimens of many species were necessary. Finally, the weevil, *Neochetina eichorniae*, was selected as the best species to control the plant. It was introduced and has become established in this country. It now has become an effective control species.

Figure 13.4. *Lake Alice, Gainesville, Florida, clogged with water hyacinth. (Courtesy of Dr. Gary Buckingham, U.S.D.A.)*

Another species, a pyralid moth, *Acigona infusella*, is also being used.

Another serious weed, the musk thistle, *Carduus* spp., a complex of three species, is found in pastures and rangeland. It greatly reduces the forage plants that would normally feed the cattle and, because of its spiny nature, is avoided by the livestock. It invades now open lands and sets seed quickly, soon taking over the entire area. As often happens, such land is overgrazed, the cattle themselves aiding the unwanted plant to survive in place of the food plant.

In Nebraska, another weevil, *Rhinocyllus conicus*, is used to control the weed. This weevil, called the musk thistle seed weevil, feeds on the thistle, the females laying their eggs on the bracts of the developing flower heads in the spring. The newly hatched larvae bore into the base of the flowers or into the stems. This infestation interferes with the proper development of seeds, or, when the infestation is large, totally prevents seed production. This cutback of the weed allows the forage plants to grow.

Programs for the biological control of weeds are costly and may take years to test and develop. Therefore, much of this type of control is possible only on public lands and cannot be applied to the normal daily control of crop weeds. Herbicides continue to be used which present problems of chemical contamination of field runoff.

Chemical control

The most widely used insect control, and at present the most practical method to use, is the application of chemicals. Stomach poisons are used for some species, contact poisons for others, while repellents and chemosterilants are useful in certain cases. Realizing that some of these insecticides affect beneficial insects, birds, fish, and mammals, including humans, has resulted in a cry of "stop" from the media-excited public. It is true that serious harm can result from the use of these nontarget chemicals. One exceedingly dramatic publication, if not entirely accurate, that made us aware of the potential problem of chemicals saturating the environment, was the book *Silent Spring* by Rachel Carlson. This book, published in several editions in the 1960s, became a best seller and proved to be the cause of a turning point in the history of entomology. It greatly benefited the field, because it caused public awareness of the science of entomology. Governmental support for entomological research of all types increased. Chemical companies were able to introduce new chemicals that were selective and new application techniques. Unfortunately, it also gave all insecticide producers a bad name, and, since neither the media nor the general public takes the time to investigate for themselves, this blanket condemnation has resulted in somewhat unscientific legislation, which is only slowly being corrected.

One result of this upheaval is the development of integrated control methods using a combination of insecticides and biological control. Another result is the intense study of the insect pest itself. For the first time, sufficient data have been gathered about major pests, so that their life cycles, distribution, indeed, every aspect of the biology of the insects, have become known. Using these data, a program of insect pest management is possible for each species. Once this is done, a wide variety of control methods are possible, including the proper application of chemicals.

Predators, parasites, and pathogens

Biological control with predators, parasites, and pathogens can be effective only after long, careful research. Once introduced, it takes some time before the results are significant. Meantime, there is always the temptation to use self-defeating chemical controls. it is, therefore, necessary for us to look closely at the natural, as well as the artificially induced, population controls brought about by predators, parasites, and pathogens.

Insects that prey on and kill other insects, thus, predatory insects, are simply termed "predators." Insects that feed externally on the bodies of other insects are ectoparasitic, and those that develop internally in the body of another insect are insect parasites. Collectively, these insects are called entomophagous. Predatory insects may be obligate insect feeders, or they may be omnivorous. As we have seen previously (table 5.1), insects, predator or nonpredator, may be monophagous, oligophagous, or most commonly, polyphagous. Parasitic insects are generally host specific, or nearly so; that is, they may parasitize either a single species or a few closely related species.

Predators occur in most orders of insects from Collembola, the springtails, through the orders to Diptera, the flies. Some are predaciousonly as immatures; others only as adults. Parasitic insects are mainly species of certain families of Hymenoptera and Diptera, and one specialized group of beetles, the Stylopiformia, sometimes called the Strepsiptera. Various species in other families and orders of insects are also parasitic.

Once again, we must tell you that we cannot treat many details about the entomophagous insects; whole books have been written on this subject, as you may well surmise, because of their importance in the biological control of insects. The following survey will give you some idea about the diversity of these insects.

Predators. Most predators consume large quantities of food, either by eating the entire body (some discard the comparatively nonnutritious wings) or by sucking out the contents of the body. They are usually larger than the prey, which permits them to grasp the victim, although bees and wasps depend on their stinger to quickly help them overcome larger insects. This, and other morphological, physiological, and behavioral adaptations fit them for this carnivorous life.

Predators may also be grouped according to the manner in which they seek their prey; random hunting, using visual acuity to seek specific kinds; ambushing; and trapping. Each of these strategies is described here.

Random hunting. Many species of carnivorous insects simply roam their habitat and will attack any suitable life form without much discrimination. They track down the prey, using a variety of chemoreceptors on the antennae, mouthparts, and legs to detect their victims. Because of their larger size, they are able to devour the prey on contact, either by grasping them and chewing them alive despite the captive's struggles or, for bloodsuckers, by inserting their proboscis, injecting saliva, which tends to subdue the prey, and sucking out the body fluids more or less at leisure. Random searching may occur in a proportionately small area, such as on a single leaf or branch, or it may range over part of a field, along a woodland path, on the bank of a stream, or in a gravel pit.

Some species of ground beetles (Carabidae) are random predators. Most of the 2,000 American species are carnivorous. Species of the genus *Calosoma*, commonly called caterpillar hunters, because they usually feed on caterpillars (or some on snails), include the European Caterpillar Hunter, *C. sycophanta*. This species is large, approximately 25 mm in length, with a brilliant golden green body with dark blue markings on the prothorax. The mandibles are sharp-hooked, in contrast to the broad blunt jaws of the many phytophagous members of this order. This beetle was imported from Europe with the hope that it would control populations of the gypsy moth, *Porthetria dispar*. Both the larvae and the adults have a fondness for the gypsy moth larvae. The adult beetles search by day, and their larvae, by night.

Other random feeders among the Coleoptera are the familiar ladybird beetles (Coccinellidae), or ladybugs, as they are often called. Larvae and adults are especially fond of aphids and scale insects. A familiar North American species is the Convergent Ladybird Beetle, *Hippodamia convergens*, found in woods, meadows, and gardens. These beetles overwinter as adults, as do most of the species of this family. They can be found under debris and leaves or, as described in chapter 10, they aggregate in large numbers in the fall and migrate to high elevations. These overwintering beetles are sometimes collected and sold to gardeners, even by mail, for release around the yard to prey on aphids.

A similar species, the Seven-spotted Ladybird Beetle, *Coccinella septempunctata* (fig. 13.5), was recently discovered along the New Jersey Turnpike in the Hackensack Meadowlands area of Bergen County, New Jersey. This insect is an excellent predator of aphids in Europe. Now that it is an established species in the United States, entomologists are collecting and rearing adults in an attempt to establish populations in the Midwest and Southwestern United States.

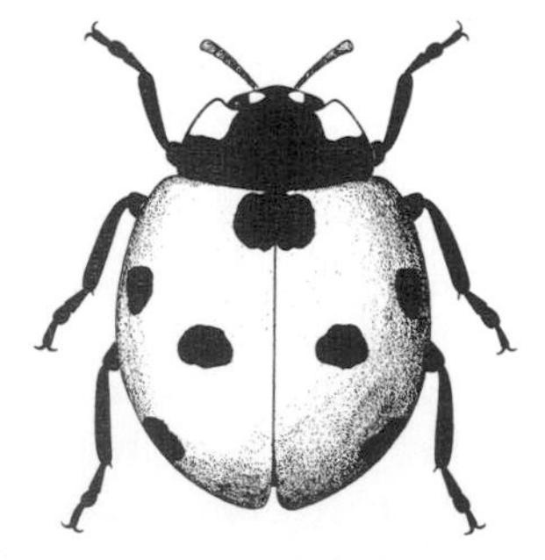

Figure 13.5.
Seven-spotted Ladybird Beetle (Coleoptera: Coccinellidae).

The beneficial nature of these predacious coccinelids has been known for centuries. During the Middle Ages, these beetles were used to rid grapevines of insect pests. In appreciation of their work, the beetles were dedicated to "Our Lady,"—hence, their common name. In the United States, however, they are frequently called "Ladybugs," probably because of the well-known nursery rhyme:

> "Lady-bug, lady-bug, fly away home,
> Your house is on fire, and your
> children will burn."

The common name, ladybird beetles, is more appropriate, since the term "bug" is applied to members of the insect order, Hemiptera, the true bugs.

Among the master hunters of insects, spiders, or indeed, most any terrestrial organism of the right size, are the many species of tiger beetles (Coleoptera: Cicindelidae). As we described in chapter 1, these insects (fig. 13.6) are brightly colored, with large eyes and swift legs. They are often seen in sunny areas along beaches, woodland glades, gravel pits, and desert sands. Their powerful jaws are capable of tearing apart and consuming their prey.

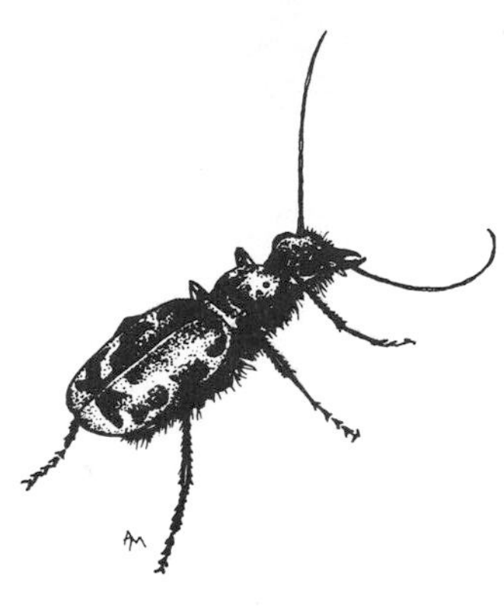

Figure 13.6. *Tiger beetle (Coleoptera: Cicindelidae).*

Still another random predator, also acknowledged for its biological control potential, are the lacewings (Chrysopidae) (fig. 13.7), members of the order Neuroptera. Lacewing larvae (fig. 13.8), known as aphislions, roam plants in search of aphids and other insects. The larvae have sickle-shaped jaws. These are formed by appressing together on each side of the head, the mandible and maxilla to create a pair of sharp, piercing-sucking tubes. With this apparatus, they are able to grasp a soft-bodied aphid, impale it on the two mandibles, and suck out the body contents of their victims. Lacewing adults are fragile-looking insects. Their two pairs of wings have many cross veins, as is characteristic of all Neuroptera. Note that the wings are held rooflike over the abdomen, an easy way to distinguish many members of this order from the dragonflies (Odonata), which they otherwise resemble.

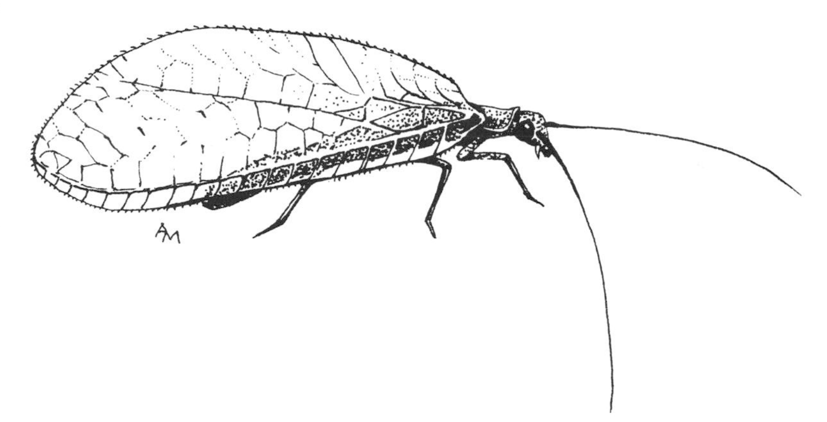

Figure 13.7. *Lacewing adult (Neuroptera: Chrysopidae).*

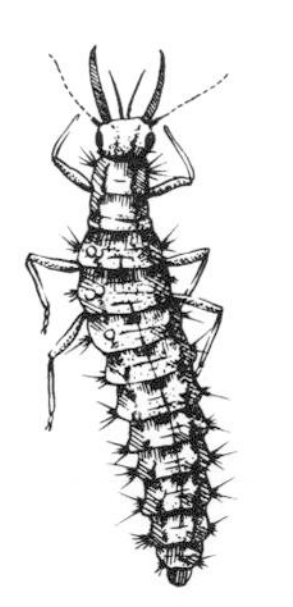

Figure 13.8. *Lacewing larva (Neuroptera: Chrysopidae).*

Hunting predators, to distinguish them from the random predators, usually search for their prey on the wing, hawklike. They have large, well-developed, compound eyes, which appear to give them sharper visual acuity. Some of the best-known hunters of the insect world are the dragonflies (Odonata) (fig.

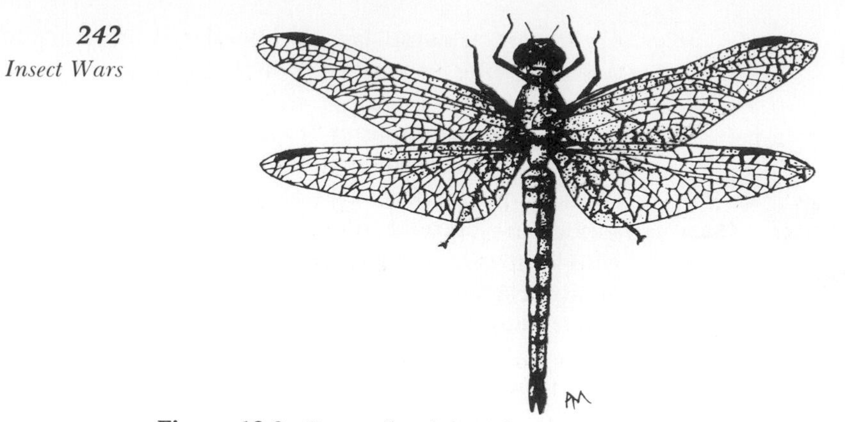

Figure 13.9. *Dragonfly adult (Odonata).*

13.9). These insects, so obvious around streams and ponds, have a variety of names. The small, fragile-looking species are called damselflies, while the larger ones, often feared by children, are variously called devil's darning needles, darners, snake doctors, and many other local names. The true darners, Aeschnidae, are among the largest (50–120 mm in length) and fastest of the flying insects. Their insect prey can be seen in all directions by their large, compound eyes that occupy most of the surface of the head. Insects are picked off in flight and carried in a jaillike basket, the bars of which are formed by the spiny legs. They may feed while still flying or alight on a cattail stem to enjoy their repast. Skimmers (Libellulidae) and other members of this order hunt in much the same manner. Dragonflies are often seen flying long distances away from water, even over parking lots, which they seem to confuse with the surfaces of ponds. They are not unknown flying over deserts, apparently in search of new bodies of water in which to lay their eggs. They have also been seen far out at sea. No one knows just how far they will fly.

Robber flies (Diptera: Asilidae) are well known for being flying hunters. These masters of the art of aerial attack pounce on their unsuspecting prey and suck out their body fluids with their short proboscis, even while on the wing. They are often seen perched on a fence, the side of a building, or a blade of grass with a feebly struggling deer fly, horse fly, grasshopper nymph, bug, or even a small butterfly or moth impaled on their merciless proboscis. Some species are known to attack even the well-protected bee. Robber flies inhabit open meadows, pastures, and gardens, and at least one species, *Promachus fitchii*, lives near beehives, where it feeds on honey bees. It is treated by the hive owners as a pest.

Perhaps the masters of flying predators are the spider wasps (Hymenoptera, Pompilidae). One member of this family is the Tarantula Hawk, *Pepsis (Dinopepsis) formosa*. These are velvety black with an orange tint to the wings. They frequent dry hillsides and plains in southwestern United States and Mexico, wherever tarantulas are found. We have even seen them enter outbuildings inhabited by these giant spiders. The female hunts down and grasps the swift-moving victim, then immobilizes it with a paralyzing sting implanted between the spider's legs (directly into the ventral nerve cord). After this, she will carry the spider to her previously constructed burrow, which she then opens, and then into a small chamber. There she lays a single egg on the still-living spider's body. She leaves the chamber, never to return. However, the egg hatches into a larva, which then feeds on the spider, kept fresh in its paralyzed state. Eventually, however, the vital organs are eaten at approximately the time the larva is ready to pupate. In due course, an adult is formed, which leaves the burrow to repeat the cycle.

Another spectacular wasp, *Sphecius speciosus*, the Cicada Killer, is found throughout most of North America, especially along the edges of woods, and even in backyards and city parks. After mating, the female sets about the task of digging a burrow. Soil or sand is loosened with the forelegs of the female wasp, and excavation is accomplished by passing the loosened soil backward with her second and third pair of legs. As the tunnel gets deeper, the wasp goes in head first, loosens some of the soil, turns around, and, with her head and forelegs, bulldozes the material out of the tunnel. Adults feed on nectar, though once the tunnel is constructed, the female turns from sipping nectar and becomes a huntress. She begins her attack on the large dogday cicadas, *Tibicen* spp. When she finds a singing male, the song turns from a mating call to a screech of terror as the two engage in a death struggle that the wasp, with her stinger, will surely win. Both wasp and cicada tumble to the ground. The fight is over in a matter of seconds as the venom from the wasp paralyzes the victim. The rest of the story is similar to that of the tarantula hawk.

The cicada is much larger, sometimes twice as large as the wasp. Once it is paralyzed, the wasp has the task of transporting it to the burrow. It turns the cicada on its back and moves astride the prey, clasping it with the midlegs. It then flies to the burrow, opens it, and drags the cicada into the internal chamber (fig. 13.10). There she deposits one egg on the body of the cicada, then leaves, closing the burrow behind her. Some burrows are up to a foot in length. After about three days the egg hatches, and the larva feeds on the paralyzed cicada.

Obviously, cicada killers and tarantula hawks are strong flyers. Their amazing ability to pick up and transport objects twice their size is comparable to a human picking up a small compact car and placing it in a parking space!

Figure 13.10. *Cicada killer (Hymenoptera) with cicada (Hemiptera). (Courtesy of* Insect World Digest)

The army ants are renowned predators. Backswimmers and waterstriders, both aquatic bugs, hydrophilid water beetles, most ground beetles, and the tiger beetles join with the ants at the secondary consumer level of the food chain. Collecting these predators is no simple task. Dragonflies, tiger beetles, and robber flies are among the most difficult insects to collect, because of their keen eyesight and swift flight. Dragonflies seem to play with the insect collector. Many an entomologist has fallen into a stream while trying to capture a prize specimen. Tiger beetles seem to torment the collector, flying just out of reach of the net each time a pass is made. One must be lightning swift with the net to capture any of these adept hunters.

Ambushing. By protective coloration and camouflage, some insects simply wait for their victims to pass by. The famous ambush predators, the mantids (Dictyoptera, Mantidae), are long slender insects reaching a length of 150 mm. Their heads are small, but they are among the few insects with a distinct neck, which permits them to rotate their head approximately 180°. Their compound eyes are proportionately large and widely separated on the head (which gives them the ability to accurately judge distances), and their raptorial forelegs, fitted with spines for grasping and holding prey, make them well adapted for their life of violence. One of the large species (60 to 80 mm in length) is the Chinese Mantid, *Tenodera aridifolia*, introduced from China in 1896 as a garden predator. These mantids overwinter in the egg stage, in cases attached to vegetation (fig. 13.11). These hatch in the spring, and the tiny nymphs immediately actively search for prey. They will eat each other if other food is not immediately

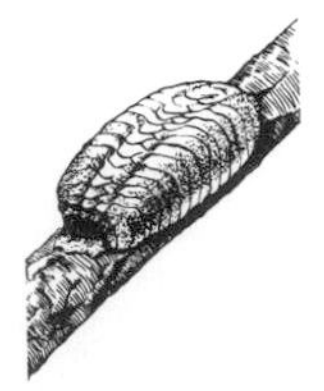

Figure 13.11. *Mantid egg case, ootheca*

available. Another species, the European Mantis (in Europe, it is simply called "the preying mantis"), *Mantis religosa* (fig. 13.12), was accidentally introduced into North America in 1899 from

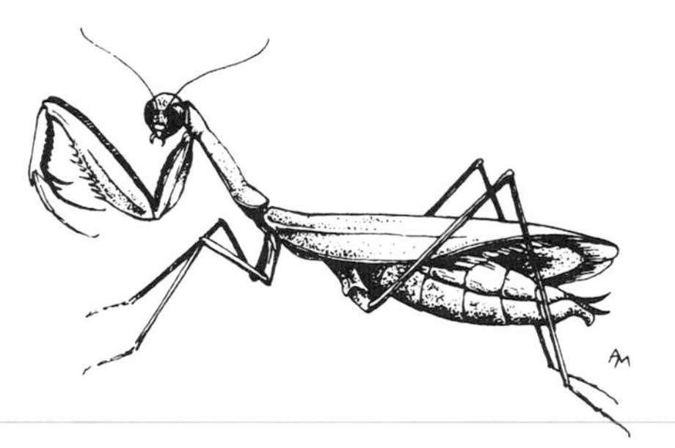

Figure 13.12. *Preying mantid (Dictyoptera).*

southern Europe. When discovered, it was hoped that this species would help control the Gypsy Moth caterpillars, but, unfortunately, mantids are rarely numerous, and, therefore, have little impact on the abundance of a pest population.

Figure 13.13. *Ambush bug (Hemiptera: Phymatidae).*

Ambush bugs (fig. 13.13) blend in with the flowers they inhabit. These hemipterans use both color and shape for their camouflage. Although relatively small (8–12 mm in length), they wait for a bee or other insects to visit the flower. The femora of the forelegs are enlarged, beset with spines, and can be used to grasp and hold prey in the same manner as those of the preying mantids. Their prey consists of bees, wasps, butterflies, and moths. Their victim is immobilized by the saliva injected through their proboscis. These species are common on garden flowers, as well as being an abundant predator in wild flowers. In general, one must consider these insects as beneficial, although it is not known whether they cause a significant toll among visiting Honey Bees.

One interesting ambush bug, recently reported from Central America, waits for termites to leave their nest in search of new places to live. The ambush bug grabs and kills them, then uses their dead bodies as camouflage. When more termites come out to inspect the dead bodies, they too soon become victims of this deceit.

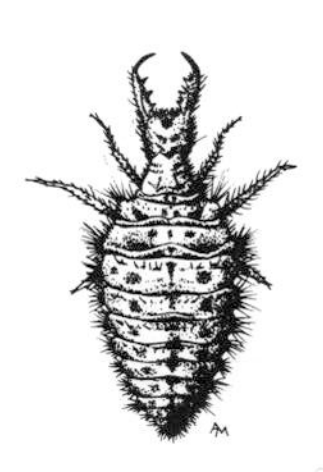

Figure 13.14. *Doodlebug, larva of an ant lion (Neuroptera).*

Trapping. A few insects construct traps for their prey. The antlions mentioned previously are well known for their sand traps. Also known as doodlebugs (fig. 13.14), these strange-looking larvae have large, sickle-shaped jaws similar to those of their cousins, the lacewing larvae. Adults resemble dragonflies, but differ by having long, clubbed antennae. Adults feed on pollen

Figure 13.15. *Aquatic net of caddis fly larva (Trichoptera) (redrawn from McCafferty, 1981).*

and nectar only. Compared to the voracious feeding habits of the larvae, the parents are docile, slow-flying insects. The "trap" set by the larva is constructed by digging a small conical pit in sand or loose soil. It then digs in at the bottom, where it lies in wait. A passing ant or other small insect walks into the trap and soon finds that it cannot escape, because of the loose sand, or, with the help of the doodlebug, more sand is loosened, keeping the insect within easy reach as it attacks and feeds.

Caddisfly larvae have been known and admired for centuries for the beautiful cases in which they enclose themselves. These aquatic larvae of the mothlike Trichoptera so resemble the bottom of a pond or stream that they pass unnoticed by fish and other predators. Besides the cases characteristic of most species, some, the net-spinning caddisflies (Hydropsychidae) are known to construct aquatic nets instead (fig. 13.15). Their nets are spread across a stone or between stones. They trap small aquatic organisms, such as crustaceans, small insect larvae, diatoms, and other algae.

Parasites. Many larvae of flies (Diptera) and wasps (Hymenoptera) are endoparasites of other insects. In these cases, eggs are laid on a single host, and they develop inside the host's body in a manner similar to those of the predatory wasps. Here, however, the host is not captured and placed in a nest. Instead, the female lays eggs on the host as it is feeding. Fly and wasp parasites are generally host specific. Many hundreds of species are involved. Only a few are mentioned here to give some idea about their various complex relationships. These are of continued interest to those concerned with the biological control of insect pests.

All species of the large hymenopterous family, Ichneumonidae, the ichenumonflies, and a closely related family, the Braconidae, the braconids, are internal parasites of the nymphs and larvae of many insects and spiders. The abdomens of these wasps are usually long and tapering. Many of them have long, needlelike ovipositors that can extend, in some species, the entire length of the body. This is used by the female to deposit eggs into the body of the host, sometime after it has been thrust into solid wood for an inch or more to reach a larva boring within. The antennae of ichneumonflies are long, threadlike, and constantly in motion. They are used by the female to locate prey. Adult wasps feed on nectar, if they feed at all. Although these wasps can sting, they seldom do so to humans.

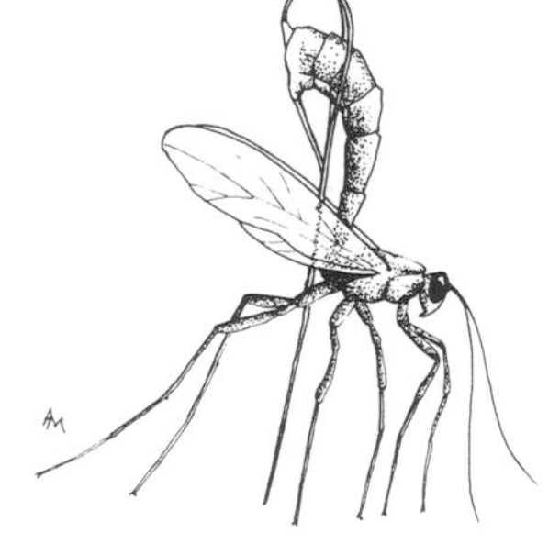

Figure 13.16. *Ichneumon fly adult (Hymenoptera) drilling into wood.*

One well-known ichneumonfly, a giant species (fig. 13.16) of the genus *Megarhyrsa*, has females that measure up to 110 mm in length, including the ovipositor. These insects occur throughout North America, almost anywhere there are trees. The gravid females search dead branches and fallen trees for the larvae of the Pigeon Tremex (or Pigeon Horntail), *Tremex columba*, and other borers. These are common in dead and dying wood. The female uses her long antennae to detect vibrations made by the tremex larvae in the tunnels. Once detected, the female curls her ovipositor up over her abdomen, curving it back down to enter the bark at right angles. The ovipositor has a sharp tip, which is able to cut deeper and deeper into the wood until the tunnel is located. The female deposits an egg on or near the tremex larva. When the ichneumon egg hatches, the larva attacks and feeds on the host larva.

Ichneumons with shorter ovipositors, such as those of species of the genus *Ophion*, parasitize free-living larvae. Females of this genus actively hunt caterpillars of tiger moths, giant silkworm moths, and various species of the noctuid moth family. Many of these species of moths are destructive crop pests, such as the well-known armyworms, cutworms, and corn earworms. Obviously, the ichneumonflies are beneficial by reducing the number of these pests.

If you wish to see these wasps in action, watch Tomato Hornworms, *Manduca quinquemaculata*, the larvae of a beautiful sphinx moth, on your garden crop of tomatoes, or, if you are in tobacco-growing areas, watch a similar species, *M. sexta*, which infests tobacco. Many of these worms will be parasitized by a small wasp, a species of *Apanteles*. The female wasp deposits her eggs on the body of the feeding larvae. These bore into the hornworms and feed on their tissues. Depending on the extent of the infection, the hornworm eventually dies, not living long enough to pupate. However, the larvae feeding inside come to the surface of the hornworm's body and spin a cocoon. These may be seen on the infected larvae. If you gather these larvae and place them in a jar, you will soon have a small collection of the adult parasites.

We mentioned previously that flies are also internal parasites of insects. They have a similar life cycle to those of the parasitic Hymenoptera. However, their females lack ovipositors capable of inserting eggs into the host. Instead, they lay their eggs on the surface of the larvae, but the result is the same.

Still another group mentioned above is the strange beetle family, Stylopidae (Coleoptera, sometimes treated as a separate order of insects, the Strepsiptera). These insects are, in one sense, the only truly endoparasitic insect species (except for a few species of fleas, the chigoe fleas), because the adult females live their entire lives within the host. Over one hundred species of Stylopidae occur in Canada and the United States. These are closely related to the beetle families Mordellidae, Meloidae, and Rhiphoridae, which also have parasitic larvae. The female stylopid is wingless, eyeless, and legless. She lives her entire life inside various species of Orthoptera, Hemiptera, Homoptera, Hymenoptera, and Thysanura, depending on her species. Males resemble somewhat tiny wasps about 5 mm in length. The male pupates within the host and emerges actively seeking a female. The latter apparently produces a sex pheromone, which attracts the males, because infected hosts confined in cages attract the males (an easy way to catch these seemingly rare insects). The mature female protrudes her cephalothorax from the host's body, usually between two abdominal sclera. The male then deposits sperm into the aperture of the brood canal of the female. The fertilized eggs mature within the female. First stage larvae, called triungulins, are dropped to the ground or on flowers, where they await a passing host. Some attach themselves to a new host, while others, such as *Stylops pacifica*, are taken into the gut of a bee when it takes nectar from a flower. When the bee returns to the host and prepares a ball of pollen mixed with nectar, the triungulins remain within this preparation. Now the larvae are deposited in the bee's cell with the bee's egg. When the bee egg hatches, the triungulins penetrate the body of the developing larva and feed on the nonvital tissues. This apparently does little harm to the bee larva, because it pupates, emerges, and takes its place in the hive with the parasite within. It is known, however, that the bee in this bizarre case of endoparasitism may have reduced vitality.

Pathogens. Besides predators and parasites used to control populations of pest species, several pathogens attack insects, some of which are now being used commercially for pest control. These pathogens are various species of viruses, bacteria, protozoa, and fungi. A few of these are described here.

Pathogenic bacteria are acquired by insects, usually by ingestion, in much the same way as they infect humans. Body contact probably plays a role in transmission (as far as we know, insects do not sneeze or kiss!). Important bacterial diseases include the milky disease of the Japanese Beetle, caused by *Bacillus popillae*. Spores of this bacterium are available commercially, and, with this preparation, soil can be inoculated. The grub (larva) of the beetle then ingests these spores while they are feeding on the roots of plants. This infection is lethal. Commercial preparations of *Bacillus thurigiensis* are also available for the control of various lepidopterous larvae. This bacterium produces a toxin known to kill at least 180 species of these caterpillars. It, too, is ingested by the larvae as they feed on vegetation. This material has been sprayed on trees in Northeastern United States in an effort to control populations of the gypsy moth.

Some bacterial diseases of insects infect beneficial species as well. A deadly disease of the honey bee, known as American Foulbrood, is caused by *Bacillus larvae* and may wipe out entire hives.

Pathogenic protozoa, including various species of microsporidians, protozoa similar to those that cause malaria, take a heavy toll of species of Lepidoptera and Diptera. These include another disease of honey bees caused by the protozoan *Nosema apis*.

Lepidoptera and sawfly (Hymenoptera) larvae seem to be most susceptible to insect viruses. The so-called polyhedrosis viruses, one of many types of viruses, are the main ones pathogenic to insects. They produce polyhedral crystals lethal to the insect.

Fungi of various species attack insects by penetrating the integument or by spores ingested into the intestine. Some fungal diseases are known to control such pestiferous insects as the Chinch Bug (Hemiptera). This insect has been controlled by the fungus *Beauveria globulifera*.

Many fungal diseases have the potential to control insect populations, but the main problems that exist with all such pathogens are their effects on nontarget species and, sometimes, contamination of the environment. Preparations of bacteria, protozoa, fungi, or viruses have the potential of attacking other organisms, including beneficial insects. This adverse effect could have serious consequences. Until a pathogen is proven to be host specific and, thereby, safe when used for control, the United States Environmental Protection Agency (EPA) will not license any commercial pathogenic preparation for use. However, research continues in an attempt to find species useful for pest control purposes.

Biological control procedures have been extremely successful. Some of these are shown in table 13.2. You will note that biological control projects have been employed for about one hundred years; hence, it is not something new. Research continues toward evening the odds in the relentless quest for the "perfect" control.

Control of pests on plants is difficult and costly. The never-ending battle is caused by the "slippery" nature of the living system. Entomologists can never predict with certainty what will happen next. Every control method will eventually fail,

Table 13.2. *Some successful biological control projects using insects as control agents.*

Pest	*Primary Control Agent and Source*	*Location and Date*
Icerya purchasi (Cottony-cushion Scale)	*Vedalia cardinalis* (Vedalia Beetle) (Australia)	California (1888–1889)
Perkinsiella saccharicida (Sugarcane Leafhopper)	*Paranagrus optabilis* (mymarid) (Australia) *Cytorhinus mundulus* (mirid) (Australia)	Hawaii (1904–1920)
Levuana irridescens (Coconut Moth)	*Ptychomyia remota* (tachinid) (Malaysia)	Fiji (1925)
Pseudococcus citriculus (Citriculus Mealybug)	*Clausenia purpurea* (encyrtid) (Japan)	Israel (1939–1940)
Dacus dorsalis (Oriental Fruit Fly)	*Opius oophilus* (braconid) (Philipines, Malaysia)	Hawaii (1947–1951)

Lepidosaphes beckii (Purple Scale)	*Aphytis lepidosaphes* (eulophid) (China)	California (1948 on) Texas, Mexico, Greece, Brazil, Peru (1952–1968)
Chrysomphalus aonidum (Florida Redscale)	*Aphytis holoxanthus* (eulophid) (Hong Kong)	Israel, Lebanon, Florida, Mexico, Peru (1956 on)
Chrysomphalus dictyospermi *(Dictyospermum Scale)*	*Aphytis melinus* (eulophid) (California)	Greece (1962)
Opuntia spp. (Prickly Pear Cactus)	*Cactoblastis cactorum* (moth) (Argentina)	Australia (1920–1925)
Hypericum perforatum (Klamath Weed)	*Chrysolina quadrigemina* (chrysomelid beetle) (Australia)	California (1944–1946), other western states

because of the changes that take place within the species itself. Resistance builds up, and new approaches are required. As much as the environmentalists may wish it, chemical control will never disappear entirely, not unless the consumer is willing to eat a wormy apple; not until the human race is willing to let a certain number of us die with malaria, yellow fever, or sleeping sickness. Meanwhile, we should encourage *all* efforts toward a sane approach to insect control.

14

Earning Their Keep

"The bee is more honored than other animals,
Not because she labors,
But because she labors for others."

St. John Chrysostom (ca. 345–407)

We usually think of insects as worthless. That is why we have been asking you to realize that insects are rarely pests. Years ago, one of us was attracted to the study of butterflies (he was in 8th grade at that time), because of an entomologist that advertised in a popular magazine, offering a list of butterflies of North America with brief descriptions and some poor black and white illustrations. That wasn't too unusual, but what was included with the description of each species was the supposed market value of each one. Descriptions of how to mount specimens, along with notations about where these species could be found, were included. Unfortunately, this book was long ago misplaced, borrowed, or wandered off, as books sometimes have a habit of doing. This *Scott's Catalogue* of the butterflies was exciting, both because of the interesting study it described and the chance it seemed to offer for selling specimens and making money. One species was listed as worth $7.50 a pair (a large sum of money in those bygone days). Unfortunately, the values placed on these specimens was more fictitious than real, especially since there was no market for these specimens. But it was thrilling to think that insects had a value, and certainly this made the cost of the book (exceedingly expensive then) worth it. Visions of an income from the sale of these specimens increased his interest in the new hobby, and friends were eagerly enlisted to aid in this new business-hobby. After the initial disappointment over lack of sales, the joy of collecting never left.

Sale of specimens

What would you think if you had a pair of rare butterflies that would bring $40,000 at an auction in Paris? This would certainly excite you enough to investigate further the value of insects. Each year such auctions take place and, on at least one occasion, a pair of rare butterflies was sold at that high price. It is not unusual for specimens to be sold at four-figure prices.

A German entomological journal, "Entomologische Zeitschrift," is issued every two weeks. In the center section are eight pages of advertisements (the section is called "Insectenboerse," meaning "insect market or exchange") offering insects, dead and alive, for sale. Many persons in Europe raise large, showy moths. This way, they get perfect specimens. Eggs and pupae are offered for sale in this publication, along with an array of large beetles and moths for pinning. The live insects may be reared on various food plants to provide a continuous stock of these insects, always useful for exchange for other specimens. Some of the large tropical beetles range in price from $25–175 a pair. Certain moths and butterflies are advertised at even higher prices.

In Costa Rica and other parts of Central and South America, there are butterfly farms—ranches with large screened-in areas in which butterflies and moths of various species are reared. The larvae of these colorful insects feed on the leaves of small trees that grow within the screened areas. When the adults emerge, they are collected, papered, and labelled. This is the stock of specimens that are offered to dealers in the United States, Europe, and Japan. The dealers buy in large numbers and then sell to collectors, just as postage stamps are bought and sold.

In Tokyo, Japan, separate specialty stores sell insect specimens and collecting equipment. Even a department store in Osaka has a section devoted to entomology. Hobbyists visit this section to buy specimens, live insects, and equipment, much the same as we buy stamps or camera equipment in this country.

Obviously, once the collecting bug bites, collectors are eager to add specimens from out-of-the-way places, much the same as stamp collectors want rare stamps. These amateur entomologists specialize in certain groups of insects, as stamp collectors specialize in certain countries. Some like butterflies; others, large moths; still others, large beetles; and so on. Of course, this kind of collecting is not what this book has discussed—it really isn't scientific, but that does not mean that it shouldn't be done. We hope that your interest will be somewhat deeper than just acquiring a collection.

As might be expected, when a group of people develops similar interests, they form a society. In Japan, numerous local entomological societies have become exceedingly large. It is told that some amateur groups hold weekend collecting trips and charter several busses to transport members from the city to the nearby countryside to collect and observe insects. Besides a large membership in these societies, there is also a wide range of books available on the insects of Japan, most of these in full color, rivalling and surpassing the bird books published in North America.

Trading specimens

The purchase of specimens may soon lead to contact with other collectors and the exchange (trading) of specimens. In times past, entomological societies were organized primarily as a way for collectors to get together to show and exchange specimens. This is seldom done today, but, perhaps, the current entomologists take their work too seriously. Exchanging specimens might make the work of insect study more enjoyable. No doubt this would lead to wider interest in insect study, which certainly would benefit the field as a whole. (Appendix I gives a list of entomological societies of interest to amateurs and some periodicals listing the names and addresses of collectors.)

Collecting for hire

Expeditions are organized by some entomologists to go to remote places for field research. Some of these trips are sponsored by museums and universities; others may be financed by research grants from the National Science Foundation; and still others, by the National Geographic Society. Rarely, collectors will go at their own expense. Some of these collectors are the dealers referred to above. It is possible to commission these individuals to collect certain groups. These arrangements are made in advance, with the understanding that a fixed sum will be paid for the specimens collected. Seldom is money raised in this manner to pay for the trip, but it may help. Gone are the days of Lord Walter Rothschild of England, who sent out hundreds of collectors to all parts of the world to capture moths, butterflies, birds, and large turtles. Let's hope that once again this will become possible before wild areas are gone and, with it, the range of species that will never be seen again. As Rothschild claimed, it is beneficial to science to collect large numbers of specimens while it is still possible. Only in this manner can the range of variation of the species be studied. (See also comments on conservation in chap. 17.)

Insects and art

Mounted specimens of butterflies and moths are for sale in boutiques, souvenir stores, and museum shops. These are primarily used as wall decorations. Until recently, jewelry made from the bright, iridescent blue wings of the giant morpho butterflies was sold in these shops. The wings were cut and pieced together into designs. Other butterfly wings were used to create pictures. So popular were these art objects that a lively trade sprang up just for the collection of morpho butterflies. Fortunately, this trade has been stopped, because of the decline of the butterflies. Some countries no longer permit the collection and shipment of these insects.

Ancient Egyptian artifacts (fig. 14.1) depict the sacred scarab beetles. All manner of facsimiles of these relics are manufactured and offered for sale. This has branched out to include other kinds of insects. We have seen bootjacks cast or molded into the form of a winged termite. You insert your foot between the antennae to pull off the boot. Other insects are molded into paper clips. The Japanese offer wind-up beetles, remarkably lifelike, that walk across the floor or table top and even come in several forms resembling different actual species. Wooden cut-out parts can be put together as mosquitoes, beetles, and so on. The list of these so-called art objects is long. It is paradoxical that these items are purchased by so many persons, who otherwise maintain that they "hate" insects.

Stamp collectors are known to specialize in stamps picturing insects. The number of these issues grows each year. One notable example of this is a new species of moth illustrated on a Cuban stamp—the stamp was issued before the actual description of the species. One might cite this as the original description of the species, except that such descriptions are disallowed by

Figure 14.1. *Carving of an Egyptian sacred scarab beetle, a widely used insect artifact.*

the International Rules of Zoological Nomenclature (see chap. 3).

Actual value of insect specimens

The real value of any material thing is entirely dependent on what is actually paid for it. Unless it can be sold, it has no value, no matter how much the price tag says it costs. Since museums and universities are interested in having collections, and they hire curators to gather and manage these collections, one way to determine the value of insect specimens is to compute how much it costs to collect, prepare, and store insect specimens. Our interest in these costs is twofold. First, if a collection is damaged or destroyed by fire, water, or earthquake, and the collection is covered by insurance, how is the value determined? Second, if a collection is donated to an institution, how much can be claimed as a tax deduction? This has been determined, and these figures have been paid by insurance companies. The U.S. Internal Revenue Service has allowed these amounts as tax deductions.

Because of a continually changing economy, there is little point in listing actual figures here. The aspects to be considered in the evaluation of a collection for either insurance or tax purposes should include the following items:

1. Count the number of specimens prepared as pinned and labelled specimens, with locality and ecological data, identification labels, and arranged in the collection. It is important to distinguish between specimens collected locally, at proportionately lower expense, and those collected through special expeditions, either on this continent or abroad. This, of course, will be important in computing the replacement costs. Specimens simply pinned, specimens glued on points, those spread with their wings set in position for display, specimens mounted on microscope slides, and those stored in alcohol (see chap. 16) have different values. Once specimens are counted in this manner, it will be possible to assign different per-specimen figures for each type.

2. Determine the cost of a collecting trip. This would include the travel expenses, food, lodging, and salaries. It is not difficult to determine the number of specimens that can be collected on such a trip. Then add the cost of materials, the time taken to mount and label the specimens, and the time to do this, that, and the other operation. The cost of identification, arranging the collection, storage, and the equipment used must also be determined. All costs are added to the per specimen price. Obviously, specimens are worth much more than a few cents each, as one might first suppose.

To appraise a collection, one must have had experience in collecting, curating, and identifying insects. Before a tax deduction can be taken for the donation of specimens, a qualified appraiser must be located. For this, a letter listing the details of the collection and the value of each group of specimens is necessary. This document should include an appraisal of the boxes, drawers, and cabinets in which the collection is stored. (Further details may be obtained from museums or the authors.)

Old collections are sometimes extremely valuable. Sotheby's in London auctioned off (in 1983) a handsome walnut-faced Gurney cabinet containing fifteen drawers of British butterfly rarities and aberrations, including three species now extinct. A private collector purchased the entire collection for about $15,000. At the same auction, approximately 100,000 butterflies were offered for sale, with a single drawer of races of the swallowtail butterfly, *Papilio dardanus*, selling for about $800. Entomological publications, often offered for sale with collections, are also valuable—a single volume often bringing several thousand dollars.

The Florida State Collection of Arthropods, Gainesville, Florida, has an unofficial membership of professional and amateurs interested in the welfare of this large collection. These persons are given the title "Research Associates." Many donate specimens to the collection. In return for their interest and help, the head curator prepares a letter and document evaluating each donation; this is useful for tax purposes. This system has resulted in the addition of many fine specimens and, sometimes, entire collections to this rapidly expanding entomological museum.

Similar donations are made to other collections. The large August Schmitt collection of butterflies was donated to the Cornell University collection, Ithaca, New York, the oldest and finest university collection in the United States. Much of the Schmitt collection was originally purchased from professional collectors.

Insects as pets

Keeping caged insects (fig. 14.2) is rarely practiced in the United States, but in Japan and elsewhere, insects as pets often replace the more traditional dog or cat. Some claim that this happens because their houses are so small that there is no room for mammal pets, but this really is not so. These people simply have a broader appreciation for living things and enjoy a wider variety of pets. One department store instituted an insect park on its roof, planting grass and small trees to harbor its stock of living insects. The markets in Tokyo offer living crickets and bamboo cages for these singing pets. Special sticky material is sold to smear on long bamboo poles used to catch cicadas to cage during their brief adult life. The Japanese claim that caring for these pets is just as easy as raising tropical fish, and much cheaper.

The practice of keeping insects is widespread in the Far East, as well as in the Mediterranean Region, where crickets and katydids have, for millennia, been esteemed for their song, according to Professor D. Keith McE. Kevan of Canada, who has made a special study of this subject. In old China, cricket-fighting was practiced. Crickets singing at night also serve as "watchdogs" in reverse; that is, when they suddenly stop singing, this is a warning that an intruder is near.

Figure 14.2. *Eighteenth-century grasshopper houses in which to cage these pets. (Courtesy of* Insect World Digest*)*

Caged grasshoppers were once sold in the busy port of Hamburg, Germany. The cages, made as cardboard cutouts shaped like miniature doll-houses, could be purchased. These were assembled (fig 14.3) with transparent windows and one side that would open to permit the insertion of the insects, as well as their fresh food. These could be purchased flat, cut out, and pasted together following the guidelines printed on the sheet of cardboard.

Figure 14.3. *Folding Hamburg "kopperhouse" to cut out and glue together as a grasshopper cage. (Courtesy of* Insect World Digest*)*

Insects as test animals

The insecticide industry uses living insects as test animals during their development of a new insecticide. Many tests must be made to establish the killing effect of the new material. The lethal dose of the chemical is established, by application in various amounts, until the LD-50 is established. This term indicates the dosage necessary to kill 50 percent of the insect population being tested.

Once this is established, the proper application rate of the insecticide can be calculated.

As indicated previously, all insecticides are closely regulated by the U.S. Environmental Protection Agency. Before a product can be marketed, the results of these tests must be submitted to the EPA, along with the exact printed label giving contents and directions for use to go on the package. Every detail must be approved before any of the insecticide is sold. The experimentation leading to the final marketing of these chemicals generally runs into the millions of dollars. After this comes the additional expense for marketing.

Insects as test animals are not limited to insecticide experimentation. Fireflies, for example, furnish valuable enzymes used in the assay of certain medical products. Mosquitoes are reared by the millions and infected with bird malaria. Then various drugs are tested as possible cures or preventatives for that worldwide disease, still a constant threat throughout the tropics. Reared mosquitoes have been used to test protective clothing for the military, as well as in the study of other mosquito-borne diseases, such as dengue and equine and human encephalitis. Of course, mosquitoes are not the only test insects. Many insect pests may be more easily reared than "wild" species. Some are used in industrial testing of many kinds and in scientific research at university and governmental laboratories.

Insects provide medicine

The famous "Spanish fly" has been the subject of much off-color discussion and laughter, because of its alleged aphrodisiac properties. True, this insect, a European beetle (Meloidae), has been used as a medicine. All members of the blister beetle family, Meloidae, as well as the false blister beetles, Oedemeridae, contain a protective chemical, cantharidin. This extremely poisonous substance is an irritant of the mucus linings, and, in fact, all body tissues. Cantharidin is secreted as a protective measure when the beetle is disturbed by flexing its legs. Drops of the material touching human skin cause blisters. These last for several days to several weeks and are exceedingly painful. In times past, the body contents of these beetles were dried and ground into a powder. This was administered in minute quantities to stimulate urination as a treatment of bladder diseases. It, of course, irritates the urinary tract in general, as well as most other body systems. The legend about its use as an aphrodisiac stemmed from oral accounts of how it was added surreptitiously to a lady's drink to make her romantic. Any such attempts would more likely have seriously injured her, if not actually caused her death. This drug is seldom used today.

In this connection, headline news recently reported the mysterious deaths of several extremely valuable race horses in Florida. After autopsies and an investigation of the food eaten by the animals, it was determined that the horses were killed from eating a few blister beetles. The beetles were trapped in the hay as it was harvested. New methods of preparing hay allow it to be cut and bundled while it is still wet with morning dew, a practice never possible in the past, because of the chance of spontaneous combustion. Wet hay will ferment. If stored in barns before it is entirely dry, it will suddenly burst into flames, once the cause of many barn fires. When the hay dries out in the field, the beetles leave. Therefore, harvesting dry hay presents little danger. Past beetle poisoning apparently went undetected. Grazing animals are, of course, in danger of eating these beetles. To prevent this poisoning, it is necessary to be certain that the hay is dry and without beetles. A veterinarian attending the race horses stated that the beetles had caused the most violent death he had ever seen.

These same beetles also cause blistering of humans. At times, the much more common beetles around lights, species of Oedemeridae, have become pests in Florida resort areas. Attracted to lights around tennis courts (they are particularly attracted to the lights from bug zappers) or open air patios, they will alight on the skin. The natural tendency is to swat it, which drips cantharidin-laden body fluids on the skin, causing blistering.

Insects as food

Honey is eaten worldwide and has long been a substantial source of sweets for humans. Details of beekeeping (fig. 14.4) is the subject of many books and is treated in dozens of scientific and trade journals available to beekeepers. Beekeeping is a small, but significant, part of American agriculture.

Bees are not the only insects producing food for man. Manna is frequently referred to in the Bible. This is a sweet material excreted by scale insects. These are pests, infesting primarily desert plants. It is an excellent energy source whenever needed. In certain parts of the American deserts, an ant, the honey ant, is eaten by Indians and others. Certain workers of these ants engorge their bodies with ant honey obtained from nectar of plants. They remain in the nest as living storage jars until the honey is needed as food. Then they regurgitate the honey to be eaten by the other ants. The storage ants hang on the ceiling of the nest dug in soil. Indians track down the ants, open the nest, and eat the ants as one would candy.

Figure 14.4. *A honey bee hive.*

Insect bodies are full of protein. They have much less fat than beef; as much as 68 percent of their body is protein, compared to the average hamburg of about 15 percent protein. Protein is a scarce item in the diet of more than 50 percent of the world's human population. It is little wonder, then, that many people eat insects.

The larvae of the maguay worm, a moth that infests the maguay plant in Mexico and elsewhere, is deep fried and eaten. These are called "fritos" in Spanish. They are eagerly sought after by children, as well as adults. Anyone with an open mind will have to admit that they are good. Similar larvae, especially large, wood-boring beetle larvae, are eaten by certain natives in Australia. They have a pleasing nutlike flavor. The well-known explorer, Jacques Cousteau, shows Amazonian natives eating alive the larvae of the dobsonfly, called hellgrammites in this country. We use them for fish bait, not realizing, of course, that the fish we catch with them grew up on these "nasty worms." Most persons will agree that seeing the wiggling abdomen of the larva protruding from another's mouth as he/she crunches the body with the teeth is hardly acceptable in our society. However, carefully prepared hellgrammite larvae could easily become a French delicacy, if served under the proper conditions.

Several books have been written about insects as human food. These point out the kinds of insects that provide a substantial amount of the needed protein in the human diet. Some of these books give recipes for the preparation and cooking of insects. Termites, for example, may be cultured in old newspapers. This practice not only recycles this waste material, but provides the ingredients for a tasty termite and rice souffle.

The eating of grasshoppers in Africa, also reported in the Bible, is so widespread that various etiquette practices have developed among various native tribes. For example, some of the more refined pull off the wings and legs before the body is eaten. These are thrown on the floor of the hut—a telltale sign of the presence of this tribe. Others eat the wings and legs, but it is reported that these are indigestible.

Not all insects should be eaten. The clue must be taken from what the insect itself eats. For example, corn earworms, the larvae of a moth pest of corn, are tasty, especially when fried with bacon. But do not eat the larvae of the monarch butterfly. These larvae feed on the poisonous milkweed, and they should be avoided. And, as tasty as a tomato worm may look, they too contain poisons. Lists of edible and nonedible species may be found in the books referred to above. Once it is established what can be eaten in the way of insect pests, we might well consider this: if you can't beat them, eat them!

Insects providing industrial products

Silk production is an ancient art and the source of fine fabrics. The larvae of the silkworm moth spins a cocoon in which the larva transforms into a pupa, and then into the adult. The pupa can be removed from the cocoon unharmed. The silk strands forming the cocoon may then be used to spin the thread woven into cloth. The cocoons are soaked in hot water to loosen the strands, which then may be unwound and spun, used much the same as the woolly hair of sheep or the cotton bolls. Silk has been produced in Japan for millenia. Trade in this commodity has had a prominent part in the development of international affairs. For many centuries, the art of silk-making was kept a secret in Japan, with the resulting monopoly placing Japan in a beneficial position. Silk was partly responsible for much of the early exploration in an attempt to find an easy route by sea to Japan and the Orient, one reason for Columbus's discovery of the New World.

The gypsy moth, already described as a severe forest and shade tree pest, arrived in Massachusetts from Europe, because it was thought that it could become a source of American silk. Alas, as we have seen, it escaped and is now a pest, instead of a beneficial insect.

Today, the silkworm moth lives only in captivity; all wild populations long ago became extinct. It can be reared by feeding the larvae on mulberry leaves, the long-held Japanese secret now known. Once revealed, the Oriental industry declined. Today little real silk is available, but plastic fabrics are taking its place. It is agreed that none of these is equal to the quality of real silk.

Natural dyes are available from various plants, but the beauty of the red obtained from the bodies of the cochineal insect, a scale insect related to many plant pests, has never been surpassed by plants. Cochineal is only available from these insects, but, of course, with the development of the coal tar industry, the trade of providing cochineal has also declined and now is used only by native artists.

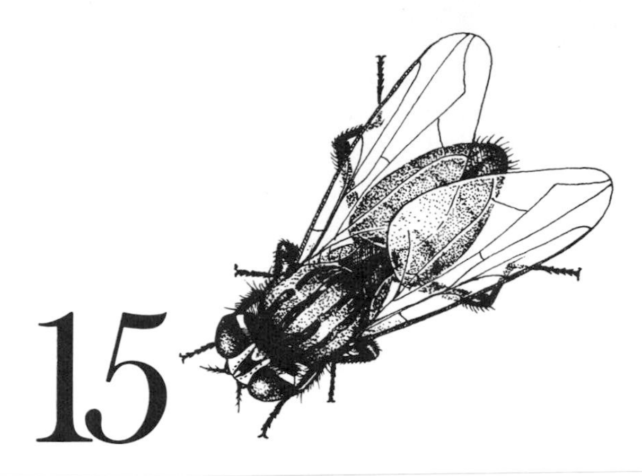

15 How to Collect Insects

"During our stay at Brazil I made a large collection of insects. . . . I may mention, as a common instance of one day's (June 23rd) collecting, when I was not attending particularly to the Coleoptera, that I caught sixty-eight species of that order."

Charles Darwin, *Journal*

Collecting, almost for the sake of accumulating "things," seems to be a part of every human, not only when growing up, but as a spare time activity of adulthood. Electric trains, tropical fish, beer cans, stamps, baseball cards, comic books, dolls, or whatever—all become collectibles and a part of someone's hobby. Some hobbies are expensive; stamps, coins, and electric trains can cost "big bucks." Bottle tops, matchboxes, and post cards are less expensive to collect and are interesting items in some collections, but really do not offer much of a challenge. Insect collecting, a hobby for thousands of persons the world over and a profession for many, is both mentally and physically stimulating. Amateurs enjoy adding more and more species to their collections, some of which can be sold, just as are stamps (see chap. 14). Insect collecting generally costs little, but as the collection grows, it becomes more valuable. How much it costs depends on how much you want to spend and how deeply involved you become. One thing is certain—insect collecting is a most enjoyable and exciting part-time hobby, because it enables you to get outside and collect during all seasons of the year and to spend long hours on cold winter nights identifying and arranging your collection for study and display.

Getting started

Certain tools are necessary to get started. To get into the swing of things, assemble the basics before the hunt begins. Table 15.1 lists the essential equipment. The following text discusses each item to help you choose the equipment most suited to your needs. Biological supply houses are listed in table 15.2. Not one carries a complete line, which makes it necessary for you to contact several. Some items can be constructed at home, and many common household items may be used, which helps to cut

Table 15.1. *Equipment needed for insect collecting.*

Equipment
Basic Equipment Needed to Get Started
Killing jars
Killing agent
Storage jars and boxes
Additional Equipment Useful for Intensive Collecting
Forceps
Hand lens
Pill boxes or cigar boxes for field storage of specimens
Cleansing tissue or cotton pads to layer specimens for storage
Note pad and pen
Aerial insect net
Sweep net (for collecting on vegetation)
Beating sheet (for collecting from shrubs and trees)
Aquatic net (for collecting in water)
Light traps and UV black lights (for night collecting)
Aspirator (for collecting small insects in confined places)
Camel's hair brush (for picking up tiny insects)
Pitfall traps (for collecting terrestrial insects)
Berlese funnel (for sorting insects found in leaf litter and debris)
Bait: sugar bait, carrion, dung (for insects attracted to these items)
Chopping tools (for removal of insects from rotting logs, branches, or roots)
Knife (for splitting stems and other plant material)

Table 15.2. *Biological supply houses.*

Write to the following for catalogs:

American Biological Supply Co. (AMBI), 1330 Dillon Heights Avenue, Baltimore, MD 21228

BioQuip Products, Inc., P.O. Box 61, Santa Monica, CA 90406

Carolina Biological Supply Co., Burlington, NC 27216

Powell Laboratories Division (of Carolina Biological Supply Co.),Gladstone, OR 97027 (for West Coast orders)

Turtox, Cambosco, MacMillan Science Co., Inc., 8200 South Hoyne Ave., Chicago, IL 60620 (They charge a fee for their catalog.)

Ward's Natural Science Establishment, Inc., P.O. Box 1712, Rochester, NY 14603, or P.O. Box 1749, Monterey, CA 93942 (for West Coast orders)

down costs. You can start with a jar, add a killing agent, and off you go. But by adding an insect net, you will greatly increase your chance, and a net makes collecting more interesting, because it will test your ability to outwit the insects. All equipment for capturing, killing and storing specimens in the field, except for the net, may be conveniently carried in a small canvas musette bag sold in Army surplus stores. The popular canvas back sack carried by bike riders is also well suited for carrying equipment. Even a folding net may be purchased from some supply houses, and it too can go into the collecting bag.

Killing jars

Glass killing jars of convenient sizes containing one of several poisons can be constructed at home, or they may be purchased from a biological supply house. Three commonly used killing agents are potassium cyanide, the best; ethyl acetate, good; and carbon tetrachloride, used only as a last resort. Only ethyl acetate may be obtained outside the university. Alcohol, both as a killing and as a storage agent, is used for certain kinds of insects. These are all poisons and must be used with great care and, by all means, keep away from children and pets. Cyanide is the most difficult to obtain—unless you are a university student or a museum associate. It generates an extremely poisonous gas, which quickly kills insects—and humans, as well. Ethyl acetate may be purchased from a biological supply house. Carbon tetrachloride (as Carbona) can no longer be purchased in drug and grocery stores. Ethyl alcohol is expensive, unless obtained through a research institution, but isopropyl alcohol works just as well, perhaps even better, and may be obtained as rubbing alcohol. At least one institution, the Florida State Collection of Arthropods, uses isopropyl alcohol for storage in such quantities that each laboratory has this material piped in.

Select jars of various sizes for various kinds of insects. At least one should be large (pint to quart size), specifically for Lepidoptera (butterflies and moths). Lepidoptera tend to lose scales from their wings, which become attached to small insects and obscure parts. Therefore, use Lepidoptera jars for Lepidoptera and other jars for other insects. Select jars with threaded lids; avoid jar tops with crimped edges, because these lids can easily work loose and fall off, releasing killing vapors and dumping out the insect catch. Beetles, bugs, and grasshoppers can be collected directly in alcohol in a separate vial or small jar. Do not place insects with a dense pile (for example, bumble bees) in alcohol. Alcohol vials are the easiest killing jars to use, and specimens may be stored in them indefinitely. When used for storage, add locality data to each one. Be sure the ink used is alcohol resistant. Do not use penciled labels, as they rapidly fade in alcohol. These vials should be closed with a neoprene stopper, not a cork, since cork allows the alcohol to evaporate within a short time, while the neoprene stopper seals the vial indefinitely.

Make killing jars in a well-ventilated room or outdoors, but away from a breeze. When using potassium cyanide, place about an inch on the bottom of the jar, press it down tightly with the end of a piece of doweling, and add about an inch of sawdust, also tapped down. Sawdust will absorb moisture from the air and from insect bodies. This will control the rate of gas generation in the jar. Place 1/2–3/4-inch layer of wet plaster of Paris on top of the sawdust to hold the sawdust and cyanide in place (fig. 15.1). Mix the plaster in an old dish, using just enough water to make it thick, but so that it will pour off the end of a spoon. If it is too wet, the cyanide gas will generate too rapidly, and if too dry, the surface will be rough and unsatisfactory. Tap the jar lightly after adding the plaster. This will eliminate any air bubbles under the plaster and smooth out the top surface. Let the plaster dry, away from direct sunlight, for several hours, before the jars are covered. (Note: always use fresh plaster of Paris. Old plaster may not set properly, and the jars are useless, because the plaster breaks.) When moisture from the plaster reaches cyanide, deadly cyanide gas begins to generate, but to work properly, the jars must "cure" for a few hours. Be careful! Avoid breathing cyanide gas; it is deadly! Label each jar "CYANIDE, POISONOUS!" Wrap adhesive tape around the bottom part of the jar, but do not cover the sides above the plaster. The tape will help to keep the glass from shattering, should you drop it. If this happens, however, be careful not to breathe the gas released, and do not cut yourself on the glass particles. Cyanide will enter the cut, and it is just as deadly as the gas. Discard old jars by placing them in a deep hole dug out of doors, well away from plants, water, and play areas. If you place the jars in trash containers, wrap well in newspapers to prevent breakage, at least until it is safely discarded in a landfill.

When using ethyl acetate in place of cyanide, which is highly recommended for use by most amateurs, add more sawdust to the jar to take up the space not used for cyanide and to provide more area for absorption of the liquid ethyl acetate. Tap it down tightly. Then add plaster of Paris as you would for the cyanide jar. You may place a cork in the center, over the sawdust, and add the plaster carefully around the cork, but do not cover the cork. You will need to remove it when the plaster is dry. Then pour in some ethyl acetate to charge the jar and replace the cork. This will prevent the sawdust from working out and small insects from working in. It is ready immediately for use (fig. 15.2.). Ethyl acetate jars have the advantage of being less deadly to humans than cyanide, while killing insects quickly, before they are ruined as specimens by their frantic attempts to escape. However, this substance changes the color of some species. For example, green grasshoppers may turn red while in the jar and remain so as specimens. An additional advantage, however, is that specimens may remain indefinitely in an atmosphere of this gas. They will not mold, and they will remain flexible for mounting at a later time.

Carbon tetrachloride may be substituted for ethyl acetate. Remember, both ethyl acetate and carbon tetrachloride are poisonous: avoid breathing; the vapors are toxic. The worst prop-

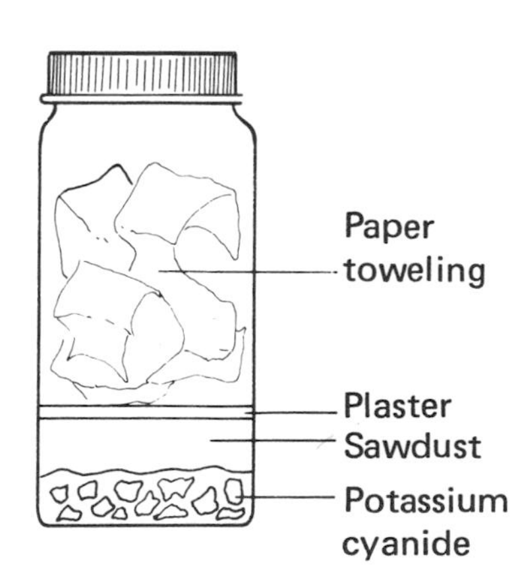

Figure 15.1. *Killing jar, cyanide.*

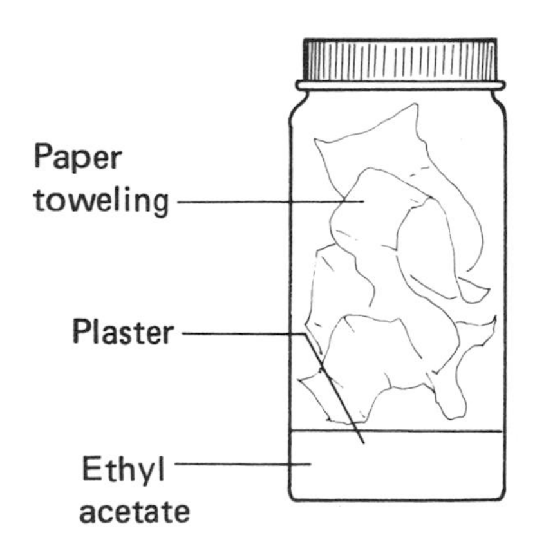

Figure 15.2. *Killing jar, ethyl acetate.*

erty of this killing agent is that specimens become rigid in the gas, and parts cannot be moved for mounting. Also the gas is extremely poisonous to humans and is a poison that accumulates in the body.

Check all jars before starting out on a collecting trip. Replace or recharge as needed.

Nets

Buy either professionally designed insect nets from biological supply houses or make your own. Aerial nets are used for collecting flying insects (fig. 15.3). To make one, sew a double thickness of muslin around the edge of the netting to protect it and to form a channel through which the ring, made of heavy steel wire, is threaded. The ring, bent to form a circle with a diameter of about 14 inches, is attached to the handle of hardwood doweling or sturdy aluminum pipe, so that it can be easily removed. Usually this is done by bending the wire ends of the steel ring parallel to the edge of the handle. Two holes, one about three inches and one about four inches from the end of the handle and on opposite sides, are drilled into the wood. The projecting ends of the ring are bent to fit into these holes. Then a brass tube about three inches long and with the inside diameter the same as the outside diameter of the handle is slipped over the ends of the ring to hold it in place. It will be necessary to gouge out a groove for the wire to sit in place beneath the brass tubing. A small screw will hold the tube in place over the wire. If an aluminum pipe is used for a handle, a flat narrow strip of steel can be used for the net ring and the ends bolted to the handle (fig. 15.4).

Aerial nets should not be used in vegetation, since they can easily be torn. Instead a heavier net, similarly constructed, with a bag of light canvas or heavy muslin, is used for beating and

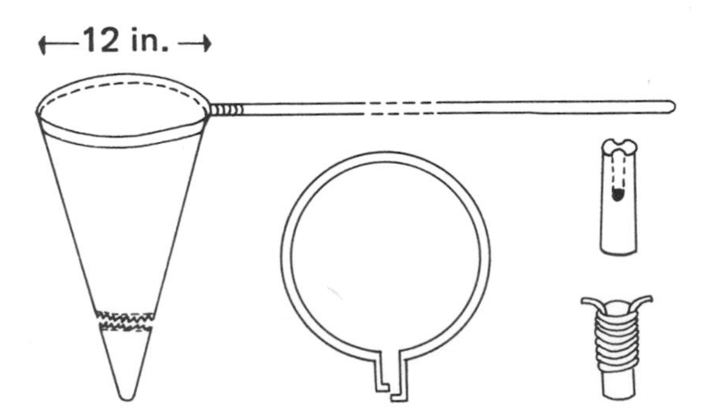

Figure 15.3. *Aerial net. left, net; right, details showing how ring is attached.*

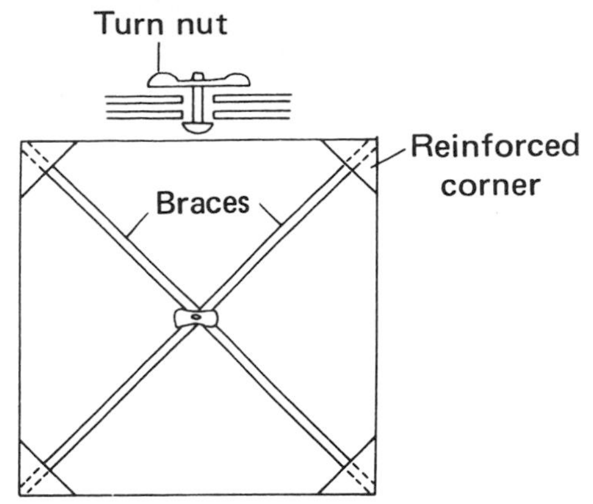

Figure 15.4. *Beating net.*

sweeping vegetation. Beating nets, when properly used, capture beetles, bugs, bees, and flies found on a wide variety of plants. Sweep nets pick up grass and weeds along with the insects. Flying insects will crawl out from under this material and fly out of the net, unless you are quick to catch them. Other insects will hide under the debris, which must be dumped out and sorted.

A beating net is made by using two small crossed handles about three feet long inserted into small pockets on the four corners of a square of heavy cloth. This forms an umbrella used to hold under branches of bushes and trees while they are beaten with a stick or net handle. Insects will drop into this umbrella and can be picked up and placed in killing jars or alcohol vials. An aspirator (described below) is useful for capturing tiny specimens. Act quickly, because insects will stay in the net for only a short time.

Collecting aquatic insects can be both challenging and fun. Aquatic nets are constructed with the ring flat on one side. These are used to scrape over the bottom of a shallow stream or the edge of a pond. Naiads of mayflies, stoneflies, dragonflies, and damselflies, as well as nymphs, larvae, and adults of other aquatic insects found in streams, lakes, ponds, and swamps, are collected in this manner. Scoop nets made of heavy hardware cloth and a mesh cover to keep out aquatic plants, but allow insects to pass through, may be purchased by the ardent collector of water insects. This is an advantage over the open net, which picks up aquatic vegetation and debris. However, aquatic debris may be dumped into a white enamel pan and sorted through for specimens. Specimens may be picked up with your fingers or forceps, but be careful. Some aquatic Hemiptera can inflict a painful bite. Material collected from aquatic habitats may be stored in small jars or vials of alcohol. Be sure to include data labels.

Aspirators (fig. 15.5) are used to suck small insects into a

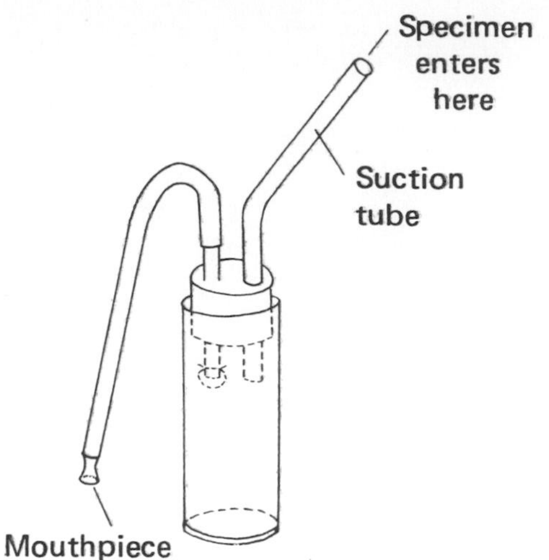

Figure 15.5. *Aspirator.*

tube without danger of swallowing them. With this, insects can be collected from windows, greenhouse walls and plants, small spaces, and beating nets. All one needs for this device is a large test tube, two bent small-diameter brass tubes, and a two-hole rubber stopper of the same diameter as the mouth of the test tube. One end of the tube used as the sucking tube is closed with fine netting inside the test tube. This prevents insects from being sucked into your mouth.

Traps

Commercially made light traps attract night-flying insects, especially moths and beetles (fig. 15.6). These traps have a UV bulb and, sometimes, a fan inside. The light attracts the insects, they hit a baffle and drop into a container attached to the bottom of the trap. Some traps will operate on a 12-volt car battery. Set the trap in an area where the light can be seen from some distance. In a field or near a pond is ideal. Light traps placed in a yard with night lights may not yield many insects. The container at the bottom of the trap is usually filled with alcohol to kill the specimens. However, if you are interested in moths and other insects that should not be stored in alcohol, a cyanide jar may be attached. Be careful if this is used, because the gas will escape into the air near the trap. If you are interested in perfect specimens of moths, a better way to collect at night is to hang a white sheet on a cord tied between two trees. Hang a UV light close to the sheet. Moths will fly to the sheet and usually settle down. You can sit nearby and collect at your leisure, removing those specimens of most interest to you. Beetles and other insects are also attracted and may be collected, without moth scales, in the same

Figure 15.6. *UV light trap.*

way. They may be killed and stored in the appropriate way according to their kind.

A Malaise trap (fig. 15.7) is a daytime collecting device made of dark netting. The trap is erected in a flyway, forming a baffle that traps day-flying insects. These tend to fly up to get away, and, instead of escaping, they fly into a killing jar at the top of the trap. Several commercial traps are available from biological supply houses, but, if not dark, they must be sprayed with a black, dull-finish, enamel paint to work best. They may be constructed at home, if you can sew, from patterns published in *Entomological News* (vol. 83, pp. 239–247, 1972).

A Berlese funnel (fig. 15.8), which can be made from sheets of aluminum, is another type of insect separator. This consists of a metal funnel held in place over a jar of alcohol and a heat source, such as an electric light bulb. (Some plastic funnels can be used.) A coarse screen is placed in the funnel to prevent debris from falling into the alcohol. Debris gathered from the forest floor (leaf litter, decaying vegetation, etc.) is placed in the funnel. The heat, by drying out the debris, drives insects and other small arthropods down into the end of the funnel and, finally, into the alcohol jar. Berlese funnels can be used all year. They are an excellent way to collect small beetles and many uncommon soil arthropods.

Pitfall traps are used for collecting terrestrial arthropods. Dig a small hole in the ground and place a one-pound-size coffee

Figure 15.7. *Malaise trap. (Courtesy of Dana A. Focks & Co.)*

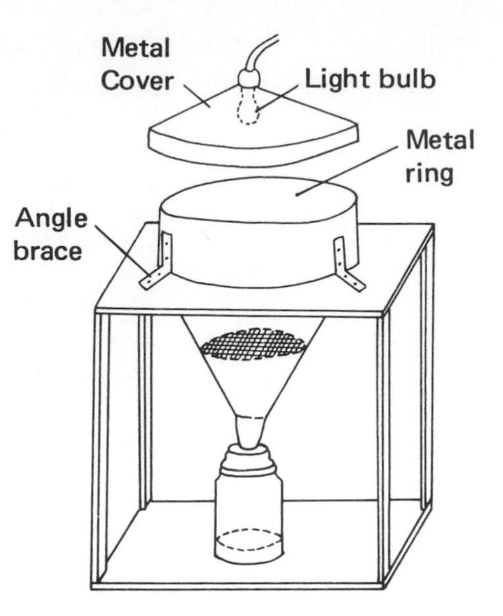

Figure 15.8. *Berlese funnel.*

can, or similar can, into the hole. Place a funnel with the top end the same diameter as the can. Cover the trap with a board held above the can as a rain baffle. Try not to disturb the environment too much, since you do not want to ward off wandering insects. Place a small amount of antifreeze, ethylene glycol, in the bottom of the can as a killing agent. Visit the trap every few days to remove the insects. These must be washed in water and placed in alcohol immediately.

Dead animals, dung, and bait of various kinds can also be placed in small containers suspended over the pitfall trap by a loop of heavy aluminum wire. The end of the loop is bent down and shoved into the ground to support the loop. This material will attract carrion and dung-feeding insects to the pitfall trap. They will fall into the antifreeze in their attempt to reach the bait.

A mixture of brown sugar, beer, rotting bananas, and molasses in proportions to make a mixture about the consistency of thick paint can be painted on the trunks of trees and bushes at night. This will attract many kinds of moths, some beetles, and a few other kinds of insects. Using a fisherman's headlight, walk around the bait trail and collect the moths attracted to the bait. This practice long used by Lepidopterists is one of the quickest ways to build a fine collection of moths.

Hand picking from vegetation is a good way to collect many kinds of insects. While many are elusive, others sit on plants and can be easily picked off. Turning over rocks and logs, and chopping into rotten logs will yield many more species, especially beetles. The serious collector always carries a collecting bottle. You just never know when that special insect will be sitting right before your eyes, so be prepared!

Collecting ectoparasites

Ectoparasites may be collected from live or freshly killed birds (you will need a Federal Bird Collecting permit either to shoot or trap birds) and mammals. Special techniques must be used to kill these pests. Place a small, freshly killed animal and a chloroform-saturated cotton ball into a plastic bag; seal it tightly. After 10–15 minutes, the ectoparasites will be dead. They may be shaken out of the animal's feathers or hair and remain in the bag after the animal is removed. This must be done before the parasites have time to leave the animal. Most ectoparasites leave when the body temperature begins to drop after death. Shake the contents of the plastic bag onto a black cloth (so that they are more easily seen) and pick them up with forceps. A fine-toothed comb may be needed to remove some of the insects from the hair of mammals. A camel's hair brush moistened in alcohol and touched to the specimen helps one pick up small specimens of any kind. Store the specimens in alcohol, until you are ready to mount them on slides, a procedure which must be done before these specimens can be identified. (Directions for slide making may be found in various technique reference books.) Be sure you put the proper collecting data, including the name of the host animal, into the vial with the specimens. It is best to wear gloves when handling dead animals, since the animal may carry diseases, such as tularemia or rabies. You should also be careful of certain ecotoparasites, such as certain mites, that may attach themselves to you, some of which are not host specific.

Storing specimens

Once you have captured and killed the insects for your collection, you must mount them, store them in a safe place, away from dust and museum pests, and then arrange the collection in the proper phylogenetic sequence. Gas-killed specimens must be mounted the same day captured, or allowed to dry and stored in boxes on layers of cellucotton. They must be perfectly dry, or they will mold and be useless as specimens. Dried specimens remain in perfect condition indefinitely, if protected from dermestid beetles, which will eat them. Once dry, specimens are extremely fragile. Before they can be mounted, place them in a relaxing chamber. They will absorb water, again become flexible, and can be mounted on pins without breaking. A suitable relaxing chamber may be made from a wide-mouth gallon jar with an opening large enough to be able to reach inside with the specimens to be relaxed. Some collectors use a metal box with a tight cover; others find that the old fashioned crock with a heavy cover works well. Fill part of the container with clean sand and enough water to saturate the sand. Wet paper towels or newspaper can be used in place of the sand. In either case, add a little carbolic acid to prevent the growth of mold. Specimens must not actually touch the water in the jar. Place them in one-half of a Petri dish or similar waterproof container. The specimens will absorb moisture from the humid atmosphere in the covered jar. Allow them to remain in the jar for at least twenty-four hours, sometimes longer, depending on the size of the insects. Once relaxed, they are ready for pinning. Beetle collectors are fortunate, because they can soak their specimens in hot (not boiling) water for a few minutes, and then mount the flexible beetles. Specimens stored in alcohol may be mounted directly from the liquid.

Storing specimens in the field (temporary storage)

Since insects should be removed from cyanide, ethyl acetate, or carbon tetrachloride jars whenever possible, the fresh specimens must be stored in a manner that will prevent any damage. They should be stored in alcohol, mounted in the field before they dry, or placed between layers of cellucotton and stored in boxes. Moths and butterflies are often placed in paper triangles, that is, oblong pieces of paper folded into a triangle with the projecting edges turned over to make a small envelope. Different size sheets are used, according to the size of the specimens. The wings are folded together above the body, before the specimen is placed in the envelope. Write all locality and habitat information on the paper triangle before the specimen is placed in it. Storage boxes should be treated to prevent the growth of mold and invasion by museum pests. The best material to use for this is paradichlorobenzene, sold under the trade name, dichlorocide. This is better than moth balls (naphthalene), since the latter is now known to be a carcinogen.

Small vials of ethyl or isopropyl alcohol serve both as killing jars and storage containers. Because of their flexibility, "wet" specimens are often easier to identify than dry, brittle ones mounted on pins or points. When there are large numbers of specimens in a vial, the alcohol should be changed after a few days; fresh specimens contain large amounts of water and will dilute the alcohol enough to prevent effectiveness as a preserving agent. Vials should be placed in a vial storage rack made of wood, plastic, or cardboard (fig. 15.9). These are offered for

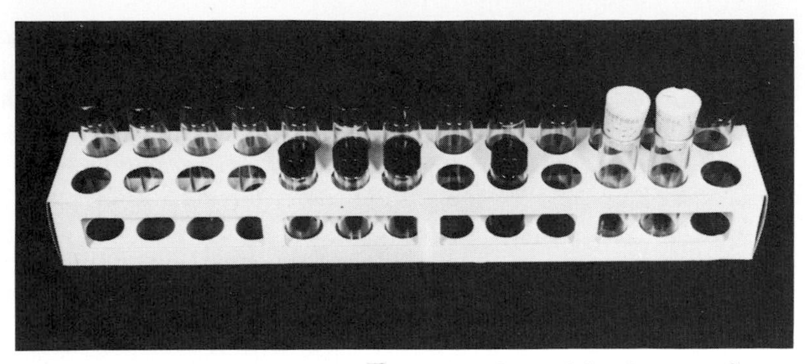

Figure 15.9. *Cardboard rack for storing four-dram vials. (Courtesy of American Biological Supply Co.)*

sale by entomological supply companies. Usually a four-dram vial is used, but for larger specimens, ten-dram vials with snap caps are available. Use a neoprene stopper for the four-dram vials. White stoppers are best, because they do not swell in alcohol. Watch vials with plastic snap caps; they tend to allow the alcohol to evaporate from the container. Cork stoppers can be used as temporary stoppers, but should be replaced when possible by neoprene stoppers. Cork absorbs alcohol, and the vials soon dry out. In either case, alcohol containers should be frequently checked to be sure the alcohol has not evaporated. By adding a small amount of glycerin (2 percent), specimens may be protected, should the alcohol evaporate. Glycerin, which does not evaporate, will keep the specimens pliable until more alcohol is added to the container.

If you are out in the field for an extended period of time, most specimens can be stored in alcohol. Specimens with scales and hair, such as moths, butterflies, flies, bees, and others, can be kept in jars containing coarse sawdust charged with ethyl acetate. They can be stored indefinitely in these jars and will remain soft and flexible, ready for mounting.

Habitat description and other recordkeeping

Specimens are of no scientific value, unless they have data labels recording the location of the collecting site, date, habitat, and the name of the collector (see chap. 16 for more details). When you collect a specimen, put the appropriate data with it, not later when there is a chance of confusing specimens. The more data recorded, the more scientific value the specimen will have. Even if you start out collecting specimens for display purposes only, keep records and label the specimens for possible future study.

Whenever a serious collector goes into the field, it is essential that a record be made of the ecological conditions at the collecting site. A mistake many beginning collectors make is to omit taking field notes, thinking the data can be recorded at the end of the day or on returning home. Experience shows that much important information is left out, because of failing memory. To overcome this, many entomologists have designed field forms. None of them are entirely satisfactory. Information to be recorded may seem endless. If too little, you will need to return to the collecting site for the missing data. Too much is a waste of time. Therefore, plan by determining objectives, then tailor recordkeeping to meet these objectives. For example, for a general collection, one needs little more than enough habitat description and locality data to enable someone to return to the area in search of additional specimens. If, however, one is attempting to learn about the habitat preferences and habits of species, details about the physical environment, as well as habit descriptions, are needed. Once these objectives are outlined, you can devise a checklist of detailed information needed, or create your own forms.

Insect photographs

Photographs of insects are difficult to take, but they provide an excellent way to record field data. Professional and amateur entomologists and photographers are tempted by the challenge of taking extreme close-ups of insects, but the results produced are among the most fascinating of photographs. Most camera lenses will focus only as close as 2½ to 3 feet. For quality insect pictures, special close-up lenses are needed. The best camera to use is a single lens reflex (SLR) camera. You will encounter many problems when filming insects and should keep several things in mind. First, try to photograph live insects only; dead insects make poor subjects and lack a natural appearance. One can slow down insect activity either by chilling or by placing them in a carbon dioxide gas chamber for a few seconds. This helps, but it is second best. Any movement during exposure is greatly magnified, so steady the camera by using a tripod.

There are ways to achieve a close-up. The simplest and least expensive way is to use auxiliary lenses. These resemble a filter and screw into the filter threads of your primary lens. Lenses of different magnification are available, usually marked as +1, +2, or +3. Various combinations enable even greater magnification. The use of extension tubes or bellows attachments is another way of taking close-ups. The first are hollow tubes that fit between the camera body and the lens to extend the lens, thereby increasing its magnifying power. These produce a fixed magnification, but bellows attachments allow greater latitude in focusing. Loss of depth-of-field occurs when you are using close-up lenses, but this can be partly compensated for by "stopping down" and using flash for additional light. The best way, of course, to photograph insects and other small subjects is with expensive macro-lenses, a good investment, if you intend to take many insect photographs. The most popular macro-lenses for 35 mm single lens reflex cameras have a focal length of 55 mm, but their focusing range permits lens extension to enlarge to about twice life size on the film. Many brands are available; you should consult a reliable dealer for more details.

Close-up photography requires a camera with slow shutter speeds, especially if you close down the lens aperture to gain depth of field or to compensate for the addition of an extension device. You must be able to use shutter speeds as low as $^1/_{25}$ of a second; anything lower will blur any movement made by the insect. However, with the recent introduction of high-speed color film, this is not as great a problem as it once was.

Good insect photographs must be in perfect focus; the specimens *must fill at least half of the picture frame*, and all legs, antennae, and wings must be in focus and inside the frame. When photographing a butterfly on a flower, it is important to keep the butterfly in focus, even if some of the flower itself is slightly out of focus. It is ridiculous to take a picture of a large flower with the insect occupying only 10 percent of the field, and don't think that this part of the picture can be enlarged to show the insect, because loss of detail always results.

Photomicrographic equipment attached to a microscope is used to take pictures of insect parts. Details of eyes, antennae, mouthparts, or genital structures useful for making identifications are photographed with this expensive equipment. For the past few years, more and more pictures taken with a scanning electron microscope (SEM) (fig. 15.10) have been published. These are expensive, but produce pictures of insect structures at high magnification and without depth-of-focus problems. All parts of the insect are in focus, no matter the magnification. These detailed pictures are not pretty, but they are extremely useful for showing details that are of great value for identification.

Various classroom projects related to the information given in this chapter will be found in chapter 18.

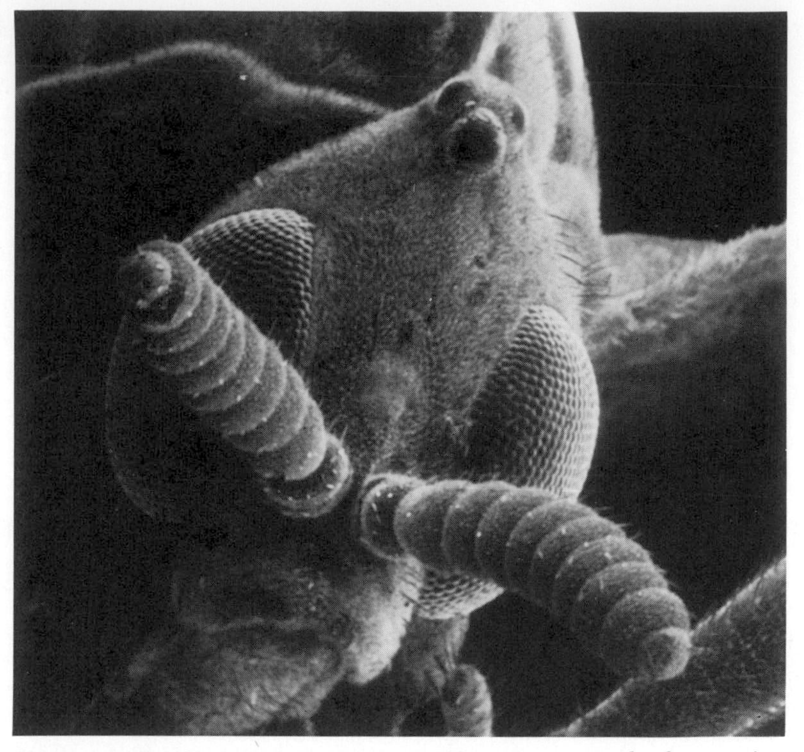

Figure 15.10. *Head of a male "lovebug" (Diptera), an example of a scanning electron photomicrograph. (Courtesy of P. Callahan, U.S.D.A.)*

Safety in the field

Collecting insects is seldom dangerous, but beginners might consider a few words of caution. Aside from being stung by wasps, ants, and bees or bitten by mosquitoes, black flies, horse flies, deer flies, and certain kinds of bugs, there is little danger from the insects themselves. Some insects transmit diseases (as discussed in chap. 13), and some others are merely annoying. Beware, however, of flying moths at night around a collecting light. These have been known to fly into people's ears and are extremely painful when they do so. It is best to use a loose, cotton earplug to prevent this from happening. Chiggers and ticks can be irritating, if they attach themselves to your skin. Use any of the good repellents sold in drug stores to ward off any of these pests. Even pet flea collars may be worn around your ankles to repel mites and ticks (and fleas, if present).

One final rule: use the "buddy" system. Never go into the field alone. Snakebites, scorpion stings, sunstroke, broken limbs, or illness may happen at any time. A collecting companion might save your life.

Be sure to carry identification with you. Not everyone is convinced that insect collecting is a harmless activity. We have been taken to police stations on several occasions while we were "checked out." Proof of identity, especially entomological society membership cards, came in mighty handy!

16

How to Mount and Preserve Insects

"What hand would crush the silken-winged fly,
The youngest of inconstant April's minions,
Because it cannot climb the purest sky,
Where the swan sings, amid the sun's dominions?
Not thine."

Shelley

"Do not mash your specimens!"

Professor Bradley

Specimens gathered and stored must be mounted and arranged for observation and study. Remember that these specimens represent a segment of nature, and to correlate the data these represent, it is necessary for you or someone to identify them. By so doing, all of the existing literature dealing with any aspect of their morphology, physiology, and life history, if anything is known about them, will become available to you. Of course, you will have to visit a large library to find some of this literature (see chap. 3 for details about identification). Entomologists usually arrange their collections in phylogenetic order, instead of by color, size, or shape. When you do this with your collection, you will notice that most groups of insects are drab. Color usually signifies something: it may be a warning saying, "I am poisonous, do not eat me." Or it may say, "Look at me, aren't I beautiful?" Male insects, like male birds, often have the brightest colors. Anyway, you will probably arrange the most colorful or the most bizarre first, just to show them off. Such arranging is a worthy goal, but soon you will understand the scientific reasons for following the arrangements given in catalogs, because these classifications are phylogenetic, that is, they reflect the evolutionary relationships of the insects. As we have pointed out previously, specimens represent samples of the population of the species.

As each species becomes better known, the less it is necessary to depend on the insect specimens in a collection, much the same as bird students now rarely capture birds for study skins. But you must remember that entomologists have a long way to go before the field guides and references are complete enough to identify insects the way one can identify birds.

How many specimens in a collection?

Time will come when you will want to limit the number of specimens you have of each species. Remember that individuals vary. Especially in insects, size range alone can be considerable. Some individuals may be twice the size of others. Unlike mammals, male insects are often smaller than females. Species of many families show great ranges in their colors and the patterns these colors form. Therefore, you can see that long series of specimens are sometimes needed to show this range of size and color. Research taxonomists do not agree on the number needed. Some say that a series of fifteen specimens is enough; others set the figure at one hundred; and still others maintain all specimens available must be studied to understand variation. It appears that you must decide. If you intend tot rade specimens, you will need trading material; hence, you will want to keep specimens in reserve for this.

If the species lends itself to culturing (see chap. 18), then variation studies will yield more certain results. Data about these species may be gathered and will fill journals with scientific articles and books with interesting facts. When this is done, specimen files become less important for study purposes, but they will always be interesting as display specimens. Unfortunately, most species are not easy to rear; hence, for the present, population samples must be filed as stored museum specimens. The method for storage is directly dependent on the ease and quality of retrievable data. Therefore, the careful preparation of all specimens is mandatory.

Although the system described here follows the standard used by most public and private collections, it may also be necessary to prepare some material for dissection. Only after an examination of the internal organs is it possible to decide where the species falls in some of the more recent classification schemes proposed. For example, the nature of the ventral nerve cord has been used to separate the primitive suborders of Lepidoptera. Almost never is this necessary for the identification of species. Nevertheless, it will be wise to keep some specimens of common species stored in alcohol for possible future use.

Mounting specimens

Most specimens are pinned and left to dry, after which they will not deteriorate externally, because no detritus bacteria or fungi will live on the dried exoskeletons of insects. The soft internal organs will dry, lose shape, and collapse, but much of them, too, remains. Muscle is barely evident. The fat body (an internal storage organ) does change, and sometimes the fat becomes liquid. This material may leak out and coat the outer part of the body. Many large specimens need to be opened, the fat removed and replaced with a wad of cotton. Smaller specimens may be soaked in ethyl ether to remove most of the fat.

Wet storage. Specimens are always stored in alcohol, never in formaldehyde (formalin). As explained previously, use 70-percent ethyl alcohol, or 80-percent isopropyl alcohol (rubbing alcohol). Specimens and their data are placed in vials and jars. The alcohol does not need to be replaced once the original water has been removed, but may have to be added to at regular intervals, due to the difficulty of finding caps tight enough to form a complete seal. Fruit jars with rims that screw down on aluminum tops with neoprene seals are the best to use. Standard-size vials and jars should be used for ease in cabinet storage. Small vials mixed with larger vials tend to tip over and are difficult to handle. Various-size storage racks are commercially available. (See chap. 15 for more details on vials, stoppers, and so on.)

Do not glue data labels to the outside of the jar. These are frequently lost, because moisture softens the glue, or light fades the label. Once lost, the specimens are scientifically worthless. Always use India ink for data labels or at least for the written part of the labels. If labels are typed, make photocopies of the labels and place the latter in the vials and jars. Caution: the toner of some copy machines does not always completely fuse and is not alcohol resistant. This problem usually develops from the use of toner other than that recommended by the manufacturer. Therefore, test photocopied labels by placing them in a jar of alcohol for at least a month before they are used in specimen jars. Also, make sure that the India ink is thoroughly dried before the label is placed in the jar. Use a good grade of rag paper for all labels.

Dry mounting. Specimens collected dry, that is, in gas-filled killing jars, may, of course, be transferred to alcohol while they are still fresh, or the specimens may be pin-mounted the same day collected. If allowed to dry, then they must be relaxed before pinning. This may be done using a relaxing chamber, as described in chapter 4, or by using a relaxing fluid. One of the best fluids is Barber's fluid (a formula follows), into which beetles, grasshoppers, bugs, and most flies may be placed. It cannot be used to relax moths, butterflies, or hairy specimens. Specimens should be left in the fluid for several minutes, removed, and pinned, if flexible.

Barber's relaxing fluid:
- 95% ethyl alcohol, 50 parts
- Water, 50 parts
- Ethyl acetate, 20 parts
- Benzene, 7 parts

(If the oily benzene separates out, add a small amount of alcohol until the solution is stable.)

Pinning insects

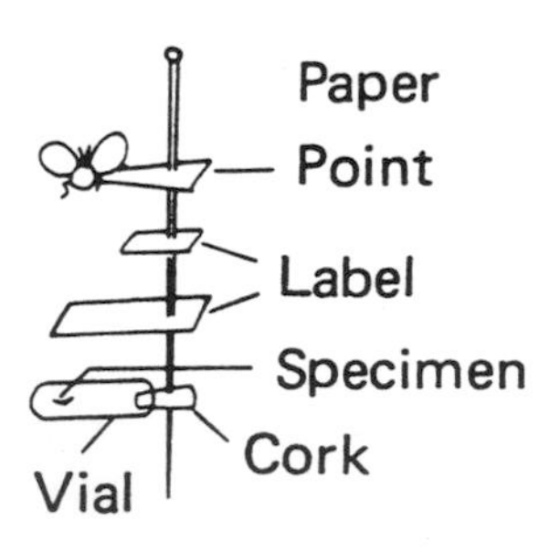

Figure 16.1. *Where to place the insect and labels on the insect mounting pin.*

When specimens are soft andpliable, they are ready to be pinned; and for Lepidoptera and some other orders, spread. Large specimens are pinned first, using standard insect pins purchased from a supply house. Seamstress (common) pins should not be used, because they are too thick, too short, and will soon rust. Insect pins, made of steel (coated, bluish in color) and stainless steel (silver), are available in a size range from 000–7. The "0" sizes are too small to use, except under unusual circumstances. Even number 1 is often too small. Most specimens are pinned on number 2 or 3 pins. The larger sizes are sometimes used for heavy-bodied insects. Specimens too small for number 2 pins are glued to points (see below) or pinned with "minuten nadeln" (described below).

Insects are always pinned vertically (fig. 16.1). The height of the insect on the pin is gauged by using a step block (fig. 16.2).

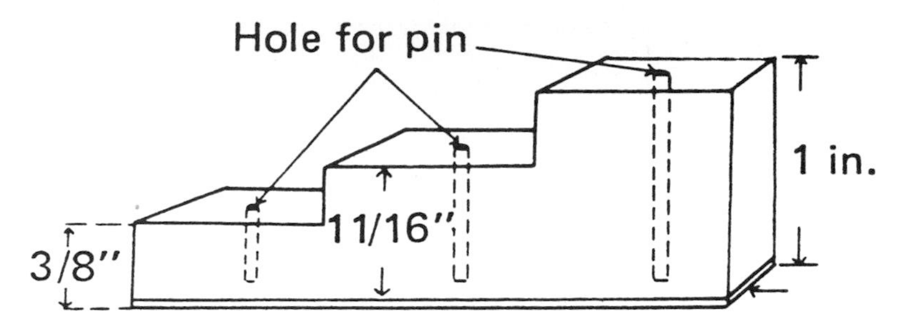

Figure 16.2. *Step block for regulating the height of insects and labels on the insect-mounting pin.*

All insects are pinned through the thorax, between the base of the front wings; or through the right wing and thorax when the wings are folded over the abdomen (fig. 16.3).

The best way to pin an insect is to cradle it between the thumb and forefinger of one hand and hold the insect pin with the thumb and forefinger of the other hand. Thrust the pin through the body until the head of the pin nearly reaches the thorax of wing. Then turn the specimen over and insert the head of the pin into the lowest step of the step block. Push the pin in as far as it will go. This will place the insect at exactly the right height on the pin. The other levels of the step block are used to regulate the height of the data labels. Not only do the specimens look better pinned at a uniform height, but the protruding pin above the insect body provides a handle for fingers or pinning forceps when moving specimens.

Never pin through the abdomen of a specimen; it is too soft and will break under the weight of the rest of the insect. Even without a pin, the abdomen tends to sag in many kinds of insects. To prevent this from happening, it is often necessary to cross two insect pins beneath the abdomen, one on each side, forming an "X," which supports the structure until it is dry. This often is needed for Lepidoptera, pinned dragonflies, damselflies, Neuroptera, walkingsticks, caddisflies, and others. While the legs are soft, extend them out to the side, and either fold the wings properly over the abdomen, or spread them for display according to the needs of the particular group. Allow the specimens to dry on "pro-tem" blocks made of cork, balsa wood, or other pinning surfaces. Do not use styrofoam for pinning, because it is affected by various fumigants. Most pinning surfaces are now made from sheets of polyethylene foam. Pinned specimens are stored in insert boxes or drawers. They should not be left out in the open, because they will collect dust, which obscures highly magnified parts, and they are subject to damage by roaches, rodents, barklice, and bunglers.

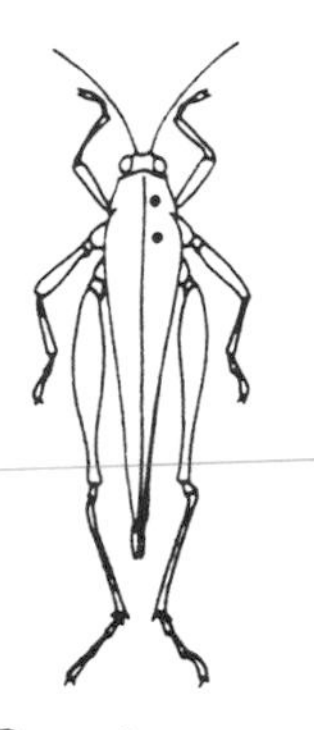

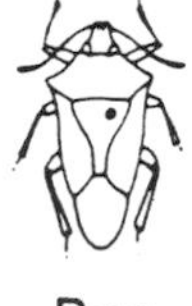

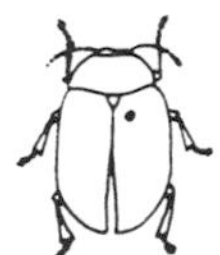

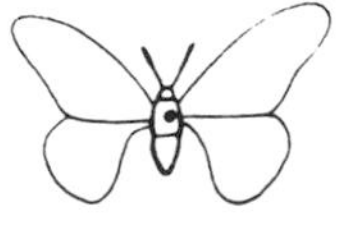

Figure 16.3. *Where to place the insect pin through the body when mounting insects (push through area indicated by the black dot).*

Spreading insects

Butterflies and moths are almost always spread—that is, the wings are extended out to the side of the body while the specimens are relaxed, pinned in place, and allowed to dry in this position. Sometimes the wings of specimens of other groups—as listed previously—are spread. These specimens are pinned, unless they are going to be displayed in a Riker mount (see fig. 16.13), a special glass case used to protect specimens while being studied by classes or hung on the wall for decoration or display. Since Riker-mounted specimens are laid on cotton in these mounts, a pin is unnecessary, but the specimens are prepared the same as pinned specimens. If already pinned, relaxing will permit the removal of the pin.

If the specimens are not relaxed, this must be done first, following the instructions given previously. To spread specimens, it is necessary to have available assorted sizes of spreading boards. Obtain these from biological supply companies, or if you are handy with woodworking tools, make them at home (fig. 16.4). Use soft pine or similar wood for the sides, and balsa wood, cork, or polyethylene foam surface for pinning. The various sizes provide for different widths of the specimens' abdomens. It is necessary to have the groove through which the pin is thrust narrow enough to permit the wing base to be pinned down and wide enough to allow the insect's abdomen to fit into the groove (see fig. 16.5). Some commercially available spreading boards have one side movable, permitting adjustment of the width of the groove. If only a few specimens are to be spread, these work well, but the seasoned collector of Lepidoptera finds that an assortment of boards is better than these adjustable types. Start with a few of the adjustable kind, note the widths

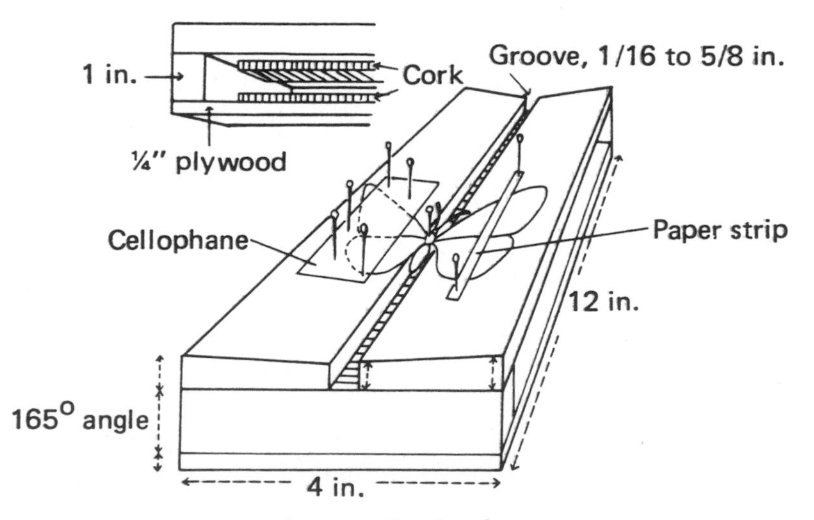

 Figure 16.4. *How to make a spreading board.*

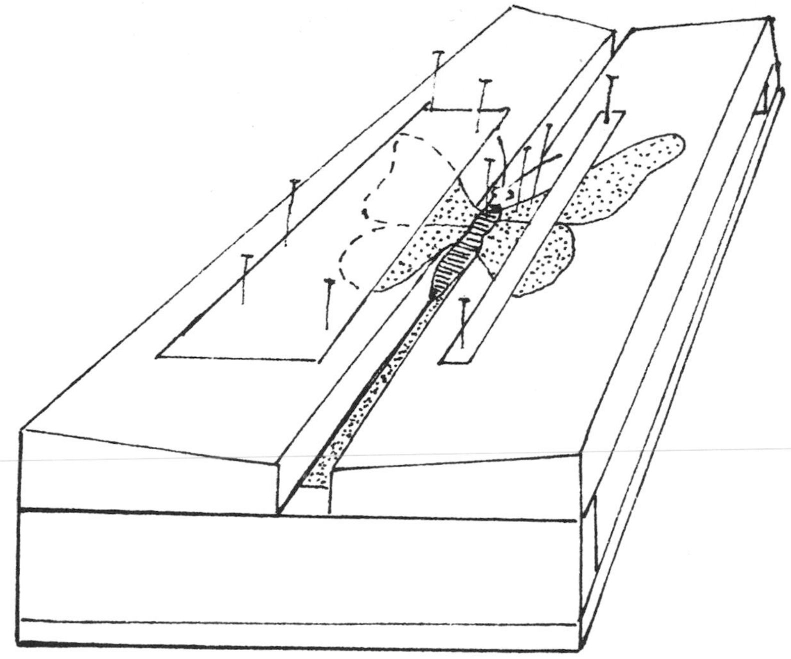

Figure 16.5. *How to spread butterflies and moths pinned on the spreading board.*

most frequently used, and then purchase or make these sizes. Probably you will need boards with groove widths of 1/16, 1/8, 1/4, and 1/2 inches, rarely wider.

To spread insects properly requires practice and some patience. Attractive specimens come after mastering the technique. Start with common species (cabbage butterflies make good practice specimens). Wait until later to spread your best specimens. Remember Lepidoptera are covered with scales, and you must keep as many of them on the specimen as possible. The best way to pin fresh specimens (which usually die with their wings folded down, instead of up) is to place forceps under the wings to hold the specimen while inserting the pin through the thorax. Then holding the specimen by the pin, carefully fold the wings upward, while pushing the pin through the pinning surface. The height of the specimen on the pin will be regulated by the sides of the spreading board and the bottom of the board that stops the pin. The wings will then lie on the sides of the board ready to be pinned. Make a final adjustment of the thorax on the pin, so that the base of the wings is exactly even with the inner side of the board. Next insert a small (number 2 or smaller) pin through the wing close to the base and just behind the large Costa vein (the anterior-most vein) of the front wing and push it forward until the hind part of the wing is exactly at right angles to the body of the insect. Repeat with the other front wing. Now bring the hind wing up into place. You will note that there is a portion of the hind wing that looks as if it should fit under the front wing, as indeed it does. Adjust the position of the hind wing accordingly. When using these pins to adjust the wings, work carefully not to tear the wing membrane. Make sure that all pressure is exerted against the heavy main vein.

There are now four pins at the base holding the wings in place. Narrowly fold strips of cellophane along one edge and place the folded edge next to the groove. The fold will prevent a sharp line from forming along the base of the wing. These strips should be wide enough to cover the wings. Insert pins around the wings to hold the wing in place, but do not pin through the wing itself. Once this is done the base pins may be removed. The wings will now dry in the set position.

It may be necessary to support the abdomen and the antennae to prevent sagging. This is done by placing two crossed pins on the sides to hold these parts in the proper places. With another insect pin, attach the locality and data labels to the spreading board near the specimen. The length of time needed to dry the wings in this position will vary with the relative humidity at the time they are set. In humid climates, it may be necessary to use heat (from a light bulb) to dry the specimens. It usually takes four or five days for specimens to set.

After the specimens have dried, carefully remove the pins and take off the paper strips. The specimens may now be removed from the spreading board and transferred to an insect box. Be sure to add locality and data labels.

Mounting small insects

Specimens too small to pin must be glued to triangular points or pinned with minuten nadeln and cork. Triangular points (fig. 16.6) are cut out of stiff, rag-content paper, preferably a heavy ledger paper. These points are made with a point punch available in two forms, one with a blunt point, the other with a sharp point. If only a few are needed, they may be cut out with scissors; the base of the triangle is about 3–4 mm, and the point, 8–10 mm long. If the small minuten nadeln pins (fig. 16.7) are used, small rectangular pieces of cork or polyporus strips cut to about the same size as the points are needed. The insect pin is pushed through one end of the cork and the minuten through the other end. Small specimens are impaled on the sharp end of the minuten, the pin inserted through the ventral surface of the thorax. Insects glued on points must be uniform. The insect pin is in-

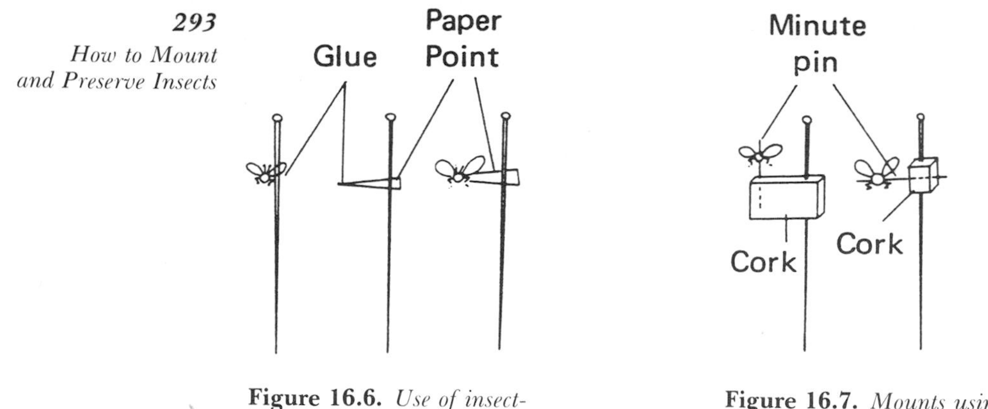

Figure 16.6. *Use of insect-mounting points.*

Figure 16.7. *Mounts using minute pins.*

serted through the base of the point, and the height on the pin is regulated with a step block. The point should be at the same position on the pin as an insect would be. Push the point nearly to the top of the pin, then turn the pin over and push the head through the lowest hole on the step block. Keep a glue pot handy. The best glue to use is polyvinyl acetate (Gelva U–25R), which is alcohol soluble and, therefore, dries quickly. Ordinary "milk" glue is second best, but readily obtainable. It is water soluble and takes longer to dry. Line up on their sides the insects to be glued to points, with their heads pointing right and their legs toward you. Place a small amount of glue on the point and touch it to the side of the insect. If the glue is too wet to hold the specimen, let it dry. Sometimes it is necessary to hold the pin horizontally until the glue dries. This can be done by inserting the pin into a pinning board held vertically. If the point is too fine to hold the specimen properly, with forceps, bend the end of the point down at right angle, or bend the end of the point to make a small shelf, which will fit against the curved thorax of the insect. It will take some practice to make good, pointed specimens. The important thing is not to use too much glue. This will hide parts that you will need to see when you are making identifications.

Broken specimens

Dried specimens are fragile and easily broken when handled. If a leg or antenna or even a wing or the abdomen is broken off, it may be glued back with the same glue used for the points. Work carefully under a magnifying glass or low power microscope. Use as little glue as possible, and try to get the part back in place.

Preservation of adults and larvae in fluid

Storage of specimens in fluids has been discussed previously under field storage. If, however, they are to be stored permanently in fluid, care must be taken to see that the container is always full of the preservative and that the labels are in good condition. Certain containers deteriorate after a few years. Selection of the proper type of cap or stopper is critical. Plastic snap tops or metal screw caps are to be avoided, if possible. The neoprene stoppers described previously are best for vials, and plastic screw tops, the best for jars. For safety's sake, add up to 5-percent glycerine to the preservative used for permanent storage. Under no circumstance, keep the specimens in the fluid used for killing larvae.

Mounting insects on microscope slides

Specimens of lice, fleas, thrips, aphids, and certain other kinds of insects must be mounted on microscope slides for permanent storage and to improve the ability to see parts when identifications are made. These insects need to be magnified with the help of a compound microscope. To do this, it is necessary to treat the specimens with certain chemicals, first for complete dehydration and, second, to clear the parts, that is, make them more transparent, so that the fine structures can be seen by transmitted light, instead of reflexed light, used with a hand lens or low-power dissecting microscope.

Many different techniques are used for this purpose, but the more or less standard system is to dehydrate the specimen completely. This is done by using small U.S.-Department-of-Agriculture-size watch glasses filled with the various reagents. Place the specimen in 70-percent ethyl alcohol, then, after a few minutes, into 95-percent alcohol. The next step is critical. If absolute alcohol is available, and the specimen is small enough, place it in this material, but remember that absolute alcohol is

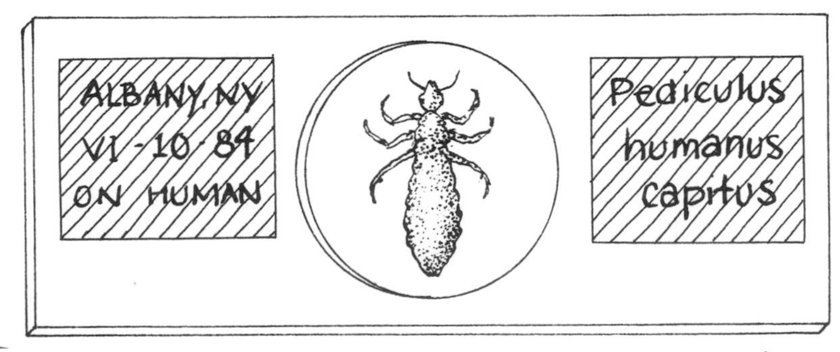

Figure 16.8. *Microscope slide mount.*

not stable. It rapidly takes on water from the atmosphere. If any water remains in the specimen, it will not clear properly. Other materials may be used in place of the absolute alcohol. A 50–50 mixture of carbolic acid and xylene may be used, or some prefer clove oil. From this step, the specimen is placed in xylene to clear. If all the water has been taken out, the specimen will clear, that is, become almost transparent; if not, it will be cloudy or milky, and the process must be repeated. Once cleared, the specimen is ready for mounting in Canada balsam on a microscope slide. A small amount of the mounting media is placed in the exact center of the slide and the specimen placed in this material. A coverslip is placed on top, and the specimen is left to dry. Sometimes a thick specimen is not easily covered with the mounting fluid. It is necessary to build a platform for the coverslip. This is done by placing small squares cut from thick plastic sheets on the slide in the mounting media. These may be stacked in three or four piles high enough to support the coverslip above the specimen. Gummed data labels showing collecting locality, and so on, are attached to the right side of the slide, and the identification label, to the left side (fig. 16.8).

Various refinements of this basic technique have been developed. For example, some specimens may be placed in potassium hydroxide solution, which will dissolve the muscle and other soft parts. Small specimens should be placed in cold hydroxide. Larger specimens may be warmed to hasten the process. Do not overheat, as it will ruin the specimen. This same technique is used for mounting dissected parts of insects, such as the genitalia, mouthparts, or small wings, for permanent storage.

Some water-base mounting media are available for quick mounts. Most of these are unsatisfactory for permanent mounts. All permanent slides, both waterbase and resin (Canada balsam and such) require time to harden. They should be stored with the cover glass up in a horizontal position.

Data labels

In chapter 15 we explained why it is necessary to have data labels (fig. 16.8). Now we need to go into some detail about the data to be added to the label and what to do with the additional information gathered with the specimens. Many labels are inadequate for a precise study of insect populations. Enjoyable as it is to go into the field for the study of insects, no single person can do all the work needed to study all the specimens collected. Therefore, it is important to record data for the future use of others. If the material collected and the recorded observations

U.S.A., New York
Albany County
Rensselaerville
E. N. Huyck Pres.
U.S.A., New York
Albany County
Rensselaerville
E. N. Huyck Pres.
U.S.A., New York
Albany County
Rensselaerville
E. N. Huyck Pres.
U.S.A., New York
Albany County
Rensselaerville
E. N. Huyck Pres.
U.S.A., New York
Albany County
Rensselaerville
E. N. Huyck Pres.
U.S.A., New York
Albany County
Rensselaerville
E. N. Huyck Pres.
U.S.A., New York
Albany County
Rensselaerville
E. N. Huyck Pres.
U.S.A., New York
Albany County
Rensselaerville
E. N. Huyck Pres.
U.S.A., New York
Albany County
Rensselaerville
E. N. Huyck Pres.
U.S.A., New York
Albany County
Rensselaerville
E. N. Huyck Pres.

Collected in
decaying veg.
Collected in
decaying veg.
Collected in
decaying veg.
Collected in
decaying veg.
Collected in
decaying veg.
Collected in
decaying veg.
Collected in
decaying veg.
Collected in
decaying veg.
Collected in
decaying veg.
Collected in
decaying veg.
Collected in
decaying veg.
Collected in
decaying veg.
Collected in
decaying veg.
Collected in
decaying veg.
Collected in
decaying veg.
Collected in
decaying veg.
Collected in
decaying veg.
Collected in
decaying veg.
Collected in
decaying veg.
Collected in
decaying veg.

Figure 16.8. *Locality labels (left) and data labels (right), full size.*

that go with this material are sufficiently detailed and reliable, great amounts of data become available to others. Mere geographical data is not sufficient, although it is basic to every label. Data are used to relocate the place where the specimen was collected; for detailed records of host, habitat, niches, and any variations noted for a single population; and to correlate weather, altitude, and latitude data with the habits and distribution of populations. Further, the purpose of the locality label is to tag a specimen in such a manner that associated field notes can be added to the information known about a species.

Actually two labels are used—one, the locality label, and the other, the data label. The locality label should give concise geographical details to help find the exact spot where the specimen was taken. First, the country, then the state, province, or similar political unit, followed by the name of the nearest permanent geographical feature (such as river or mountain range), the nearest village, town, or city, with directions from this location or similar directions from the given locality. The examples shown in figure 16.9 use these data. This information should be kept to four lines, if possible, and the lines kept short to make a compact label to place under the specimen.

The data label is correlated with a lot-record book. All observations on the insect, as well as the habitat and environmental factors, are recorded in this permanent record book while still in the field. Now it is time to index this with the specimens by assigning lot numbers and adding these numbers to the specimens. Although some biologists object to this system, because of the danger of the lot-record book's becoming lost or destroyed, rendering the lot numbers meaningless, this objection can be overcome easily by exercising two precautions: 1) the data label should contain enough information so that the specimen will be useful even if the lot record book is unavailable; 2) the lot-record book should be photocopied and copies made available with the collection. The data label is also a four-line label recording the following information: 1) the lot number; 2) the date the collection was made; 3) the name of the collector; and 4) a simple statement about the habitat or conditions under which the specimen was taken. For example, the statement could read, "taken at UV black light" or "on carrion." Finally, the data and the collector are recorded on the label. Dates should be written either by abbreviating the name of the month or by using a Roman numeral for the month. Generally, in scientific records, the day is given first, followed by the month, and then the year: for example, 2–V–1984, never 2–5–1984, which is ambiguous.

A new lot number is assigned to each separate collection as it is made. Use the lot number to label the specimens in the field, but also add the locality data while still in the field.

Most collectors now have their labels printed, but if only a few are needed, they may be hand printed using India ink and a fine-dip pen. Locality and data labels are typewritten, using a carbon ribbon, in strips of ten identical labels. Many labels may be typed in this manner and then mounted on a large sheet for camera reduction. Consult with a local printer about the best way to go about this, or order them from an entomological supply company. Labels must last for hundreds of years. Therefore, they should be printed on 100-percent rag-content, 32-point, ledger paper. Nothing else should be used. Once printed, the labels are cut out ready for mounting on the insect pin beneath the specimen. Labels may be made up in advance, leaving a blank space for the date and the lot record number. The position of the label is regulated by the step block, the locality label at the second level beneath the specimen, and the data label at the third level. The label should run lengthwise with the specimen with the top line of the label to the right.

Vial labels contain the same data as pin labels. However, these may be typed and photocopied without reduction, cut, date and lot numbers added with alcohol-resistant ink, and inserted into the container. Whenever possible, add dates and lot numbers to the label before photocopying. This will assure that the added data is permanent and will not wash off in the fluid.

Housing the collection

Once the specimens have been mounted and labeled, they are ready for sorting for identification. This is usually done using "pro-tem" blocks, which consist of a pinning surface, usually polyethylene foam glued to a thin piece of wood about 4-by-7 inches. These can be placed inside a drawer or insect box and the specimens kept covered when not permanently stored. The specimens are sorted to order and, if possible, to family. Then comes the fun—identification (for this, see chap. 3). Once identified, they must be stored in a container that will protect them from harm. Cigar boxes and similar cardboard boxes are useful for temporary storage. Eventually, the specimens must be placed in standard insect boxes or drawers. These are expensive, but they may be made at home, if you have the proper woodworking tools. The best thing to do, if you are going to make boxes, is to purchase one professionally made to use as a pattern, or purchase drawer kits to put together. Most of us must depend on the commercially manufactured ones. Therefore, study the dealers' catalogs carefully and consult with other collectors, before you make your final decision about the type of storage you will use. By all means, use a standard system. The time will come when you will want to combine other collections with yours, or yours with other collections. Using a standard system will save time repinning specimens. The best system, and the most expensive, are the storage cabinets (fig. 16.9) for glass top drawers, with unit trays (fig. 16.10) of various sizes inside. A single tray is used for a single species. The number of specimens to store determines the size of the tray. Usually an assortment of four sizes is used.

Figure 16.9. *Cornell-type insect cabinet with drawers. (Courtesy of American Biological Supply Co.)*

Figure 16.10. *Unit trays for use in Cornell-type insect drawer.*

Temporary storage and shipping boxes may be made from cigar boxes by gluing polyethylene foam to the bottom. Remember, however, that these boxes do not have tight-fitting tops and, therefore, are not pest proof. Fumigant may be added and the top sealed with tape, if these are needed for longer storage. When shipping specimens, place a piece of cardboard over the top of the pins to hold them in place. Paper toweling placed on top of the cardboard will hold it in place during shipment. Pack the box in a larger carton filled with plastic "peanuts." (Note: the Postal Service offers a special rate, the library rate, for the shipment of specimens for scientific study. Ask for rates at the post office.)

Many private collections are housed in cardboard boxes (fig. 16.11) of the Schmitt-box type, a standard-size box that measures 9-by-14-by-2¾ inches. These boxes have tightly fitting tops and polyethylene foam bottoms.

Schmitt boxes and glass-topped drawers may be housed in special cabinets of wood or metal. These cabinets offer more protection from museum pests, but, of course, this is an added expense. Schmitt boxes are easily stored on their sides on book shelves.

Specimens are arranged in the boxes phylogenetically, as explained previously (see chap. 3, table 3.4 for a list of the orders

Figure 16.11. *Standard cardboard insect box. (Courtesy of American Biological Supply Co.)*

of insects arranged phylogenetically). The beginner will start by arranging the collection to order, usually a single box for each order. Of course, it doesn't take long to catch more than one box full of moths, many boxes of beetles, and so on. Before these can be arranged properly, they must be sorted to family. This will require some reference books, some of which are listed in appendix II. As the collection grows still more, you may decide that collecting every kind of insect will take up more space than you wish to devote to the collection. Now is the time to specialize. Certain kinds of insects will interest you more than others. Follow your own inclination to specialize on one or a few groups. But keep on collecting all kinds of insects, because this additional material will be of interest to other collectors and may be used for trade, or it will have scientific value to specialists at universities and museums.

Displays for classrooms, camps, and home entertainment

Glass-topped drawers with foam bottoms are generally used for classroom or camp displays. They can be arranged on a table or hung on the wall. When hanging these drawers, be sure that the pins are pushed deeply into the foam. Also, hinge or tape the top to keep it secure. It is a good idea to tape closed all display cases, because not everyone realizes how fragile dried insects really are and that drawers should not be opened except by those competent to handle specimens.

One of the most popular ways to display insects is to place them in the Riker mounts (fig. 16.13) mentioned previously. Riker mounts are shallow cardboard boxes filled with cotton and covered with a glass-topped lid. These are available in assorted sizes from supply houses. Similar display boxes may be made by using the display boxes that wallets and similar objects come in. The transparent plastic cover makes a "riker mount" when the box is filled with cotton. Or if you prefer to make your own, take any suitable, thin box and cut out most of the top with a sharp knife, after marking the edges on the inside. Glue or tape glass or any transparent plastic inside the top to form the cover. These mounts may be painted black or any other suitable color for uniformity. A hook or fastener may be taped to the back, permitting it to be hung on the wall.

Carefully place the insects in the mount. They must be dry, or they will mold once covered, and, being dry, they are fragile. Remove some of the cotton to accommodate the thick bodies of large moths, beetles, and grasshoppers. Place an identification label next to the specimen where it can be read. Keep the other data labels out of sight, but with the specimen. Always add

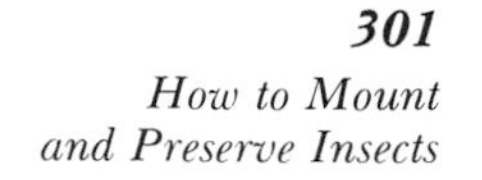

Figure 16.12. *Riker mounts. (Courtesy of American Biological Supply Co.)*

fumigant. The completed displays, artistically arranged, are attractive, as well as scientifically instructive (see fig. 16.13). Obviously these mounts are not suitable for the display of anything other than moderate-to-large-sized insects. Usually the collector arranges the mounts to show some particular theme, for example, mimicry, predation, color variation, or just plain beauty.

Care and protection of the collection

All insect collections have problems that haunt them—museum pests. Mold and other insects like dry insects even more than collectors do. They are excellent food for these organisms. These museum pests are usually beetles or psocids—the book lice. Proper drying will prevent mold from starting, and the fumigant used to repel insects will also deter mold growth. But all collections must be kept fumigated at all times. For some reason, whenever fresh specimens are introduced into the collection, museum beetles seem to follow. It is doubly important to fumigate all new specimens. We mentioned previously the type of fumigant to use. Naphthalene flakes, long used to repel these pests, are more dangerous to use than paradichlorobenzene crystals, because of possible carcinogenic properties. Whatever you use, place the material in a small box (if the storage containers are to be kept flat) or in a bag of cheesecloth. Pin these in the corner of the box or drawer. Evidence of pests, usually first spotted by the presence of a pile of chewings under the pin, or by the caste skin of the larva, indicates the need for more drastic fumigation. The best way is to place the collection in a cabinet with the boxes or drawers open. Place a dish of carbon tetrachloride in the cabinet and close it tightly. Remember that this material is deadly to you, too, so do not inhale it. Fumigate for at least forty-eight hours. This will kill all of the active pests, but usually it will not kill the eggs. Therefore, the process may need to be repeated within a week or two to kill any newly hatched larvae.

Finally, if you should ever grow tired of the collection, do not just abandon all the work that you put into it. Offer it to a museum and take a tax deduction for your donation (see chap. 14)!

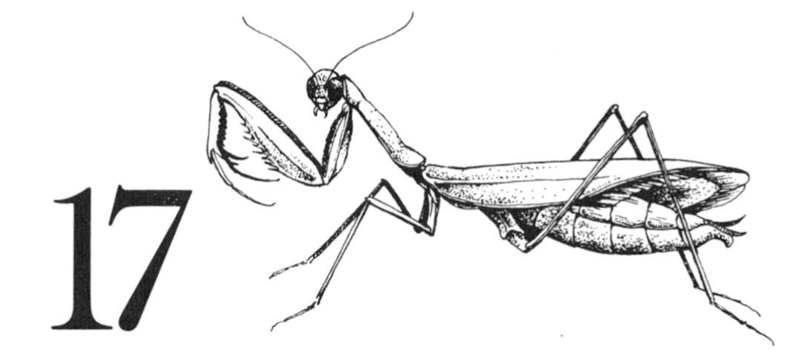

17

Saving Our Insect Environments

"The Subject is small, but not inglorious. The Ant, as the Prince of Wisdom is pleased to inform us, is exceeding wise. In this Light it may, without Vanity, boast of its being related to you, and therefore by right of Kindred merits your Protection."

William Gould (1747)

"I can't believe you are serious about keeping insects alive! Why would you want more of these pests?" These words have been repeated to both authors year after year, as we have talked to our students about insects. Students and readers deserve to know why insects should, indeed, must be, conserved. Before this can be really comprehended, the nature of ecosystems must be understood, because this is basic to all conservation attempts. Conservation practice based on anything else is and will be faulty, even disastrous. One cannot conserve those attractive plants and animals we love so much and destroy those "creepy, crawly things." One ardent nature lover once remarked: "I hate those horrible greenbriers!" Little did she know that some species of the genus *Smilax*, the greenbriers, have remarkable meiotic (cell division) processes, and, without these plants to study, we would never know that cell division of this sort takes place. Who knows, an understanding of these obscure phenomena may lead to more knowledge about cancer.

Ecosystems

Up to this point, we have discussed in some detail the nature of insect populations. We have pointed out where these populations occur on this earth, but so far we have discussed little about how closely tied each population is to other populations of its own species, other species, and the physical environment. Ecology, the study of these relationships, may be approached from two directions: the interrelationships of each species to its environment (autecology) and the intertwining of all species and their physical environment to form a community (synecology). Synecology, or the ecology of communities, has been broadened to include the interrelationships of communities, particularly as they form parts of a food chain, a concept termed an ecosystem. These ecosystems control all life as we know it, penetrating every community, including those composed of humans. In fact, it is because we cannot escape these ecosystems that we are concerned.

Many of us have experimented with a terrarium (see chap. 18). We gathered small plants (some bloodroot, violets, and, perhaps, a jack-in-the-pulpit), moss, and lichens from the woods, planted them in the soil in the bottom of an aquarium, watered them, laid a piece of glass over the top, and waited. Our two-cubic-foot, isolated world did nothing. It needed energy, so we placed it in the sun, and it became alive; we had wound the clock spring and learned that no system can run by perpetual motion. No completely closed (isolated) system occurs on this earth. All living systems are dependent on energy as light and heat from the sun.

So far, the terrarium contains only plants and probably several microorganisms, including fungus spores, water bears (Tardigrada), springtails (Collembola), maybe a wood louse (Crustacea), and even an earthworm (Annelida). The system seems balanced, but monotonous. Soon, however, we find green shoots coming up all over the soil. Some grass seeds have sprouted and threaten to take over—so we pull them out. We want a more interesting terrarium. About this time, the jack-in-the-pulpit dies—why? We figure it received too much sun, so we move the terrarium into the shade. Then the violets stop flowering. Apparently things are not as well balanced as we thought. Anyway, we will liven things up a bit by adding a couple of young preying mantids that we noticed in our garden. But they soon die. We will try again and add a rotting apple with some fruit flies flying around it. Soon thereafter, all fruit flies are gone, the new preying mantids are dead, and the apple is covered with mold. The more we tamper, the worse things become. Finally, we decide to raise geraniums.

What is the problem in balancing a terrarium? Simple; we cannot isolate a community. It must be tied to other ecosystems to function. This same struggle to balance a community occurs every year in our lawn. We fight weeds. We add fertilizer, then we mow to overcome the effects of the fertilizer. The battle is constant, because every community is dynamic; it cannot be made static.

Ecosystems, it is easy to see, depend first on an energy source, ultimately light energy from the sun. This is converted into carbohydrates, through the process photosynthesis, by combining water, carbon dioxide, and certain minerals in the presence of chlorophyll, a substance found only in green plants. A by-product of photosynthesis is oxygen. The only source of oxygen, this vitally necessary element on this earth, is from photosynthesis. Many of us think that oxygen is a substance that is endlessly present in our atmosphere—not so. It is produced by our green plants, and, without it, nothing would live on this planet. If our terrarium animals and plants were sealed in the aquarium, it would only be a matter of days before everything died from lack of oxygen, if the animal life used more oxygen than the plants could produce. That is why an air pump is necessary for an aquarium tank of fish. The fish and the plants both use oxygen at night when photosynthesis is not taking place. New York City and Los Angeles both need air pumps when they are covered by a blanket of smog.

An important part of every ecosystem is, therefore, a balance of green plants. Fortunately for us, plants produce an abundance of oxygen, extra leaves to eat, fruit, seeds, and fiber, nearly as fast as we can use it, but not everywhere. If this happened everywhere, there would be no starvation. Briefly, then, apparently all animals, being without chlorophyll, must depend on plants as their energy source, hence, their food, because only plants are capable of synthesizing inorganic material into the organic materials, carbohydrates, protein, and sterols that animals must have. So why did the preying mantids die in the terrarium? They are second generation (predators) in the energy-transfer chain. But we provided them with a first-generation (phytophagous) food source, the fruit flies. Alas, nature played a trick on carnivores!

Look at fig. 17.1, a simple ecosystem. We start with the corn plant, which uses a certain amount of the sun's energy to produce leaves and ears of corn. Along comes a grasshopper and chews on the leaves and ears of the corn. The corn is the one and only primary producer in this ecosystem. It must produce for itself, that is, the carbohydrate it synthesized is used for its own growth and for the formation of seeds eventually to produce more corn plants. It has a seed shadow, just as insects have egg shadows. More seeds are produced than are needed to keep the corn species in existence. The bulk of the seed shadow (at least 80 percent) is eaten and never germinates. Even if 20 percent reaches fertile soil, most of that is eaten by grasshoppers and other pests. The grasshoppers are primary consumers. Grasshoppers are eaten by skunks, secondary consumers. It takes, however, ten times as much energy to feed meat (protein) to skunks as it does to feed plants to grasshoppers. A great energy loss occurs byinterposing a protein converter (carnivore) between the primary producers and the secondary consumers. Skunks are eaten by hawks, tertiary consumers, which increases the energy cost another ten times, or now a hundred fold. Or, just to put it closer to us, if we get our protein from beef, it takes ten times as much food to create a steak for our energy source, as it would if we ate the corn directly. How complex energy flow becomes! What a trick of nature, and that is not the end!

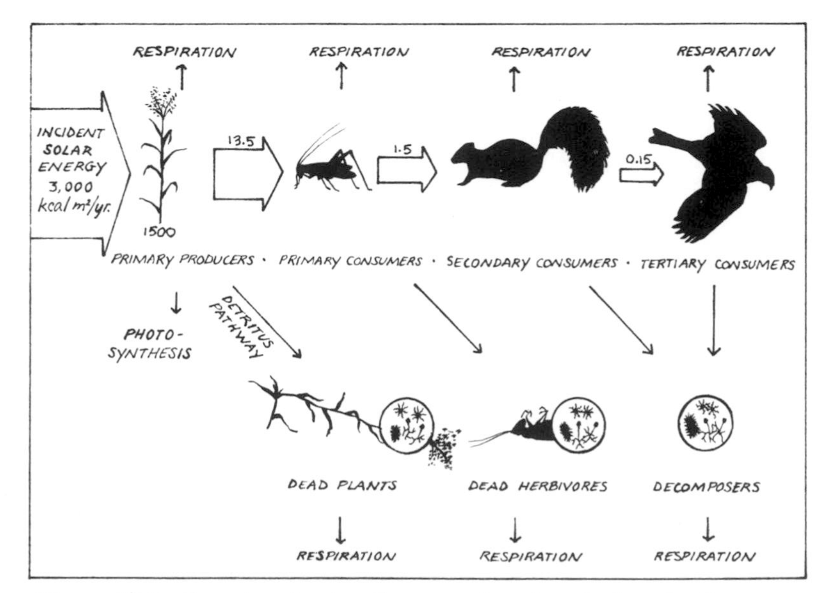

Figure 17.1. *Ecosystem showing the energy drop from herbs to herbivores to predators, and the recycling of the energy through the detritus cycle.*

All living organisms die, certainly the one characteristic feature of all life. But what happens on death? There lies the true cost of energy. Sunlight is stored in living tissues, and it must be released. Conversion of protoplasm back to its original form, carbon dioxide, water, and, most important, minerals used by green plants to store this energy, is the biological process known as oxidation. This takes place in our cells as we eat, grow, and move. This is the way we get the energy to move. Sunlight reaches our cells, keeps us warm, and supplies us with the force to contract muscles. Exactly the same process takes place whenever we turn the ignition key in our automobile. Fossil plants converted into gasoline are combined with oxygen to release sunlight to turn the wheels of our car at a great expenditure of energy stored finitely by plant life of long ago. How far-reaching these ecosystems are! Dinosaurs and automobiles are contemporaries when it comes to their energy source. But what of our immediate needs to recycle this energy?

Dead bodies do not decay by themselves. Camels in the desert and mammoths in ice remain unchanged, because the action of decay bacteria is stopped. Embalming prevents decay and selfishly stops the release of energy and nutrients (nitrates and other minerals) loaned during the life of the body. This recycling is the result of a chain of detritus organisms. Bacteria and fungi feast on nonliving flesh. These are joined by insects of many kinds. Fly larvae, dermestid, staphylinid, and silphid beetles play out their role in the chain. Eventually the corpse returns through decay to the dust from whence it came. This "dust" contains the essential chemicals needed by plants to store energy again. Particularly important is nitrogen as nitrates, nitrogen/oxygen compounds, the basis of all protein.

Plants must have nitrogen to make protein. Nitrogen is so abundant as a gas that it forms most of our atmosphere; oxygen is a small part of "air." But it is as hard to get nitrogen to form water-soluble nitrates, as it is to get carbon dioxide and water to form carbohydrates from the energy of a lightning bolt; yet, this is the way some nitrates are formed. Certain bacteria specialize in creating those nitrogen compounds by acting on the protein of dead bodies, or, more directly, by fixing atmospheric nitrogen in the soil. The latter bacteria live in little nodules on the roots of clover and other leguminous plants. So, you see, plants can be independent of animals, but animals cannot exist without plants.

Once we have established the existence of these complex food (energy) chains, we are ready to consider the role of insects in these ecosystems. First, however, let's sum up these principles in the five laws, four of which were first proposed by Barry Commoner in his book, *The Closing Circle*. These principles, unscientifically written, state that:

1. Everything is connected to everything else (ecosystems)
2. Everything must go somewhere (detritus cycle)
3. Nature knows best (all biological processes operate regardless of man's attempt to alter them)
4. There is no such thing as a free lunch (energy robbed from one place must be made up elsewhere)

We would like to add another:

5. What a tangled web we weave once we try to outwit nature (or, the best way to conserve nature is to leave it alone)

It is said that if a tree is cut down on the top of Mount Washington in New Hampshire, it will affect the lobster fishing off the coast, about 60 miles away. Cutting down a tree lets sunlight reach shade plants. These die, as did those in our sunny terrarium experiment. Loss of ground cover permits erosion, which, in turn, increases the silt reaching the ocean, and this, in turn, covers marine organisms. These marine deserts force the lobsters farther out to sea. Can you imagine lobster fishermen complaining to lumbermen that they are destroying their fishing industry? Can you deny that it is true (Commoner's Law no. 1)?

Some pages back, we discussed termites and how these pests first nest in wood in the soil and then penetrate the wood of our buildings. How do these pieces of wood get into the soil? Watch the builders in a new development. They build the houses, cut pieces of wood to fit, dropping the cutting onto the ground. Later a bulldozer will come to landscape the grounds, cover the wood scraps with soil (Commoner's Law no. 2). These seed beds for termites soon make work for the local extermination company.

Then study the case of the citrus growers in California. Some years ago, they were heavily spraying with organic chemicals to control certain scale insects. Shortly thereafter, plant-feeding mite pests took over as the major pest, these being unaffected by the chemical spray (Commoner's Law no. 3). You can't beat the laws of nature. Ecosystems were once unbalanced by the improper overuse of these sprays. They kill natural mite predators at the same time they kill the other pests. This bitter lesson has now been learned, and we are happy to say that attention is being given to the balancing of natural controls with the proper use of newly developed insecticides. (It is obvious that cultivation itself is a disruption of a natural balance of plants; therefore, if we are to eat, we must continue to use control measures for crop production.)

Four hundred years ago, much of southern New Mexico and eastern Arizona, if we are to believe the great explorer, Cortez, was covered with dense stands of grass and foraging buffalo. He reported grass up to the bellies of his horses. Today, due solely to the overgrazing of the imported European cattle, the same areas are deserts. Buffalo are sustained by native grasses. European cattle need more succulent plants, which they soon exterminated in vast areas of the southwest. It now takes one hundred acres to feed a single cow and her calf in the same area that one acre might have fed a buffalo (Commoner's Law no. 4). What a price to pay for our needs as secondary consumers, especially to plug our arteries with excessive cholesterol.

So, we go about to correct these past errors by enforcing conservation practices. Remember what happened when we experimented with the terrarium? Adjust something here, and something else goes out of adjustment. Try to save the elephant from extermination, and they eat and kill all the Baobab trees (fig. 17.2). African Baobab trees are reported to be over a thousand years old. Elephants living when these trees first grew in Kenya did not destroy them for all the many generations that they have lived. It was not until man came along, first killing the elephants, and then trying to save them from extinction (Commoner's Law no. 5) that the trees became endangered.

Figure 17.2. *A thousand-year-old Boabab tree (Kenya) damaged by elephants feeding on the bark.*

Habitat preservation

Our knowledge of the breeding populations of insects is poor. Probably the breeding habits of not over 1 percent of the described species of insects is known. This is particularly true of the "rare" species, the exact ones we are concerned about saving. What we do know about most of the species involved is the *kind* of habitat in which they breed. It seems obvious then, if we are to save the species to study it, we must save the habitat. Little can be done if the species has no place to breed.

Several years ago, the United States Congress (1973) passed a law permitting the designation of certain species as threatened or endangered. The exact difference between these two categories is not clear, but the responsibility for these designations is given to the United States Department of the Interior, Fish and Wildlife Service, Endangered Species Program. Roughly, a threatened species is one that is known to be scarce and needs to be protected to continue its population levels. These species are usually those with a distribution range limited to a few square miles. An endangered species is one that has been reduced by habitat destruction to the point where the organism faces almost immediate extinction. Obviously, most threatened or endangered species are wild flowers, birds, mammals, and a few lower vertebrates, groups that have been studied in considerable detail. Their exact distribution range is well known, and, sometimes, the exact number of individuals is known. For these plants and animals, the system works fine.

As might be expected, few insects have been designated as threatened or endangered. At this writing, less than a dozen species have been placed on either list. Some of these species are listed in table 17.1.

Besides the list of species, the species to be protected as endangered, the known habitats are also designated as a "critical habitat." These protected areas are set aside and specimens cannot be collected there.

Various publications published by the Endangered Species Program list the species that are protected, additions approved for the lists, and the areas harboring these species. These lists are circulated, and it is presumed that law enforcement agencies are also provided with this information and that they are capable of enforcing these laws.

Several factors must be considered before placing insects (or any organism) on either of the lists. Some of the species of butterflies considered for the list have not been native species, but, rather, those strays filtering up from Mexico or over from the West Indies. Obviously, the first thing to consider before placing a species on the list is whether it is a native species. This

Table 17.1. *Threatened and endangered species.*

Scientific Name	*Common Name*	*Distribution*	*Status*
Coleoptera	(Ground beetles)		
Carabidae	(Delta green ground beetle)		
Elaphrus viridis	(Longhorned beetle)		T*
Cerambycidae	(Valley elderberry longhorned		
Desmocerus californicus dimorphus	beetle)		T
Lepidoptera	(Swallowtail butterflies)		
Papilionidae	(Bahaman swallowtail)		
Papilio andraemon bonhotei	(Schaus swallowtail)	Strays in s. FL	T
Papilio aristodemus ponceanus	(Hairstreak and blue butterflies)	Common in s. FL	T
Lycaenidae	(Hessel's hairstreak)		
Mitoura hesseli	(Palos Verdes blue butterfly)	NH, NY, PA, NJ, MD, VA, NC	T
Glaucopsyche lygdamus palosverdesensis	(Karner blue butterfly)	Los Angeles Cty., CA	T
Lycaeides melissa samuelis	(Metalmark butterflies)	n.e. NA	E
Riodinidae	(Lange's metalmark)		T
Apodemia mormo langei	(Brush-footed butterflies)	Contra Costa Cty., CA	
Nymphalidae	(Oregon silverspot)		E
Speyeria zerene hippolyta	(Callippe silverspot)	Salt marshes of n. CA, OR, WA	
Speyeria callippe callippe	(Nymphs, satyrs, and others)	San Francisco area, CA	T
Satyridae	(Mitchell's satyr)		
Neonympha mitchellii		MI, IN, OH, PA, NJ	E

T = Threatened
E = Endangered

merely emphasizes the need to know more about the organisms in question before action is taken.

Second, as pointed out in our discussion of dispersal, some of the "endangered" species are not really established in the locality where they are "rare." Protecting them will be a futile effort. They may breed for a short time in the area protected, but die out during a severe winter.

Third, if a species is actually confined to a small area, the restricted gene pool will probably hasten the extinction of the species through natural causes. A species depends on variation to survive. Populations with wide latitudes in such gene-controlled features as cold resistance, heat resistance, food selection, and so on will be able to survive extremes in environmental changes. Small populations in restricted areas can be wiped out by an extremely cold winter or other extremes.

We must understand that not all species are adapted to survive beyond a certain time. Species also grow old and die. They will become extinct no matter how hard we try to save them. Such is the history of all attempts to restore a species once its numbers are reduced to a certain level (determined by the genetics of the species). This genetic function is much the same as we see in mechanical equipment. Machines are designed to do a certain job. Once conditions change, and there is no longer a need for it, it is no longer manufactured and it becomes obsolete.

Endangered and threatened species and the collector

Many species of insects have been proposed for the list of endangered and threatened species. Only a few have been selected for the list (see table 17.1). There seems little doubt that these are selected because of public interest, rather than for any real scientific reason. The species are only a few of many thousands of known species that might be on the list. But most of the species that should be here are poorly known. It is impossible to document the reason for their inclusion. Scientific logic is in conflict with bureaucracy.

No evidence exists to show that collectors have even seriously depleted a population of insects. All evidence points to habitat destruction as the cause of the extinction or near extinction of a species or subspecies. The widely publicized extinction of the Xerces Blue Butterfly *(Glaucopsyche xerces)* (fig. 17.3) was caused by a military post in 1943 on the upper San Francisco Peninsula near the city of San Francisco. The species was in decline at the time because of habitat destruction elsewhere. Its former range extended from about North Beach to Presidio and

Figure 17.3. *Xerces Blue Butterfly, an extinct species (Lepidoptera).*

south along the coast to the Lake Merced district. Specimens are in many insect collections, but there is little doubt that it is now extinct. A society has been formed, the "Xerces Society," with worldwide membership, devoted to the conservation of rare insects and other invertebrates. They endorse the need to recognize and protect as many unique habitats as practical. The work they are doing is helpful and should be supported, along with the several other organizations of a similar nature. For example, the Audubon Society, Nature Conservancy, and the Sierra Club are all concerned with land preservation as a basic conservation practice.

Unfortunately, one governmental agency may be in conflict with other agencies. Many government projects involving stream control have resulted in habitat destruction of many of our fresh water habitats. Almost thirty orders of insects are represented in or near wild water. Of course, when these bodies of water are disturbed as part of "flood control," these habitats are destroyed. Stream bank clearing is one such practice. When this is done to "prevent" flooding (which it seldom does), the streams are rendered sterile of all plant and animal life. To counter this, the fish and game people stock the same streams with game fish, and then, if any succeed in surviving the fisherman's lure, food will also have to be imported to keep the fish alive.

As you have no doubt been taught, any abandoned field is an "eyesore," and all attempts to "improve" the property will destroy it as a biological study area. University officials seem to be "death" on such areas. They find every means possible to do away with these fields (and then turn around and pay the bill for the purchase of specimens for classroom study, many of which could be collected at no cost right on the campus). County engineers believe that the best thing to do with a vacant area is to turn it into a "landfill." Eventually the public must learn that such areas harbor, and are the breeding grounds for, many predators and parasities of insect pests. Man needs these reservoirs. Hence, when these areas remain wild, they make excellent collecting grounds and preserve many "endangered" species.

Each of us interested in insects, as well as other animals and plants, should be alert to habitat destruction and determine whether alternatives are possible. Further, we can encourage the propagation of butterflies (but remember, some are pests) and beneficial insects by planting flowers that will attract them.

Inviting insects into your garden

Why are many gardens alive from sunrise to sunset all summer long with butterflies and bees? Much of this activity is due to the species of plants in the garden, not the quantity. Entomophilous plants, that is, plants that require insects for pollination, offer the insects sweet nectar in return for their services as pollinators. Wild and cultivated plants may be grown. You may already have some of them in your garden. Increase the number and variety by studying table 17.2 and planting some of these and you will be rewarded with a wonderful variety of interesting insects.

The destruction of many wild flower habitats throughout the United States is due to the growth of housing expansion and the building of new shopping malls, and corporate industrial parks. Added to this is the tremendous amount of lawn mowing on the interstate highways and state roads, and the destruction of habitats by the building of dams and the stream-channeling mentioned previously. This renders vast areas completely without these plants where they once grew in abundance. As this number decreases, so does the number of insects; especially obvious is the lack of butterflies. You can do something about this by planting those garden flowers that attract insects. Select one small corner of your yard for some wildflowers, such as milkweeds, clover, dandelions, and goldenrod. If you select the right species of plants, you will have visits by butterflies all season long. Try to encourage local officials in your community to leave some ground available for wildflowers. A small section of a park, an area adjacent to a railroad track, and the island between superhighways are all areas where wildflowers and their insects will flourish.

The National Council of State Garden Clubs, Deep-south Region, has made butterfly conservation a project. They even

Table 17.2. *Entomophilous plants for your garden.*

(These plants attract butterflies, bees, and other pollinators.)

Plant Name	*Perennial = P Annual = A*	*Cultivated = C Wild = W*	*Blooming Time*	*Comments*
Lilacs	P	C	May	Shrub, fragrant blooms of white and lavender
Butterfly bush	P	C	June till frost	Shrub, purple, white, and yellow flowers
Rose of Sharon	P	C	July till frost	Shrub, white, purple, and pink flowers
Mimosa tree	P	C	July–Aug.	Tree, pink tassles
Zinnias	A	C	All summer–fall	Popular garden flower, easy to grow, all colors
Marigolds	A	C	All summer–fall	Popular border garden flower; also an excellent plant to deter insects
Garden phlox	P	C	July to Sept.	Pink, white, and red flower
Spearmint	P	C	Summer	White flowers, fragrant plant, common herb
Chive	P	C	Summer	Pink flower
Lantana	P	C-W	Summer	Superior butterfly lure; at least 6 colors available; may grow wild in the south
Dandelions	A	W	Spring till frost	Not a weed to many! Very attractive to butterflies; good to eat in a fresh salad
Clovers	A	W	All summer	Flowers of white, red, and purple; common in lawns along with dandelion
*Butterfly weed (*Asclepias tuberosa*)	P	W	Summer	Sold by seed growers and catalogs; DO NOT REMOVE FROM WILD LOCATIONS! Brilliant orange flowers
Bee balm	A	C	Summer	Leaves with lemonlike flavor used in seasoning

*Butterfly weed is but one of a number of plants in the milkweed family that is highly attractive to insects, especially butterflies. Many of these plants are considered weeds, but actually they can be attractive border plants—it is all according to your perspective. Included in this group are common milkweed, purple milkweed, red milkweed, swamp milkweed, white milkweed, whorled milkweed, four-leaved milkweed, and blunt-leaved milkweed.

Other "weeds" that prove attractive to butterflies are ironweed, Joe-Pye weed, goldenrod, and thistles.

have bumper stickers which read: "Take a butterfly to lunch—plant a butterfly garden."

Beneficial insects to conserve

Various predators and parasites are discussed in some detail in chapter 13. But to relate this information to insect conservation, one must realize that there must always be a reserve of these insects. If these reserves are killed by insecticides, there is no way that enough will become available to have much effect on the control of pest populations. Remember, if the parasite or predator succeeds in killing an entire population, it also kills itself, because it will no longer have anything to eat—unless there is a wild area in which to retreat and rebalance its population.

18

Insect Zoos and Other Insect Study Projects

"Constant alertness is required to see that fit conditions are maintained, and a large measure of ingenuity is often necessary to adapt places and circumstances to keeping conditions fit."

James G. Needham (1868–1957)

Years ago the books of William Beebe, a famous naturalist at the Bronx Zoological Gardens, were as popular as those of Thor Heyderdahl, of Kon-tiki fame, or the voyages of Jacques Cousteau. Beebe, too, wrote about undersea life, but he was also famous for his work in a tropical rain forest in Venezuela. He tells of the massive tropical beetles, including vivid descriptions of the "battle of the beetles." These large beetles reach a length of six inches. Their head and thorax horns are used to fight with other males, and they are also neatly curved to carry off the female, should they win her hand in battle. Sad to say, more often than not, while two males are doing battle to win the fickle female, a bystander male is as likely as not to run off with the prize!

Reading such tales in our youth instilled a lifelong desire to see these things firsthand. And this one of us was able to do for two and a half years in Panama. During that stay, close attention was paid to how insects live, what they do, and how long it takes them to do it. We have combined into this chapter information on "laboratory" studies, as well as hints on what kind of data one must look for while in the field. Alas, there is not space enough in this book to tell you about all the wonderful things to be learned, but, instead, we will tell you how you can do it yourself. You need not go to Panama or Venezuela to do it, because you can duplicate some of these marvels right at your own home.

Not many years ago the idea of an insect zoo was considered to be of little interest to the public. People like to watch monkeys; we have to admit to a certain kinship. Elephants are wonders; and all the range of mammals, birds, reptiles, amphibians, and fish are brought to us in zoos and aquaria. Plants are grown and exhibited in botanical gardens, but where are the invertebrates, by far the majority of species? That question was asked by some of the staff at the Smithsonian Institution in Washington, with the net result that an insect zoo was started. It enjoyed almost instant success, even outdoing the entomologists themselves, who, as it happened, were having an international meeting right in Washington at the same time the zoo opened.

Up until this time, the only insects seen alive in captivity were honey bees in observation hives at a county fair or, rarely, at a museum. But since the Smithsonian's insect zoo opened, several others have been built and opened to the public. If there is one near you, by all means visit it and talk to the director. Directors are proud of their success and will be glad to tell you all about how to start a zoo of your own. If you can't see for yourself, read the way to get started on the following pages.

The Smithsonian zoo has numerous habitat exhibits (fig. 18.1), showing the natural environment of the species that they are able to keep in captivity. These natural habitats are easier to construct than those for elephants or monkeys, because they can be smaller. The species are local, as well as exotic (fig. 18.2). Since so little is known about local species by most people, these species seem to be of as much interest as the foreign species. One cage is a large glassed-in area with small trees and shrubs. Butterflies, bees, and other flying insects are free to roam. A small stream flows into a pond in this exhibit. The pond contains water striders and other water insects. Various species of ants, beetles, and bugs are in smaller exhibits. One cage has a microphone in it. Inside crickets stridulate, and their song is amplified throughout the room.

As it happened, one of us worked behind the zoo doors in the Lepidoptera range for several months while working on

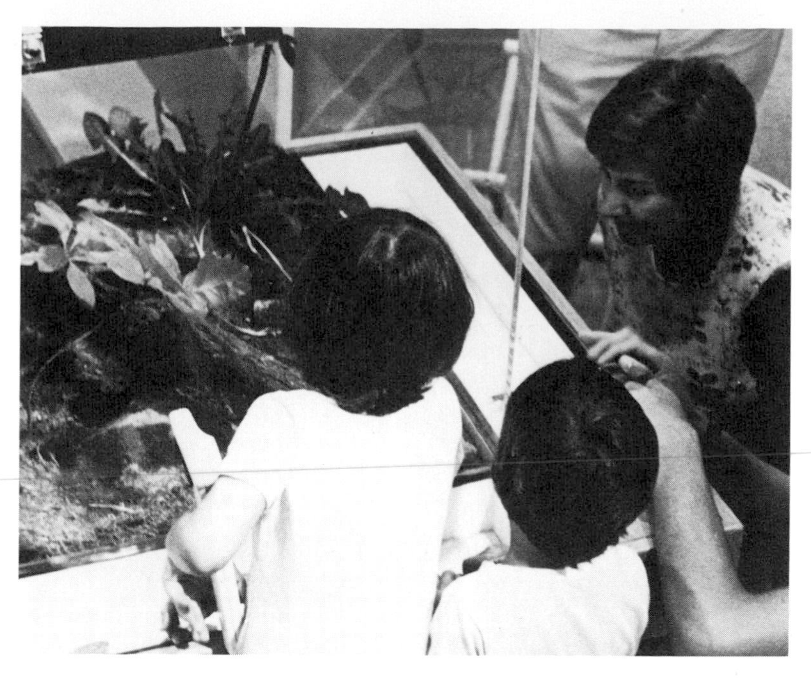

Figure 18.1. *Plexiglass insect cage for public viewing of an insect-zoo display of living insects. (Courtesy of the Smithsonian Institution)*

another book. Nearby were the rearing cages that supplied the stock for the zoo cages. This gave us a chance to see what could be easily raised and what gave trouble. Some of this information is passed on to you here.

Insect zoos should become popular in any neighborhood.

Figure 18.2. *A rhinoceras beetle from Panama, an example of a giant insect that can be displayed in an insect zoo.*

Once you can convince people that no diseases will be spread by these insects, and that they can be contained and controlled in homes and schools, you will find not only pleasure for yourself, but for those who come to view your work.

Insect zoos have some distinct advantages over ordinary zoos. One can house a fine display in a small area. A three-foot-square plexiglass case makes a beautiful display for an entire colony of ants. Another advantage is the abundance of local species for the display and the ease of acquiring specimens. As you know, it takes zoos many years to acquire a rounded-out exhibit of tigers, lions, bears, deer, and so on. A trip to a wild meadow in nearby countryside will supply you in an hour with far more than you can hope to install in your zoo. Actually, what you have to do is decide just how much space you can devote to this living menagerie of insects, and then limit your collecting to the species that are easy to raise at first. Then gradually work into the more difficult species and those that are exotic and bizarre. Once established, insects reproduce rapidly, and they are fully capable of eating you "out of house and home," as they say. A colony may be kept self-reproducing for as long as you wish to attend to it.

Initial costs of setting up a zoo of this type are minor—certainly nothing like trying to provide for a polar bear den or an alligator pit. Since you will not have to take these pets to the veterinarian for shots, they will cost less than even a stray cat or dog. Insects may require some special environmental features for their housing, but their food will probably be free—they either eat what you do or what you don't want! For some of these creatures, you may have to invest in a few magnifying glasses for your viewers to use. Once you have learned to "think small," you will have your own "Lilliputian World of Insects."

Cages

The construction of a series of insect cages, terraria, and aquaria will be needed for these rearing projects. Each cage may be used to depict a different habitat or a different kind of insect. Some will be cages for "tigers," others for "elephants," and still others for "birds." These cages need to be ready for the insects when they are captured. If you build your own cages, they may be started in the winter months in anticipation of a "wild animal" hunt in the spring.

Cages will vary in size according to the insects displayed.

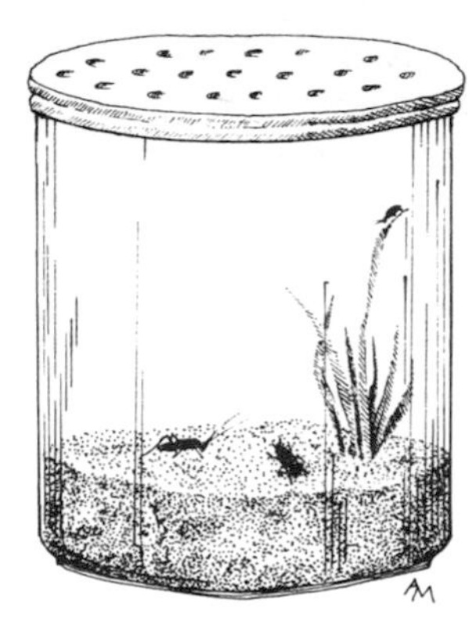

Figure 18.3. *Jars can be used as cages for home or classroom insect zoos.*

Small jars (fig. 18.3) are useful for a single or a few specimens. Larger terraria the size of 10- or 20-gallon tropical fish tanks will be needed for other insects (fig. 18.4). A convenient size for some insect rearings is the gallon jar used by restaurants. These widemouthed mayonnaise jars will permit easy access for the preparation of the habitat and are ideal for many uses. Be sure that they are clean, especially the screw-top metal lid. For certain species, nail holes punched in the top will allow enough air to enter, but generally it will be necessary to cut away most of the center of the lid and replace it with fly screen. Pint and quart fruit jars with two-part lids are also good. The center of the lid may be removed, and plastic fly screen placed on the top of the jar and the ring screwed in place.

The inside of the jar or terrarium should be provided with sand, soil, or debris, according to the natural habitat of the insects that are to live there. Aquatic displays are usually aquaria of assorted sizes. Terraria may have all glass sides, or some sides may be constructed of wood. Even fish tanks may be used, but these are more expensive and probably should be reserved for aquatics. The top or cover of the terrarium must fit tightly enough to prevent the inhabitants from escaping. Sometimes, it is necessary to place a rock or similar weight on top to hold down the lid.

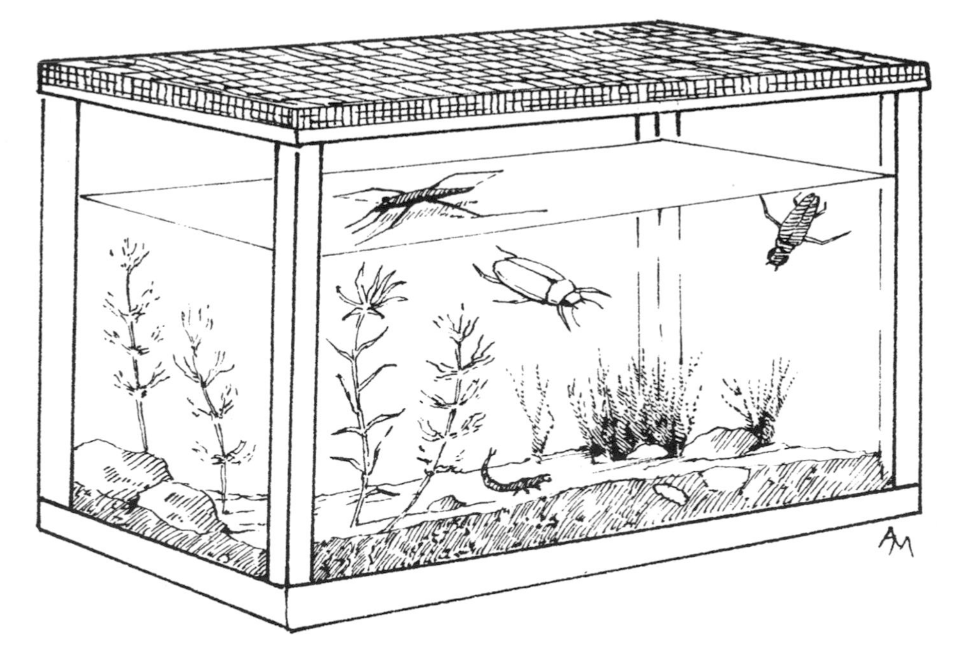

 Figure 18.4. *Fish tanks may be used as cages or for aquatic insects.*

Aquaria may be stocked with a wide variety of insects, particularly, water boatmen, water striders, backswimmers, diving beetles, water bugs, and the naiads of dragonflies and damselflies. Be sure the tank is clean to start with, and use the pond or stream water in which the insects lived, instead of tap water. Avoid tap water, as it may be too hard, too soft, or poisoned with chlorine. Place some gravel, sand, or small rocks on the bottom of the tank. Aquatic plants may be added as available in the original habitat. Since many of these aquatic insects are predacious, great care must be taken to properly balance the tank. Interesting prey–predator relationships can soon get out of hand, and as we have seen in our previous discussion of trying to balance a terrarium, it will be next to impossible to balance the aquaria, as well. You will need to regulate the light for the plants and the air for both the plants and the animals. The ambient temperature of your locality is suitable for these tanks. But, if you keep them over the winter with living naiads, nymphs, and adults, the tanks will need to be warmed.

Collecting live insects

Collecting insects for the zoo is much simpler than trying to find additions for your pinned collection. It is also inexpensive, because wild insects are free! Avoid damage to the environment, however. A good rule to follow is "take nothing, but insects; leave nothing, but footprints." Some of the species to look for, and the way to handle them, follow.

Hand-picking specimens is the quickest and most selective way to capture insects for your zoo. Such insects as mantids, grasshoppers, and crickets may be found in the late summer. Some of these will remain active, especially crickets, throughout the fall and winter. Mantids and grasshoppers will lay eggs, and the young can be reared in the spring. Beetles, certain species of bugs, and, best of all, for showy specimens, moths and butterflies, may be gathered. Be sure to have several containers with you to keep specimens unharmed.

Figure 18.5.
UV light trap with collecting cage, instead of killing jar, attached to bottom for collecting live insects.

Night collecting at UV lights is best for the large, nocturnal moths that are always popular in a zoo. If you use a light trap, a small screened cage can be fitted to the bottom of the trap (fig. 18.5). It is wise to monitor this cage closely. You will need to remove large moths and some of the predators that will enter the cage. Another way is to hang a UV light from a limb in front of a white sheet suspended from another branch. You will soon find a supermarket of lively visitors to select from, many of which will be gravid females ready to lay eggs in your cages. Before you go too far, you must figure out what the hatching larvae will eat.

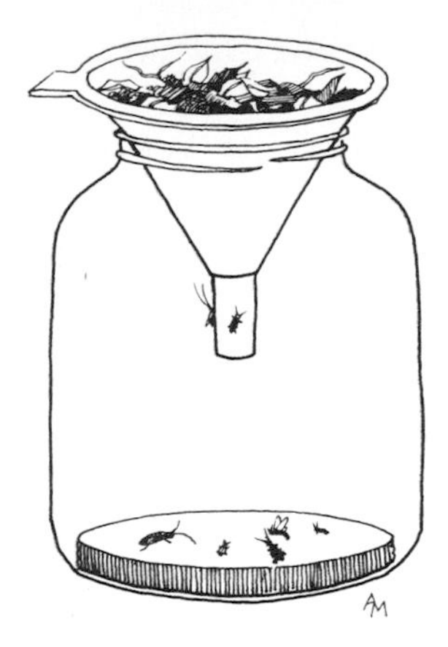

Figure 18.6. *Berlese funnel.*

One way to add unusual species to the zoo is to collect by using a Berlese funnel (fig. 18.6). Insects from leaf litter, humus, moss, and fungi can be captured alive with this device. Microarthropods include the members of class Entognatha (see table 3.2), the Collembola, Protura, and Entotrophi (also known as Diplura). Other noninsect arthropods often collected include mites, pseudoscorpions, symphylids, and pauropods, if you include soil in the material added to the funnel. Along with everything else, the first to appear in the container under the funnel will be beetles. "Why are they there?" you may ask. Because they are eating the smaller animals. This gives you another reason for collecting in this manner, because it will provide food for these larger insects that you may wish to display.

A pitfall trap is another method of collecting (fig. 18.7) larger insects, such as ground beetles. Instead of using a killing agent in the bottom, smear a little petroleum jelly around the rim of the jar to prevent trapped insects from climbing out. Details about the construction of these traps are given in chapter 15.

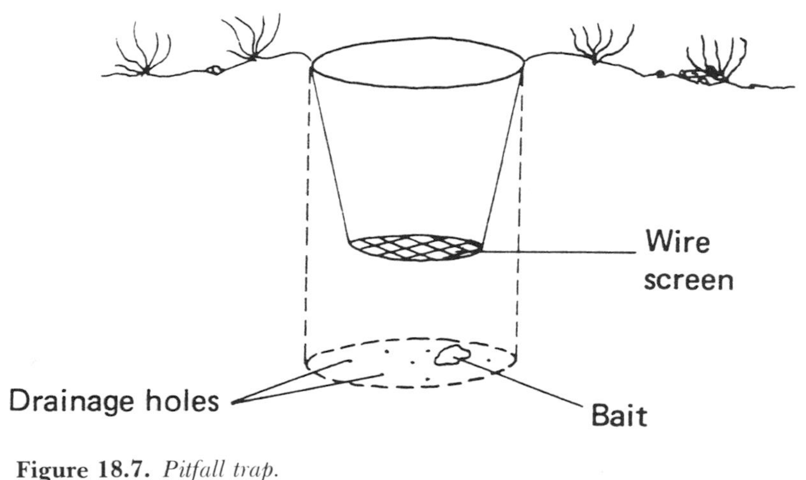

Figure 18.7. *Pitfall trap.*

Keeping and rearing insects

Now that you have assembled, by a variety of methods, living insects, you must immediately place them in their new homes. A few general rules should be followed, rules needed for the survival of the individuals for their normal lifespan.

1. Don't overcrowd; better to start with a few insects.
2. Keep cages, terraria, and aquaria clean.
3. Provide adequate light and air.

4. Provide the right amount of food and water. Never overfeed.
5. Do not let them escape, unless you are releasing them back in their natural habitat. (Obviously, if you are raising pest species, these must be destroyed, if no longer being used.)

Keeping insects alive for short periods of time for display is relatively easy. Fresh material may be added from the field during their normal flight period. Rearing insects to stock the home zoo is a different matter, but this is, of course, the challenge that makes it interesting. Either at home or in the classroom, there are several species that will provide excellent examples of their life history, habits, and behavior.

Some of the species may be purchased from biological supply houses, usually any time during the year (table 18.1).

Cockroaches or earwigs, both potential pests, must be handled in a way to be sure that small nymphs cannot escape. Petroleum jelly can be placed around the top of the jar or fish tank to prevent the insects from crawling out. A fine mesh screen should be placed over the top. For aquatics, small jars or fish bowls can be used. Dragonfly nymphs are predacious and should be kept in separate containers, or they will eat each other. They may be fed various aquatic arthropods, including *Daphnia* spp.; mosquitoes or even house fly larvae are satisfactory food.

Table 18.1. *Easy-to-rear insects.*

Name	*Order*	*Type of Metamorphosis*	*Length of Life Cycle***
Dragonflies (naiads)*	Odonata	Hemimetabolous	Several years
Field crickets	Orthoptera	Paurometabolous	About 35 days
Mantids	Dictyoptera	Paurometabolous	45–65 days
Cockroaches	Dictyoptera	Paurometabolous	Some about 30 days
Earwigs	Dermaptera	Paurometabolous	About 90 days
Milkweed bugs	Hemiptera	Paurometabolous	About 40 days
Mealworms	Coleoptera	Holometabolous	About 50 days
Hornworms	Lepidoptera	Holometabolous	About 30 days
Mosquitoes	Diptera	Holometabolous	About 30 days
House flies	Diptera	Holometabolous	About 14 days

*Dragonfly naiads are carnivorous and remain in the naiad stage for many months to years, depending on the species and the instar they are in when captured.

**Time in days from egg to emergence of the adult; temperature and food availability can change these times.

When rearing crickets, a small container of fine sand should beplaced in the cage for egg laying. Peat cups or egg cartons provide ideal hiding places for nymphs and adults. The female cricket has a long ovipositor, which she uses to place her eggs in the sand. Eggs are laid singly and at room temperature. They hatch in two to three weeks. To prevent adults from eating the eggs, place a wire mesh screen over the container of sand after the eggs are laid. Food, such as dry dog food for protein and fresh apples, pears, and lettuce for a balanced diet, should be provided. If crickets are kept at 27° C, they will mature in about two months; but, if it is warmer, up to 32° C, the maturity time will be cut in half when they have the proper food.

Rearing mantids the year around is tricky. The main problem is to keep living prey available. One mantid ootheca may contain as many as 250 eggs. In warm weather, the ootheca will hatch two to three weeks after laid. If the ootheca and the nymphs hatch in a fish tank or terrarium, it will be necessary for you to separate them, or they will soon eat each other. This is, of course, one way to provide food for awhile, at least until the number you want to rear is reached. The best time to rear mantids is in the late spring, when you can find plenty of live insects to feed them. Therefore, if you find ootheca in the fall or winter, keep it in the refrigerator to prevent hatching until food is available. However, if you want nymphs and adults around all the time, you can also raise a variety of other insects to feed to them. For example, young nymphs will eat fruit flies; larger nymphs and adults will eat cockroaches, house flies, blowflies, and even mealworms.

Earwigs must be placed in a container with loose potting soil, damp, but not wet. Provide them with dog food, carrots, and potatoes, plus a supply of water. The female earwig will deposit about fifty eggs in a small "cave" that she has constructed in the soil. She will display some maternal care by guarding these eggs until they hatch. At 21° C, the eggs will hatch in five weeks. Earwigs may be pests in some parts of the United States, because they feed on the blossoms of garden flowers. Therefore, extra care must be taken to prevent their escape. (Incidentally, the posterior forceps at the end of the abdomen are harmless to humans, and they are not known to bite, despite their common name.)

Milkweed bugs are widely used in entomological-physiological research. They are abundant on several species of milkweed (*Asclepias* spp.), on which they may be easily seen and captured. The life cycle is ideal for this showy species. It is a favorite for classroom rearing. The bright orange and black bugs are utterly harmless and provide good study specimens. At

29° C, the egg stage lasts only four days. The five nymphal instars each last for six days, and the adults live up to a month. All stages may be seen at any one time, which is ideal for demonstrating paurometabolous metamorphosis. Milkweed pods and seeds may be stored for long periods of time, and these are easily collected in the fall. Milkweed pods are the bugs' sole food, but the seeds of sunflowers, watermelon, squash, cashews, and almonds may be substituted.

Mealworms are the larval stage of the large, block tenebrionid beetle, *Tenebrio molitor.* These have long been used as food for turtles, lizards, and frogs, and sometimes as fish bait. To rear them in quantity, place bran meal in the bottom of a container—a covered plastic dish or a gallon pickle jar will do. Add a potato slice for moisture and some shredded paper on top of the bran. The adults will climb on the paper. Purchase about twenty-five adults to start the colony. The first larvae will appear in ten to fourteen days. Clean out the pans periodically by changing the food, or start new jars by removing newly emerged adults. This method will give you an endless supply of these insects to watch and to feed to other insects. Remember, these beetles are household pests and will live in cereal boxes, as easily as in your jars.

Hornworms are the larval stage of the tobacco hornworm sphinx moth, *Manduca sexta*. This species is a destructive pest of tobacco, tomato, pepper, and eggplant. Be careful not to let hornworms escape into your garden. Their large size makes them ideal for observing the stages characteristic of the holometabolous insects. It is possible to carry the hornworm through its entire life cycle in thirty days, but only if the critical factors, light and temperature, are carefully controlled. The larval stage must have constant light from a 100-watt bulb, and the temperature in the cage must be kept at 27° C. Larvae are fed on most any of the solanaceous plants mentioned previously as their hosts. The plants must be grown in pots placed inside the rearing cage. The larval stage lasts about fourteen days. After they complete their feeding stage, they will crawl into the soil in the pot to pupate. This lasts another fourteen days. When the adults emerge, they will crawl up the stem of the plant and then fly around in the cage. After a few days, mating will take place, and the cycle will start over again. These insects are somewhat more difficult to raise than the others we have talked about, but with care, beautiful moths will emerge. Using the same method described here and providing the correct host plant, many different kinds of moths, butterflies, and phytophagous beetles can be reared.

House flies and mosquitoes are both easy to rear, but they have different habitats. House flies have a fast life cycle, as do certain species of mosquitoes. From egg to adult can take only about two weeks at 27° C or more. House flies may be reared in gallon containers with about 10 cm of wood chips or wood shavings on the bottom of the container. Avoid cedar, redwood, or pine chips, since they are toxic to the fly larvae. Mix one part powdered milk with two parts water and pour over the wood chips. Place house fly eggs on the wood chips, but not in the milk mixture. Eggs may be obtained by attracting house flies to some substance (such as garbage), and then capturing them in a jar. The females will lay eggs in a few days. The larvae will feed on the milk mixture for a few days, then leave the wood chips and crawl on the sides of the jar. These are ready to pupate. Special measures need to be taken for the pupae and adults. Place the mature larvae in another container lined with paper towels where they will pupate. The pupae may then be removed to a cage, and then the adults will emerge. Adults can be fed a solution of granulated sugar and powdered milk.

Mosquito eggs may be purchased from biological supply houses or gathered from pans of water placed outside in the summer. The eggs are transferred to small, covered aquaria or to pans in a small, screened cage. The larvae are fed small quantities of brewer's yeast. It is best to float a clean microscope slide, supported by a cork raft, on the water. Place the yeast on the slide and slightly dampen. The larvae will swim to the edge of the slide and feed on the yeast. If the water becomes contaminated by the yeast, it must be changed. After a few days the larvae will mature and transform into swimming pupae. The pupae may be removed to pans in separate cages, or left where they are. Adults will emerge, mate, and lay eggs. Cultures of various species of mosquitoes may be kept indefinitely. But one thing you should know: the female must have a blood meal. Either you provide the blood by placing your arm in the cage, or you get a rabbit or guinea pig to volunteer to do it for you!

Predacious insects, such as lacewings and ladybird beetles, are more difficult to rear than preying mantids. However, lacewing eggs and moth eggs may be purchased. Themoth eggs laid on squares of paper are supplied commercially. The lacewing eggs, laid at the end of a long stalk, may be obtained cut from their stalk. They will hatch and feed on the moth eggs. Each lacewing larva eats less than one hundred eggs. It pupates, and adults soon emerge. From then on, you can keep colonies of these beneficial insects, if you provide them with moth eggs.

Ladybird beetles feed on scale insects and aphids. To rear them, it is necessary either to rear aphids on potted plants or to collect leaves and stems infested with scale insects. Some species are host specific, making it necessary to capture adults that are feeding to know what species of prey to provide.

Predacious aquatics, such as dragonflies or predacious beetles, will feed on the larvae of mosquitoes, mayflies, crane flies, and other aquatics. The larger species will eat guppies and other small fish. Some water beetle larvae are known as "tigers of aquaria" and, once introduced, will provide exciting "sea" battles.

Parasitic insects, too, may be reared. The parasitic wasp, *Mormoniella vitripennis*, or jewel wasp is one that may be reared. (This wasp does not sting humans.) Females will oviposit in the puparia of house flies and other species. They have a short life cycle. Infested flies will produce adults in a few days. Because they enter diapause, they may be used to study this phenomenon (see chap. 6).

Most terrestrial insects will require, as we have stated above, small amounts of water. This may be provided by filling a glass vial with water and plugging it with cotton. Attach the vial, inverted, on the side of the cage. The adults will fly to it to obtain this moisture. Larvae will also go to it, if the vial is placed down low where it can be reached. Sometimes fresh fruit and vegetables can be a sufficient source of moisture.

Hours of fun and learning may be had observing these tiny zoo animals. Rearing a variety of species, the work shared by various friends or classmates, will provide instruction for all. Learning the details of rearing, applying the principles of insect behavior, and devising ways to overcome various problems is an exciting way to learn. Insects are plentiful, cheap, and, most of all, enjoyable to watch.

Field studies

You do not need to confine your studies of insect behavior to laboratory zoos. Carefully planned field work is exceedingly important, however, if field observations are to yield data comparable to those gathered by laboratory studies. Field biologists are usually concerned with natural populations of organisms. A population is an open system, in contrast to those closed systems that we attempt to create in our laboratory zoos. Therefore, you should try to circumscribe the system affecting each species observed, as suggested in chapter 2. A species cannot be fully studied without some knowledge of its life cycle or life system. A life system is composed of the subject population and its effective environment, or, in other words, the totality of external agencies influencing the population, be it insect or man. The various genotypes (the individual's genetics) that compose the population, coupled with environmental factors, determine the characteristics of the population. Genes determine the time of birth, death, and type of dispersal of the individuals of the unit. The whole complex study of population ecology is based on knowledge of genetic factors. You as a student of entomology should be interested in those aspects of the study that contribute data useful for the formation of a hypothesis to explain the origin and evolution of each species.

Practical study

To illustrate the points outlined in the preceding paragraph, as well as throughout this book, we have listed here some projects which, if undertaken, will help you to gather more data on insect life. Significant data about natural populations gathered in the field will help you learn enough about insects for you to be able to hypothesize about isolating mechanisms (the factors that keep the species separate) affecting the species (see chap. 4). It may lead to the gathering of data helpful for the discovery of previously unknown information about a species. These data will be useful as a supplement to the morphological descriptions of species.

Selecting a study area. Spend some time walking through the collecting site to determine which insect species are dominant. Then, with note pad, record as much information as can be gathered by observation. This should include data on the plants, soil, temperature, moisture, light, wind, and the physical nature of the habitat. Collect a few specimens of the abundant species for later identification. Record all the data needed for locality and ecological data labels.

Ecology data. These field data are grouped under information on the ecology, life history, distribution, and behavior (especially reproductive and feeding behavior). To help you organize your data, answer the questions in the following checklist, recording your observations as they are made. (Write down only observations made at the moment and omit any biased assumptions based on previous knowledge.)

Gathering Ecological Data

1. Record date and time of observations. How does this contribute data on the ecology of the species?
2. What are the weather conditions, including temperature, relative humidity, light intensity, and wind velocity?
3. What other organisms are present that might affect the life system of the species?

4. Record the altitude, along with the locality data. Is there any evidence that altitude affects this species?
5. Is there any evidence that soil influences the distribution of the population?

Gathering Life History Data

1. What stages of the species are present?
2. Where are these stages found?
3. What is their relative abundance, that is, adults to nymphs, larvae, etc.?
4. What are the obvious "spot" characteristics of each stage present? (Field records of appearance are important supplements to the description of dead specimens.)

Gathering Data on Distribution

1. What is the distribution pattern for each stage of each species observed in relation to the limits of the habitat?
2. What is the general trend of movement of the individuals in the population?
3. Is there any evidence of a dispersal stage? What is the means of dispersal?

Gathering Data on Behavior

1. Is there evidence that the act of your observation affects the behavior of the individuals?
2. What are the individuals doing: feeding, courting, etc.?
3. How is this being done (details)?
4. Is there any evidence of negative or positive reaction toward other species?

On returning home, write your notes into a permanent record book (see chap. 16), and correlate those notes with the specimens collected.

GOOD HUNTING!

Appendix I
Entomological Societies

Jules Verne once said that Americans form a society whenever there are enough people to elect a president, a secretary, and a treasurer. Entomologists certainly fit that pattern, because there are dozens of societies devoted to the study of insects in the United States. The rest of the world is no exception, however, because there are well over three hundred such societies that regularly publish a journal or newsletter and, probably, several hundred local societies that meet regularly without publishing a journal. Moreover, many of these organizations or groups published special newsletters or even a journal just for a single group. We personally have a list of sixty-four publications devoted to a single genus (for example, "Cicindela," devoted to tiger beetles only), a single family, or a single order, some of which are listed below. Those organizations listed below are widely known and generally interested in having both professional and amateur members. A complete list of entomological societies may be found in the *Naturalists' Directory and Almanac (International)*, 44th edition.

AMERICAN ENTOMOLOGICAL SOCIETY
1900 Race Street
Philadelphia, PA 19103
The oldest entomological society in the United States; publishes *Entomological News*, *Transactions*, and *Memoirs*.

CAMBRIDGE ENTOMOLOGICAL CLUB
16 Divinity Ave.
Cambridge, MA 02138
Local and international membership; publishes *Psyche*.

COLEOPTERISTS' SOCIETY
c/o Department of Entomology
Smithsonian Institution
Washington, DC 20560
Publishes the *Coleopterists' Bulletin*; originally founded for amateurs, but has since become a formal journal primarily for museum-based systematists.

ENTOMOLOGICAL SOCIETY OF AMERICA
4603 Calvert Road
College Park, MD 20740
The leading national professional society; publishes several journals devoted mainly to applied entomology.

ENTOMOLOGICAL SOCIETY OF WASHINGTON
c/o Department of Entomology
Smithsonian Institution
Washington, DC 20560
An active local society with an international membership; publishes the *Proceedings* and a series of memoirs; many useful papers on taxonomy published each year. Amateurs welcomed.

INSECTA MUNDI, INSECT WORLD DIGEST, and HANDBOOK SERIES
4300 NW 23rd Ave., Suite 100
Gainesville, FL 32606
An irregular journal, an annual, and a handbook series; devoted to the publication of new taxonomic data, biological information, and useful works for the identification of insects; worldwide in scope.

LEPIDOPTERISTS' SOCIETY
1041 New Hampshire Street
Lawrence, KS 66044
The society for those interested in moths, skippers, and butterflies; publishes a journal and a newsletter; amateurs take an active part in all society affairs; annual meeting and collecting trip.

YOUNG ENTOMOLOGISTS' SOCIETY
c/o Gary Dunn
Department of Entomology
Michigan State University
East Lansing, MI 48824
Publishes a quarterly journal devoted to general articles on entomology of interest to the beginner.

Appendix II
Entomological Literature

During the past thirty years, more and more books and pamphlets dealing with entomological subjects have become available for the general public. The gap between the highly technical literature used by the advanced amateur and the professional systematist is narrowing. This book, we hope, helps to close this gap. The following list of readily available publications includes some of the better beginning field guides, works useful for the preliminary identification of insects, and some of the more advanced works that can be purchased for your own use or by your local library, if not already available there. The books are arranged by author for ease in finding in the library card or computer file. The title, date of publication, and name of the publisher or dealer are cited. This is followed by a brief comment to guide you in the selection of what you may wish to refer to for more information than that given in this book.

Selected field guides

Arnett, R. H., Jr.; Downie, N. M.; and Jaques, H. E. *How to Know the Beetles*, 2nd ed. Dubuque, IA: W. C. Brown Co., 1980.

Keys to the families of beetles of United States and Canada and to hundreds of common species; many species illustrated.

Arnett, R. H., Jr., and Jacques, R. L., Jr. *Guide to Insects.* NY: Simon & Schuster Co., 1981.
Full color photographs of 350 common insects found in the United States, each species with symbols, showing habitat, habits, and ecological significance; with a picture key to the orders of insects.

Bland, R. G. *How to Know the Insects*, 3rd ed. Dubuque, IA: W. C. Brown, Co., 1978.
Orders and common families of insects are treated in key form with many illustrations of the most common species; very useful for the beginner.

Borror, D. J., and White, R. E. *A Field Guide to the Insects.* Boston, MA: Houghton Mifflin Co., 1970.
Treatment of the major families of insects of the United States and Canada with many illustrations in black and white and color; no keys, but each group with important family characteristics highlighted and well illustrated.

Helfer, J. R. *How to Know the Grasshoppers, Cockroaches, and Their Allies.* Dubuque, IA: W. C. Brown, Co., 1953.
Picture keys to most of the species of Orthoptera, Dictyoptera, Dermaptera, and Phasmatodea of United States and Canada; well illustrated.

Klots, A. B. *A Field Guide to the Butterflies.* Boston, MA: Houghton Mifflin Co., 1951.
All of the butterfly and skipper species east of the Great Plains are described and illustrated. Somewhat out of date now.

Otte, D. *The North American Grasshoppers.* Vol. I: *Acrididae: Gomphocerinae and Acridinae.* Cambridge, MA: Harvard University Press, 1981.
Although monographic in nature, the excellent illustrations and relatively simple identification keys make this a useful field work in the sense that it will enable the user to make accurate identifications of the species involved.

Slater, J. A., and Baranowski, R. M. *How to Know the True Bugs.* Dubuque, IA: W. C. Brown, Co., 1978.
Picture keys to the families and common species of bugs of the United States and Canada; well illustrated.

White, R. E. *A Field Guide to the Beetles of North America.* Boston, MA: Houghton Mifflin Co., 1983.
Most of the families of beetles of the United States and Canada are treated, with brief descriptions, but no keys; well-illustrated work for nonspecialists.

HANDBOOKS (more advanced works for identification and reference)

Arnett, R. H., Jr. *American Insects, A Handbook of the Insects of America North of Mexico.* New York, NY: Van Nostrand Reinhold Co., 1985.
All of the orders, all families, subfamilies, tribes, and genera are treated, with the number of species of each genus and hundreds of species illustrated and described; keys to the orders and families, and some subfamilies, tribes, and genera; extensive bibliography; a standard reference work for amateurs and professionals.

Arnett, R. H., Jr. *The Beetle Fauna of North America.* Vol. I: *The Families and Genera*; Vol. II: *Checklist of the Species.* Gainesville, FL: Flora and Fauna Publications, (in press).
Keys to the families and genera of the beetles of the United States and Canada, and a list of all the species, giving their distribution; extensive bibliography; hundreds of species illustrated.

Howe, W. H. *The Butterflies of North America.* Garden City, NY: Doubleday & Co., 1975.

The best book on butterflies and skippers since Holland's famous butterfly book of the early part of this century. Each species and most subspecies are described, illustrated, with details about food plants, distribution, and migration. Currently the standard work on the subject.

McAlpine, J. F. *et al. Manual of Nearctic Diptera.* Vol. 1. Ottawa, ON: Research Branch, Agriculture Canada, 1981.

Keys to the families and genera of the flies of the United States and Canada, with extensive discussions of the anatomy of each group; extensive bibliography; superior illustrations; when the work is completed (probably in 1986), it will be the standard work on the flies for this region.

NOTE: These are the major works that are readily available in print at the present time. Other works for other groups are in press. More extensive lists of publications useful for identification are given in Arnett, 1985, listed above.

Entomological information

Arnett, R. H., Jr. *Entomological Information Storage and Retrieval.* Gainesville, FL: Flora and Fauna Publications, 1970.

A mine of useful information about entomological literature; it is not intended as an exhaustive study of the subject, but is useful as a reference to sources of entomological supplies, dealers, and specialized booksellers; it includes descriptions of reference resources, discussion of literature, how to search the literature, how to index, and how to cite literature.

Blackwelder, R. E., *Taxonomy: A Text and Reference Book.* New York, NY: John Wiley, Inc., 1967.

The most complete guide to the methods of taxonomy, the use of the International Rules of Zoological Nomenclature, and the practice of taxonomy.

Dice, L. D. *The Biotic Provinces of North America.* Ann Arbor, MI: University of Michigan Press, 1943.

Provinces based on dominant plant and animal species, named and mapped for all North America, including Mexico.

Gilbert, P., and Hamilton, C. J. *Entomology: A Guide to Information Sources.* London, Eng.: Mansell Publishing, Ltd., 1983.

Packed full of reference work citations dealing with all aspects of entomology, including history, naming and identification, specimens and collections, how to search the literature, and keeping up with current events in entomology; very useful to amateurs and professionals.

Kuechler, A. W. *Potential Natural Vegetation of the Conterminous United States* (with colored map). New York, NY: American Geological Society, Special Publication no. 36, 1964.

A description, illustrations and map of 116 vegetation types of the United States.

McCafferty, W. P. *Aquatic Entomology: The Fishermens' and Ecologists' Illustrated Guide to Insects and their Relatives.* Boston, MA: Science Books International, 1981.

The best-illustrated, most up-to-date discussion of aquatic insects now available; useful to fishermen, as well as ecologists.

Michner, C. D. *The Social Behavior of the Bees.* Cambridge, MA: Harvard University Press, 1974.

A thorough discussion of all aspects of the biology and behavior of wild bees.

Taylor, R. L., and Carter, B. J. *Entertaining with Insects.* Santa Barbara, CA: Woodbridge Press Publishing Co., 1976.
How to prepare insects for appetizers, lunches, dinners, and celebrations, with basic recipes, sources of insects for food, and how to raise your own insects for food; also includes earthworm recipes.

Wilson, E. O. *The Insect Societies.* Cambridge, MA: Harvard University Press, 1971.
A thorough discussion of all aspects of social life among the insects.

Directories

Arnett, R. H., and Arnett, M. E. *The Naturalists' Directory and Almanac (International)*, 44th ed., Part I: *Insect Collectors and Identifiers.* Gainesville, FL: Flora and Fauna Publications, 1985.

An up-to-date list of entomologists of the world interested in collecting and identifying insects; each person listed with complete mailing address, nature of interests, and exchange desires; geographical and subject index, and articles on entomology; of particular interest to amateur entomologists.

Arnett, R. H., and Samuelson, A. *Insect and Spider Collections of the World.* Gainesville, FL: Flora & Fauna Publications, 1985.
A complete list of the insect and spider collections of the world with names and addresses of curators, descriptions of the collections, size, and emphasis or specialization of the collection.

Glossary

Most of the words in the following glossary have to do with parts of insects or with special phenomena. Generally these words are defined in the text where first used. Some are more formally defined here.

Abdomen the posterior-most body region of the three basic regions of insects.

Adult the sexually mature stage of Hexapoda; except for certain primitive species, this stage no longer molts; usually the winged stage. Note: only adult insects have wings.

Ametabolous a type of metamorphosis characterized by no or little external change except size; characteristic of primitive Hexapoda and the lice.

Anaphylaxis an unusual or exaggerated reaction to foreign protein or other substances.

Androconia special scent scales located on the wings of some butterflies.

Antenna a sensory organ, always paired, on the head of almost all Hexapoda, consisting of one or more segments, each with sensory setae or spines.

Apical pertaining to the apex or outer end of a structure.

Apterous without wings.

Auditory organs special chambers covered with a stretched membrane on the anterior tibia of some Orthoptera, or the abdomen of these and other insects; function, as an ear.

Beak the long proboscis or sucking mouthparts of some insects (see also snout).

Bifid in two branches.

Biome a major ecological division of a continent characterized by certain dominant plants.

Biparental with a male and female parent, as opposed to hermaphroditic or parthenogenetic reproduction.

Capitate a type of antenna with the terminal segments enlarged to form a spherical mass.

Carnivorous feeding on animals.

Category a class group or hierarchy to which a taxon (q.v.) belongs; e.g., Noctuidae represents the taxa of the family hierarchy.

Caudal tail or posterior position of an organism.

Caudal filament a thin, usually segmented median posterior abdominal process.

Cell an area of the wing of insects enclosed by veins.

Cercus (pl. cerci) a lateral posterior abdominal process usually slender, which may be filamentous and segmented.

Chelicera (pl. chelicerae) the pincherlike first pair of appendages of certain Arthropoda.

Chitin a complex nitrogenous carbohydrate forming the main skeletal substance of the arthropods.

Clasper a clasping organ at the apex of the male abdomen.

Class a category used as a division of phylum.

Classification the arrangement of species in the hierarcial system of categories and taxa.

Clavate gradually thickened toward the apex; clubbed.

Claw a hollow, sharp organ, usually paired at the end of the tarsus.

Cline a gradual series of morphological or physiological differences exhibited by a group of a species, generally geographically related.

Clubbed or **club-shaped** shaped like a club with apex or distal portion enlarged (see also capitate).

Clypeus part of the insect head below the front to which the labrum is attached.

Cocoon a pupal covering composed in part or entirely of silk spun by a larva.

Compound eye an aggregation of separate ommatidia or visual elements, usually paired, on the head of most adult insects.

Compressed flattened from side to side.

Coxa (pl. coxae) the basal segment of the leg, articulating to the body.

Cryptic coloration color patterns of insects and other animals that resemble part of the habitat and hence, tend to "hide" the animal by making it difficult to see when not moving (camouflaged).

Depressed flattened dorsally–ventrally.

Determination the identification of a specimen.

Detritus cycle the part of the ecological process that involves the feeding on and processing of dead plant and animal material.

Diapause a period of metabolic inactivity between active periods.

Dichotomous divided into two parts.

Distal away from the base or point of attachment.

Diversity different body types and systems present in a taxon.

Ectoparasite an insect parasite that feeds on the surface of the host, usually bloodsucking.

Egg the first free-living stage of the life cycle of most insects, usually within a flexible shell.

Elytron a modified thickened and hardened front wing of an insect, particularly the Coleoptera.

Endoparasite a parasite living internally in the host.

Exoskeleton the chitinous, sclerotized covering of the insect body.

Exuvium (pl. exuvia) the cast skin of larvae or nymphs at metamorphosis.

Family a category in the hierarchy of classification, a division of an order.

Femur (pl. femora) the segment of the leg of an insect beyond the trochanter before the tibia.

Filament a long, slender, threadlike, usually segmented process.

File stationary part of the sound-producing surface.

Filiform threadlike, slender, and of nearly equal diameter throughout.

Forceps hook or pincerlike processes at the apex of the abdomen in Dermaptera.

Form sometimes used as a subdivision of a species, but not as a recognized taxon; also refers to a species or a group of species in general terms.

Galea the outer lobes of the maxilla.

Gaster abdomen of Hymenoptera beyond narrowed first and/or second segments, i.e., beyond petiole.

Genitalia the modified apical abdominal segments used in copulation.

Genus the category in the classification hierarchy to which species are assigned.

Gill a thin-walled structure with tracheae used for the absorption of oxygen in aquatic insects.

Glossa (pl. glossae) the paired inner lobes of the labium, particularly the coiled sucking tube of most Lepidoptera.

Gradual metamorphosis (see hemimetabolous), usually with egg, nymph, and adult stages.

Group used to refer to a taxon without using the Latin name, e.g., "This group (referring to the preceding taxon) is large and widely distributed."

Hairlike seta (pl. setae) a fine seta, often referred to as a hair on the surface of the insect body; a term used in insect anatomy in place of "hair," a structure characteristic of the mammals.

Halter (pl. halteres) the modified, vestigial hind wings of Diptera, sensory balancing organs.

Head the anterior-most body region of insects.

Hemelytron (pl. hemeytra) the basal thickened portion of the wing, typical of Hemiptera.

Hemimetabolous metamorphosis incomplete metamorphosis with eggs, aquatic naiads (aquatic nymphs) dissimilar to the adults.

Herbivorous plant feeders.

Hibernation a period of lethargy, or suspension of most body activities, with a greatly reduced respiration rate, occurring particularly during periods of low temperature.

Holometabolous with complete metamorphosis, with egg, larva, pupa, and adult stages.

Holotype the single specimen on which the description of the species is based and so-designated at the time of description; also known as the type specimen.

Homoiothermic warmblooded animals, as opposed to poikilothermic, or coldblooded animals; insects are poikilothermic, not homoiothermic, because their body temperature varies with the ambient.

Hypermetamorphosis a form in which an insect passes through more than the normal number of stages, particularly a stage other than the pupa after the transformation of the full-grown larva.

Identification the process of finding the correct name of an organism by keys and other pertinent literature or similar processes (see text).

Imago the adult stage.

Inquline a socially parasitic species living its entire life in the nests of host species.

Instar the stages between molts of nymphs or larvae; also called statium.

Isolating mechanism physical, genetic, or chemical barrier to reproduction.

Joint an articulation; the area of flexion between sections of an appendage.

Labellum (pl. labella) the sensitive tip of the mouth structures of certain Diptera (for example, the house fly).

Labium the lower lip of the mouthparts of insects.

Labrum the upper lip or flap of the mouthparts of insects.

Law of priority a rule of zoological nomenclature, which states that the oldest validly proposed (published) scientific name of an animal must be used, and later names proposed for the same animal must be treated as junior (invalid) synonyms.

Larva (pl. larvae) the growing stage of the life cycle of holometabolous insects, hatches from the egg, feeds, and eventually changes into the pupal stage, except in rare incidences.

Larviform adult resembling a larva, but sexually mature.

Lateral, laterad at or toward the side.

Malpighian tubes long and slender blind tubes lying in the haemocoel (blood cavity), opening into the beginning of the hind intestine of insects, used for excretory functions.

Mandible one of the pair of jaws of insects, stout and toothlike.

Mandibulate having biting jaws.

Maxilla (pl. maxillae) the second pair of jaws of a mandibulate insect.

Maxillary palpus (pl. palpi) segmented sensory structures located on the maxilla.

Median in or at, or pertaining to the middle.

Melanin an organic pigment in the cuticle of insects producing amber, brown, and black coloration.

Mesal, mesad pertaining to or toward the median plane of the body.

Mesothorax the second, or middle, thoracic segment, which bears the middle pair of legs and, in winged insects, the anterior pair of legs.

Metamorphosis the series of changes through which an insect passes from egg to adult.

Metathorax the third, or last, thoracic segment, which bears the third pair of legs and, in winged insects, the second pair of wings.

Migration a mass movement, usually seasonal, away from or toward the breeding area; in insect migration, the movement may be random or, more rarely, in one direction.

Molt to cast off or shed the outer covering of the integument periodically during growth.

Mutualism Symbiosis benefiting both species involved in the relationship.

Myiasis a condition caused by the infestation of the body by fly larvae.

Naiad the nymph of aquatic hemimetabolous insects.

Nasute a soldier caste of termites with a snout on the head, through which is extruded an irritating defensive substance.

Nearctic one zoogeographic region of the earth, which includes Canada, Alaska, Greenland, the United States and the temperate, northern part of Mexico.

Nomenclature the scientific names of animals and plants and the application of these names.

Nymph the immature stage of insects with gradual (paurometabolous) metamorphosis.

Ocellus (pl. ocelli) a simple eye in some adult insects consisting of a single, beadlike lens.

Ootheca an egg packet, eggs covered with a hardened case.

Order the division of a class in the hierarchy of categories.

Osmeterium V-shaped scent organs behind the head in the larvae of Papilionidae (swallowtail butterflies), Lepidoptera.

Ovipositor the tubular or valved structure with which the eggs are placed.

Parasite any animal or plant that lives in and at the expense of another animal or plant.

Parasitoid a parasite of insects that completely consumes the host.

Parthenogenesis reproduction from unfertilized eggs.

Paurometabolous having a metamorphosis in which the change from the immature stage (nymph) to the adult is more or less gradual, i.e., the nymph resembles the adult, but is without functional wings and is sexually immature.

Pectinate comblike.

Petiole the narrowed portion of the abdomen of certain Hymenoptera, typical of ants and wasps.

Phosphorescent organ a light-producing structure found on the adults and larvae of certain insects, particularly some beetles.

Phylogeny the arrangement of organisms in a series to show their evolutionary relationships.

Phylum a major division of the animal kingdom, a category.

Phytophagous feeding on plants.

Pile a covering of fine, hairlike setae.

Poikilothermic coldblooded, or body temperature at or close to that of the ambient.

Population a group of individuals of a single species living in a particular region and capable of reproducing by mating with each other.

Prepupa an active, but nonfeeding, stage of nymphs of Thysanoptera and some Homoptera.

Proboscis the elongated mouthparts of many insects, used for sucking or piercing-sucking.

Pronotum the upper or dorsal surface of the prothorax.

Prothorax the first thoracic segment; bears the first or prothoracic legs.

Proximal the part of an appendage nearest the body.

Pubescence setae, usually fine, hairlike.

Pulvillus (pl. pulvilli) soft, padlike structures between the tarsal claws.

Punctate set with impressed punctures or small impressions, pits, or deep depressions on the body surface.

Pupa (pl. pupae) the transformation stage between larva and adult in holometabolous insects.

Puparium (pl. puparia) in certain Diptera, the thickened, hardened, larval skin within which the pupa is formed.

Race a term usually applied in zoology to populations showing geographical differences; races are no longer given Latin names and have no validity as named taxa under the International Rules of Zoological Nomenclature.

Raptorial adapted for seizing prey.

Retractile capable of being drawn back (after being extended).

Saltatorial capable of jumping.

Saprophagous feeding on dead or decaying animal or vegetable tissue.

Scale a modified, broad seta; the covering of certain Homoptera, particularly the scale insects.

Scalelike resembling a scale.

Scientific name the Latin or Latinized name used for an organism, as opposed to the common or vernacular name.

Scraper moving part of the sound-producing surface of insects.

Scutellum the posterior division of the notum of the thorax; in Coleoptera and Hemiptera, the triangular piece between the front wings.

Segment a ring or subdivision of the body or of an appendage between areas of flexibility with muscles attached for movement.

Series a major group of superfamilies within an order; also several specimens of a species.

Serrate sawlike.

Seta (pl. setae) a hollow structure developed as an extension of the epidermal layer of the body wall.

sexual dimorphism morphological differences between the males and females of a species.

Snout the prolongation of the head, at the end of which the mouthparts are situated.

Species (pl. also species) a group of individuals or populations that are similar in structure and physiology and are capable of interbreeding and producing fertile offspring.

Specific name the two-part name (generic name and trivial name) in Latin or Latinized words used to refer to a species.

Spermatophore a packet of sperm.

Spiracles external openings of the tracheae, usually with valves capable of opening and closing the opening.

Spine a more-or-less thornlike process or outgrowth of the body wall, not separated from it by a joint; a large seta.

Stage, stadium (pl. stadia) the period between molts of the developing insect.

Stria (pl. striae) a groove or depressed line, usually on the elytra.

Striate with striae.

Stridulation production of sound by rubbing one part of the insect body against another part, usually as a file and a scraper.

Stylet a needlelike structure, especially the elongated parts of the piercing-sucking type of insect mouthparts.

Stylus a short, slender process.

Subfamily a division of the family category in the classification hierarchy.

Subgenus (pl. subgenera) the subdivision of the genus category in the classification hierarchy.

Subspecies the division of the species category in the classification hierarchy, the only division of the species that is validly given a Latin or Latinized name.

Sulcus a furrow or groove.

Superclass a division of a phylum above the category of class in the classification hierarchy.

Superfamily a division of an order above the category family in the classification hierarchy.

Suture a seam or impressed line between parts of the body wall; also in Coleoptera, the line of junction of the inner or posterior edges of the closed elytra.

Synonym another name for a species or genus, invalid either because it is a younger name, or invalidly proposed.

Systematics the science of the taxonomy and evolution of species of plants and animals.

Tarsus (pl. tarsi) the portion of the leg of insects distal to the tibia.

Taxon a named group of organisms arranged in a category in the classification hierarchy.

Taxonomy the naming and arranging of species and groups into a system of classification.

Thorax the middle body region of insects.

Tibia (pl. tibiae) the section of the insect leg between the femur and the tarsus.

Tomentose covered with matted, woolly, hairlike setae.

Trachea (pl. tracheae) an internal respiratory tube in insects.

Tribe a division of the subfamily, a category in the classification hierarchy.

Trochanter the portion of the leg between the coxa and the femur.

Tympanal organ the ear or auditory organ of an insect.

Tympanic membrane the membrane stretched across the surface of the tympanic cavity as a part of the tympanum.

Tympanum a cavity (the ear) on the metathorax or leg of certain insects.

Type specimen usually the holotype, but may be another specimen designated to replace a lost holotype, termed the neotype, or a later designated type, the lectotype.

Variation individual differences found in populations and species.

Variety a term applied to a specimen showing some variation from the majority of the members of a species; a named variety no longer has validity in zoological nomenclature.

Vein a tube or tubes running through the membrane of the wings of insects.

Vouchered specimen a specimen whose identification is verified by comparison with the type specimen or specimens used when writing a review, revision, or monograph of the species.

Wing the paired membranous flight organs located on the mesothorax and usually the metathorax of insects.

Wing pad the encased, undeveloped wings of the nymphs and pupae of insects.

Wing scales the modified, broad, flattened setae found on the wings of certain insects, particularly the Lepidoptera; the stumps of wings of winged termites after the wings have been shed.

Index